Single Particle Tracking and Single Molecule Energy Transfer

Edited by
Christoph Bräuchle, Don C. Lamb, and Jens Michaelis

Further Reading

M. Sauer, J. Hofkens, J. Enderlein

Handbook of Fluorescence Spectroscopy and Imaging

From Ensemble to Single Molecules

2010

ISBN: 978-3-527-31669-4

C.F. Matta (Ed.)

Quantum Biochemistry

2 Volumes

2010

ISBN: 978-3-527-32322-7

H. Fukumura, M. Irie, Y. Iwasawa, H. Masuhara, K. Uosaki (Eds.)

Molecular Nano Dynamics

2 Volumes

2009

ISBN: 978-3-527-32017-2

G. Lanzani (Ed.)

Photophysics of Molecular Materials

From Single Molecules to Single Crystals

2006

ISBN: 978-3-527-40456-8

Single Particle Tracking and Single Molecule Energy Transfer

Edited by
Christoph Bräuchle, Don C. Lamb,
and Jens Michaelis

WILEY-VCH Verlag GmbH & Co. KGaA

The Editors

Prof. Dr. Christoph Bräuchle
Ludwig-Maximilians-Universität München
Department Chemie und Biochemie
Butenandtstrasse 11
81377 München
Germany

Prof. Dr. Don C. Lamb
Ludwig-Maximilians-Universität München
Department Chemie und Biochemie
Butenandtstrasse 11
81377 München
Germany

Prof. Dr. Jens Michaelis
Ludwig-Maximilians-Universität München
Department Chemie und Biochemie
Butenandtstrasse 11
81377 München
Germany

Library of Congress Card No.: applied for

British Library Cataloguing-in-Publication Data
A catalogue record for this book is available from the British Library.

Bibliographic information published by the Deutsche Nationalbibliothek
The Deutsche Nationalbibliothek lists this publication in the Deutsche Nationalbibliografie; detailed bibliographic data are available on the Internet at <http://dnb.d-nb.de>.

Printed in the Federal Republic of Germany
Printed on acid-free paper

Cover design Formgeber, Eppelheim
Typesetting Toppan Best-set Premedia Limited, Hong Kong
Printing betz-druck GmbH, Darmstadt
Bookbinding Litges & Dopf Buchbinderei GmbH, Heppenheim

ISBN: 978-3-527-32296-1

Contents

Single Particle Tracking and Single Molecule Energy Transfer
Edited by Christoph Bräuchle, Don C. Lamb, and Jens Michaelis

ISBN: 978-3-527-32296-1

Preface

Approximately 20 years ago, the first single fluorophores were detected using low-temperature fluorescence spectroscopy. Nowadays, it is possible not only to detect single particles and fluorophores at room temperature on a routine basis, but also to perform experiments with single molecules. Just as atomic theory has altered our way of thinking about matter, single-molecule methods are beginning to change the way in which we experimentally investigate the natural world. We are no longer limited to ensemble averaging, but can easily detect subpopulations, rare events and measure the dynamics of individual biomolecules, making synchronization no longer necessary for investigating the kinetics of dynamic processes.

Among the areas being impacted by single-molecule methods, the influence on the biological sciences is perhaps the most apparent. For example, the kinetics of individual enzymes has been followed, conformational changes in proteins have been characterized using single-pair FRET, and the entry pathway of single viruses into live cells has been elucidated. Yet, the application of single-molecule methods to the nanosciences is equally important, as these techniques are well suited to measuring interactions and dynamics on the nanoscale. One excellent example of this is the investigation of the dynamics of catalysis.

In this volume, we highlight three realms where single-molecule methods are making a major impact, namely single-particle imaging and tracking, energy transfer on the nanoscale, and single molecules in nanosystems. To help in assembling such an overview, we have accumulated contributions from a number of expert research groups, with each chapter incorporating a review of the application of single-molecule methods in the bio- and nanosciences. While preparing these chapters, the authors had to make difficult decisions regarding which scientific works to include; consequently, it was impossible to cite – much less highlight – all of those contributions in the field of single molecules that are worthy of recognition. Regardless of whether you are an expert in the field of single molecules, a novice entering the field, or simply have a casual interest in current research topics, we hope that you benefit from the overview of single-molecule methods presented in this volume, and wish you as much enjoyment in the reading of this book as we had during its preparation.

Single Particle Tracking and Single Molecule Energy Transfer
Edited by Christoph Bräuchle, Don C. Lamb, and Jens Michaelis

ISBN: 978-3-527-32296-1

We kindly acknowledge the many people who contributed to the work included in this volume. First, we thank the current and former members of our research groups, as well as our coauthors and colleagues who were responsible for obtaining the results reviewed in the book. We gratefully acknowledge the contributions of many people at Wiley-VCH for ensuring that the publication was performed smoothly and rapidly. Finally, we give a special thanks to Ms Silke Steger, who has greatly assisted in the administration and organization of the entire project.

Munich, September 2009

Christoph Bräuchle
Don C. Lamb
Jens Michaelis

List of Contributors

Paul Alivisatos
University of California Berkeley
Department of Chemistry
B-62 Hildebrand Hall
Berkeley, CA 94720-1460
USA

Christoph Bräuchle
Ludwig-Maximilians-Universität
München
Department of Chemistry and
Biochemistry
Center for Nanoscience (CeNS)
Butenandtstrasse 11
81377 München
Germany

Gert De Cremer
Katholieke Universiteit Leuven
Department of Microbial and
Molecular Systems
Kasteelpark Arenberg 23
3001 Heverlee
Belgium

Maxime Dahan
Université Pierre et Marie Curie,
Paris 6
Laboratoire Kastler Brossel
CNRS UMR 8552
Physics and Biology Department
Ecole Normale Supérieure
46 rue d'Ulm
75005 Paris
France

Enrico Gratton
University of California at Irvine
Laboratory for Fluorescence Dynamics
3210 Natural Sciences II
Irvine, CA 92697-2715
USA

Jeffery A. Hanson
Princeton University
Department of Chemistry
Princeton, NJ 08544
USA

Johan Hofkens
Katholieke Universiteit Leuven
Department of Chemistry
Celestijnenlaan 200F
3001 Heverlee
Belgium

Single Particle Tracking and Single Molecule Energy Transfer
Edited by Christoph Bräuchle, Don C. Lamb, and Jens Michaelis

ISBN: 978-3-527-32296-1

Seamus J. Holden
University of Oxford
Department of Physics and IRC
in Bionanotechnology
Clarendon Laboratory
Parks Road
Oxford, OX1 3PU
UK

Laurent Holtzer
Leiden University
Physics of Life Processes
Leiden Institute of Physics
Niels-Bohr-Weg 2
2333 CA Leiden
The Netherlands

Christophe Jung
Ludwig-Maximilians-Universität
München
Department of Chemistry and
Biochemistry
Center for Nanoscience (CeNS)
Butenandtstrasse 11
81377 München
Germany

Achillefs N. Kapanidis
University of Oxford
Department of Physics and IRC
in Bionanotechnology
Clarendon Laboratory
Parks Road
Oxford, OX1 3PU
UK

Ulrich Kubitscheck
Rheinische Friedrich-Wilhelms
University Bonn
Institute for Physical and Theoretical
Chemistry
Department of Biophysical Chemistry
Wegelerstrasse 12
53115 Bonn
Germany

Don C. Lamb
Ludwig-Maximilians-Universität
München
Department of Chemistry and
Biochemistry
Center for Nanoscience (CeNS) and
Munich Center for Integrated Protein
Science (CiPSM)
Butenandtstrasse 11
81377 München
Germany
and
University of Illinois
Department of Physics
Urbana, IL 61801
USA

Valeria Levi
University of Buenos Aires
Departments of Physics and
Biological Chemistry
Intendente Guiraldes 2160
Pabellón I – Ciudad Universitaria
Buenos Aires C1428EHA
Argentina

John M. Lupton
University of Utah
Department of Physics
115 South 1400 East
Salt Lake City, UT 84112-0830
USA

Jens Michaelis
Ludwig-Maximilians-Universität
München
Department of Chemistry and
Biochemistry
Butenandtstrasse 11
81377 München
Germany

Wolfgang J. Parak
Philipps Universität Marburg
Fachbereich Physik
Renthof 7
35037 Marburg
Germany

Jörg Ritter
Rheinische Friedrich-Wilhelms
University Bonn
Institute for Physical and
Theoretical Chemistry
Department of Biophysical
Chemistry
Wegelerstrasse 12
53115 Bonn
Germany

Maarten B.J. Roeffaers
Katholieke Universiteit Leuven
Department of Chemistry
Celestijnenlaan 200F
3001 Heverlee
Belgium
and
Katholieke Universiteit Leuven
Department of Microbial and
Molecular Systems
Kasteelpark Arenberg 23
3001 Heverlee
Belgium

Vahid Sandoghdar
ETH Zürich
Laboratorium für Physikalische
Chemie
Wolfgang-Pauli-Strasse 10
8093 Zürich
Switzerland

Thomas Schmidt
Leiden University
Physics of Life Processes
Leiden Institute of Physics
Niels-Bohr-Weg 2
2333 CA Leiden
The Netherlands

Bert F. Sels
Katholieke Universiteit Leuven
Department of Microbial and
Molecular Systems
Kasteelpark Arenberg 23
3001 Heverlee
Belgium

Jan-Peter Siebrasse
Rheinische Friedrich-Wilhelms
University Bonn
Institute for Physical and Theoretical
Chemistry
Department of Biophysical Chemistry
Wegelerstrasse 12
53115 Bonn
Germany

Yan-Wen Tan
Princeton University
Department of Chemistry
Princeton, NJ 08544
USA

Roman Veith
Rheinische Friedrich-Wilhelms
University Bonn
Institute for Physical and
Theoretical Chemistry
Department of Biophysical
Chemistry
Wegelerstrasse 12
53115 Bonn
Germany

Dirk E. De Vos
Katholieke Universiteit Leuven
Department of Microbial and
Molecular Systems
Kasteelpark Arenberg 23
3001 Heverlee
Belgium

Manfred J. Walter
University of Utah
Department of Physics
115 South 1400 East
Salt Lake City, UT 84112-0830
USA

Haw Yang
Princeton University
Department of Chemistry
Princeton, NJ 08544
USA

Part I
Single-Particle Imaging and Tracking

Single Particle Tracking and Single Molecule Energy Transfer
Edited by Christoph Bräuchle, Don C. Lamb, and Jens Michaelis

ISBN: 978-3-527-32296-1

1
Three-Dimensional Particle Tracking in a Laser Scanning Fluorescence Microscope

Valeria Levi and Enrico Gratton

1.1
Introduction

In recent years, single-particle and single-molecule techniques have each become essential tools in the fields of biophysics and cell biology [1]. One of the main reasons for the strong impact of these techniques is that they provide crucial information that is averaged out in traditional ensemble methods. Among these new techniques, single-particle tracking (SPT) has constituted a remarkable new tool for the study of dynamics in biological processes.

Today, several fluorescence microscopy techniques have been developed to measure the motion of molecules, the two most widely used being fluorescence recovery after photobleaching (FRAP), which was developed during the 1970s by Axelrod *et al.* [2], and fluorescence fluctuation spectroscopy (FFS), which was established during the same decade by Magde *et al.* [3]. Some of the most important characteristics of these techniques are represented schematically in Figure 1.1. While FRAP averages, in time and space, the behavior of a large ensemble of molecules, FCS averages the behavior of a small number of molecules within the observation volume. In both cases, the mobility properties determined in these experiments correspond to the average behavior of the observed molecules. Such averaging may, however, be problematic. For example, one problem may occur in the complex environment of the cell, where particles can interact with multiple targets and result in populations with different mobility properties. Moreover, the dynamics of each population may change in time and/or space. In such cases, both FRAP and FCS will provide only limited dynamical information.

Although SPT was first applied in biophysics during the 1980s and 1990s [16–19], the number of applications of these techniques has since grown significantly, based mainly on advances in microscopy and labeling techniques which have led to significant improvements in the accuracy and speed of these methods. Such advances have also presented the possibility to study more complex processes, with better spatial and temporal resolution.

Single Particle Tracking and Single Molecule Energy Transfer
Edited by Christoph Bräuchle, Don C. Lamb, and Jens Michaelis

ISBN: 978-3-527-32296-1

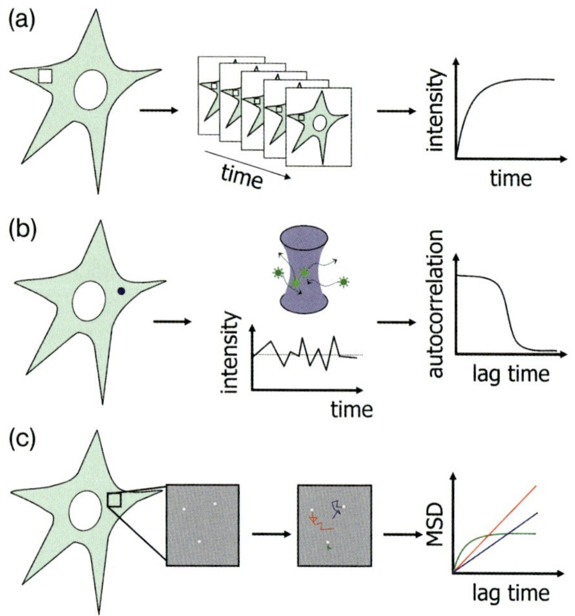

Figure 1.1 Studying dynamics with fluorescence microscopy. (a) Fluorescence recovery after photobleaching (FRAP) (for a review, see Ref. [4]) involves the selective photobleaching of fluorescent molecules in a given region (shown as a white square) of a sample with a high-intensity laser, followed by measurement of the recovery of fluorescence intensity in the bleached region as a function of time. The recovery kinetics depend on the speed at which the fluorescent particles move from other regions of the sample to the photobleached area, and provides information regarding the mobility of the labeled molecules; (b) Fluorescence fluctuation spectroscopy (FCS) and related techniques [5–7] measure fluctuations in fluorescence intensity in the femtoliter observation volume (blue) obtained in a confocal or two-photon microscope. These fluctuations are due to fluorescent molecules moving in and out of the observation volume (green stars), and their quantitative analysis through calculation of the autocorrelation function provides information regarding the average residence time, and therefore the mobility of the labeled molecules in that volume; (c) Single-particle tracking (SPT) determines the trajectories of single particles within the sample. These trajectories are analyzed to extract quantitative information regarding the motion mechanism of the particles [8]. The most common approach to analyze single-particle trajectories is to calculate the mean-square-displacement (MSD) as a function of the lag time. This parameter indicates how far a particle has traveled, and its dependence with τ is related to the motion properties of the particle. Consequently, a possible mechanism of motion can be obtained by comparing the experimental MSD plot with predictions from theoretical motion models [8–15].

These fluorescence microscopy techniques were developed to follow the position of individual particles in time. Provided that the spatial and temporal resolution of the method is adequate, these trajectories can be analyzed statistically to extract quantitative information regarding the mechanism involved in the motion of the particle (for recent reviews, see Refs [8, 20]). Since the properties of the particles are not averaged (as in bulk measurements), SPT represents an appealing technique to achieve the ultimate goal of understanding dynamics within cells.

In this chapter, we briefly describe the techniques that are used most often for tracking particles, and focus on recent advances in microscopy that have led to improvements in these methods. We discuss the different strategies employed to obtain information with regards to the axial position of the particle in image-based tracking approaches, and describe in detail a routine designed to achieve three-dimensional (3-D) tracking, using laser scanning microscopy. Finally, applications of the technique to the study of chromatin dynamics in interphase cells are demonstrated, the aim being to highlight possible applications for this new tracking procedure.

1.2 Image-Based Single-Particle Tracking Methods

The methods used most often to track fluorescent particles are based on recording images of the sample in a widefield or confocal fluorescence microscope as a function of time, and then locating the particle of interest in every recorded frame of the stack by using a specific algorithm (Figure 1.2).

In an optical microscope, a point-like particle forms a diffraction-limited image of width approximately equal to $\lambda/(2\text{NA})$, where λ is the wavelength of the light and NA is the numerical aperture of the objective. This diffraction limit implies that the image of the particle would have a diameter of ~200 nm for visible light. Thus, if particles are close to each other within this diffraction limit, it is not possible to determine their individual positions.

When the distances among particles exceed this limit, however, the position of each particle will correspond to the center of the intensity distribution of its image. In such cases, this position can be determined with high precision by using an algorithm that identifies the center of the distribution [21, 22]. For example, Yildiz *et al.* located molecular motors that had been labeled with single fluorophores, with an error of 1.5 nm, by fitting a Gaussian function to the intensity distribution of the fluorophore image [23, 24].

Recently, the present authors have designed a pattern recognition algorithm with 2 nm accuracy and 10 ms temporal resolution, which does not require the assumption of an intensity distribution function for the particle as it uses the particle's own intensity profile [25]. This approach improves the precision with respect to other techniques when tracking particles of finite size. Moreover, the algorithm makes it possible to correct for an inhomogeneous background, which is ideal for tracking experiments in living cells.

Importantly, these approaches only provide information regarding the two-dimensional (2-D) motion of particles at the focal plane. Those particles that move away from this plane will change their intensity profile until they completely disappear from the image. Thus, these techniques have been mainly used to study

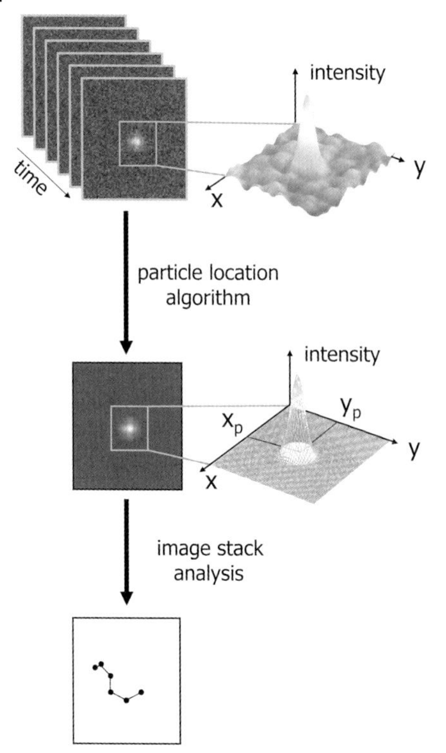

Figure 1.2 Image-based methods for single-particle tracking. The image of a fluorescent point-like particle observed through a fluorescence microscope shows an intensity distribution, the center of which corresponds to the position of the particle. This position (x_p, y_p) can be recovered by using a deconvolution algorithm which, for example, fits the experimental intensity of the image with a known distribution function allowing determination of the center of the image. This routine is repeated in every image of an image-stack acquired as a function of time to calculate the particle trajectory.

processes that occur in two dimensions, such as the diffusion of membrane components [26–28] and their transport along cytoskeleton filaments positioned within the focal plane [16, 23].

When using the SPT routine, it is important to analyze the factors that determine the precision of the particle position calculation. Thompson *et al.* [29] studied the theoretical sources of noise in the determination of particle position when fitting the image of a single fluorophore with a Gaussian distribution function and obtained the following expression:

$$\left\langle (\Delta x)^2 \right\rangle = \frac{s^2}{N} + \frac{a^2}{12N} + \frac{8\pi s^4 b^2}{a^2 N^2} \tag{1.1}$$

where Δx is the error in the particle position, s is the standard deviation of the point spread function (PSF), a is the pixel size, N is the number of photons collected, and b is the background noise, which includes fluorescence coming from the background and the noise of the detector.

The first term arises from the photon noise, which results from the fact that photon emission is a random process that follows a Poissonian distribution. Thus, this error represents the fluctuations in the number of photons collected in a given temporal window. The second term is the pixelation noise, and is due to the finite size of the pixels. This noise arises from the uncertainty as to where the photon arrived in the pixel, and thus increases the apparent size of the image spot. Thompson *et al.* [29] demonstrated that, in practice, the pixel size should be in the region of the standard deviation of the PSF. The final term is related to the background noise, and represents the error introduced on the position determination by photons coming from sources other than the particle. Common sources of background noise include readout error, dark current noise, and autofluorescence of the sample.

1.3 Advanced Fluorescence Microscopy Techniques for Single-Particle Tracking

One of the main problems encountered with SPT approaches based on standard epifluorescence microscopy is that excitation occurs throughout the entire depth of the specimen. Consequently, fluorescence derived from regions far away from the focus will increase the background intensity, an in turn decrease the effective signal-to-noise ratio (SNR) of single particles at the focal plane. These out-of focus regions are also unnecessarily photobleached and photodamaged. For single-particle tracking experiments, these factors will lead to reductions not only in the accuracy of the tracking but also in the effective duration of the tracking experiments.

Confocal microscopy (for a review, see Ref. [30]) constituted an improvement in widefield approaches due to its ability to reject out-of-focus light. In most commonly used confocal microscopes set-ups, the sample is excited by epi-illumination with a laser, while the objective collects the fluorescence emitted at the diffraction-limited volume. The emission passes through the dichroic mirror and an emission filter, is focused at an aligned confocal aperture or pinhole exactly in the primary image plane of the objective, and is then detected with a photomultiplier tube or avalanche photodiode detectors. Fluorescent light from above and below the primary image plane come to a focus elsewhere, and so must be defocused at the confocal aperture. Consequently, they will not reach the detector, which in turn will thus eliminate the blurring of images caused by out-of-focus light, as is observed in standard epifluorescence microscopes.

This simple set-up allows detection of fluorescence from a single point in the sample. To reconstruct an image, it is necessary to move either the sample or the excitation laser in such a way that the laser can scan a given region of interest. Generally, the excitation laser is raster-scanned by using two galvanometer scanning mirrors. In this conventional confocal set-up, the image acquisition frequency is normally less than 10 frames per second, because not only must the laser scan the sample, but it is also necessary to integrate the intensity signal during at least 10 μs per pixel. By contrast, the image acquisition time in widefield set-ups normally ranges from 1 to 30 ms [31–35], but may be as short as 25 μs [27]. Thus, conventional confocal microscopy is useful only for single-particle tracking studies of processes with very slow dynamics.

Modifications of this original set-up have been carried out for the observation of fast dynamics. In the *spinning-disk confocal microscope*, a rapidly rotating disk with an array of pinholes is used to generate an array of beams that are focused on the sample. The pinholes on the disk are arranged in a pattern that allows every location of an image to be covered when the disk is rotated. The fluorescence is collected through the same set of pinholes, and focused in a charge-coupled device (CCD) camera. Instead of scanning one point at a time, a spinning disk confocal microscope can measure the intensity of a thousand points of the sample simultaneously; consequently, the image acquisition frequency can be increased to approximately 300 frames per second (for a review, see Ref. [36]). Recent examples of the use of spinning disk confocal microscopy include the tracking of cytoplasmic and nuclear HIV-1 complexes [37], and the monitoring of mRNA transport [38].

Total internal reflection fluorescence (TIRF) microscopy represents an interesting alternative to standard epifluorescence microscopy for SPT experiments (for recent reviews, see Refs [39, 40]). In a TIRF microscope, excitation is achieved with an evanescent field generated by total reflection of the excitation laser. As this evanescent field decays exponentially with the distance normal to the surface, excitation of the fluorophores can only occur within 100 nm of the specimen surface. This presents a significant minimal reduction in out-of-focus background with respect to widefield and confocal approaches. Also, as no excitation occurs in regions far from the thin evanescent field, both photodamage and photobleaching are significantly lower than in standard fluorescence microscopy. Because of its strong dependence with the distance, SPT-TIRF microscopy has been used successfully to study the dynamics of biomolecules nearby, or inserted into, biological membranes [41], and the processes occurring at the coverslip surface [23, 42].

During the past decade, super-resolution, far-field microscopy techniques have successfully overcome the diffraction limit [43–50]. For example, stimulated emission depletion (STED) microscopy generates fluorescent focal spots that are smaller than what might be predicted according to the diffraction limit. In this type of microscope, the focal spot of an excitation laser is overlapped with doughnut-shaped spot of another laser of lower photon energy that quenches excited molecules in the excitation spot periphery by stimulated emission. As a result of the stimulated emission, the excitation volume is effectively reduced in size, such that resolution down to 20 nm can be achieved [43]. Imaging with these techniques usually takes longer than with conventional widefield techniques; consequently they are mainly applied to the study of fixed specimens [51–54]. However, in a recent study reported by the group of Hell, the STED-imaging of synaptic vesicles was demonstrated at video-rate in living neurons [55]. Clearly, a growing number of applications of these exciting new techniques for SPT studies in living cells is to be expected in future.

1.4 Two-Photon Excitation Microscopy

Although two-photon absorption was originally predicted by Maria Göppert-Mayer in 1931, this important principle could not actually be demonstrated until the development of high-intensity lasers [56]. In fact, several more years were to pass until two-photon excitation was first applied to laser scanning microscopy [57], some excellent reviews of which are available [58–61].

Two-photon excitation (TPE) is a nonlinear process that involves the almost simultaneous absorption of two photons ($\sim 10^{-16}$ s) that, in the simplest case, will have half of the energy required for transition to the excited state (Figure 1.3a). Two-photon excitation is typically achieved with near-infrared (NIR) excitation sources, because most common dyes used in fluorescence microscopy absorb in the ultraviolet (UV) or visible spectral region.

As the molecular cross-sections for two-photon absorption are very small, ($\sim 10^{-56}$ cm^4 s per photon; [59]), high photon fluxes are required to achieve significant levels of fluorophore excitation. For this reason, TPE requires high-power continuous-wave lasers or femtosecond-pulse lasers that provide pulses of high intensity during brief periods of time, but have a low average excitation power, which limits the heating of the sample.

The TPE probability is proportional to the square of the laser intensity, which implies that TPE will only occur at the focus of the laser, where the photon flux is highest [60]. Figure 1.3b shows the emission volume obtained in a solution of a fluorescent probe under two-photon or one-photon excitation. While two-photon excitation occurs only in the femtoliter volume in which the laser was focused, one-photon excitation is also observed in out-of-focus regions.

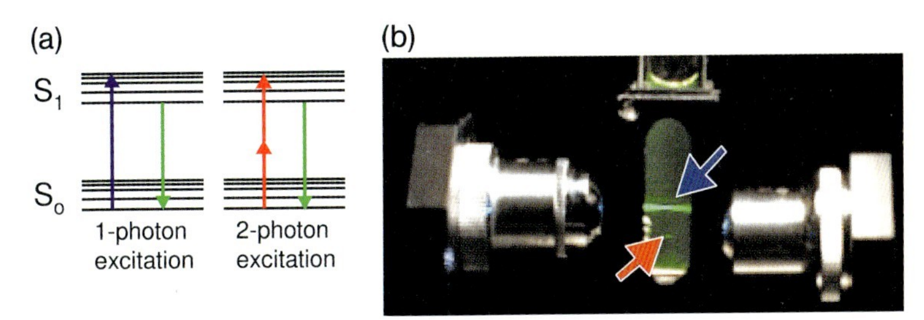

Figure 1.3 Two-photon excitation microscopy. (a) Energy diagrams showing the electronic transitions occurring during one-photon (blue) and two-photon (red) excitation; (b) Emission volumes obtained in a fluorescein solution excited under one-photon (blue arrow) and two-photon (red arrow) conditions. The numerical apertures of the objectives used in this experiment were the same. Reproduced with permission from Ref. [60]; © 2000, Annual Reviews (www.annualreviews.org).

This intrinsic, optical 3-D sectioning capability is the most appealing characteristic of TPE for fluorescence microscopy, as it is possible to perform 3-D imaging as in a confocal microscope, without the need for a confocal pinhole. Moreover, as most biological samples do not absorb significantly in the IR range, the out-of-focus regions will not contribute to the background and so will not be photodamaged.

The background fluorescence in two-photon microscopy is also reduced by the fact that the excitation is significantly red-shifted with respect to the emission. Thus, it is easier to separate emission light from the excitation beam than in a one-photon experiment, where they may be spectrally close together. Another advantage of two-photon microscopy is that it can be used for imaging thick specimens; this is because the NIR light employed for the excitation is less scattered and absorbed than the UV or visible excitation light required for the excitation of common fluorophores in one-photon microscopy.

The main drawback of TPE fluorescence microscopy is that photobleaching within the focal volume, with laser power levels typically used in biological imaging, is usually much more extensive than would be expected for one-photon excitation, and may be attributable to higher-order resonance absorption [59].

1.5
3-D Tracking in Image-Based SPT Approaches

The initial methods developed for 3-D tracking utilized confocal or two-photon microscopy to achieve a z-sectioning of the specimen, and consisted of collecting z-stacks as a function of time. The resultant stacks of images were analyzed using a deconvolution algorithm to locate the particle of interest with high precision [37, 62]. However, these methods required the sample to be scanned several times, which made the tracking very slow for most applications.

One interesting possibility of obtaining the z-position of a particle simply by analyzing the 2-D images was proposed by Speidel *et al.* [63]. This group made use of the fact that the image of a point-like particle changes when it is defocused. The spot diameter increases and, for large z-displacements from the focus, ring intensity patterns are formed. Thus, the distance of the object to the imaging focal plane is encoded in the intensity pattern. Speidel's group observed that the radius of the outermost ring was linearly related to the actual z-position, and used this relation to calculate the axial position of the particle (Figure 1.4). This method can be used to follow particles moving within a range of 3 μm from the focal plane, with an accuracy of approximately 1 nm and a time resolution of about 100 ms. However, the accuracy in the radial position was seen to decrease with the distance from the focal plane, as the intensity of the off-focus images was spread over a higher number of pixels.

In order to improve the radial position accuracy, different groups have used multifocal imaging methods in which two or more focal planes about the

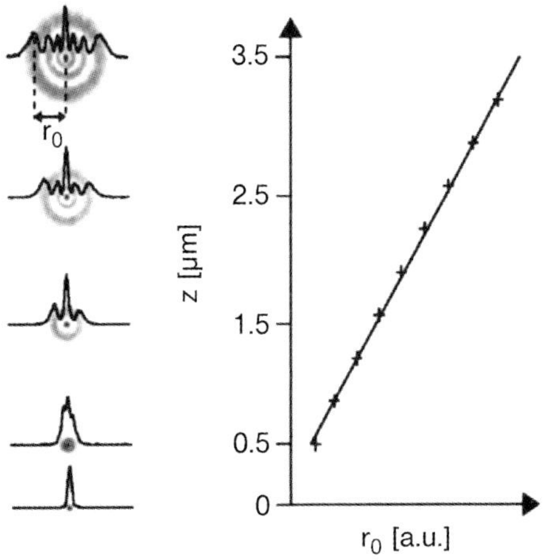

Figure 1.4 Three-dimensional defocusing-based SPT techniques. Images of a fluorescent bead in an agarose gel at different *z*-distances from the focal plane (left). The radius r_0 of the outermost ring in the images scaled linearly with *z* (right). Reproduced with permission from Ref. [63]; © Optical Society of America.

selected particles of interest were recorded, either with different cameras or on separate halves of a single CCD camera [64, 65]. The radial and axial positions of the particle which were obtained by analyzing the focused and defocused planes, respectively, could be determined with 2–5 nm accuracy and 2–50 ms time resolution [65].

A similar defocusing-based approach was previously followed by Kao *et al.* [66], who introduced a cylindrical lens into the detection optical path of a widefield fluorescence microscope, the aim being to generate axial astigmatism in the collected image. As a result, the images of point-like or spherical fluorescent particles were circular in focus but ellipsoidal above and below the focus, with the major axis of the ellipsoid shifted by 90° in going through the focus. Thus, the absolute z-position of the particle can be determined from the image shape and orientation. Holtzer *et al.* [67] further improved this method to track single quantum dots (QDs) in cells, and attained 6 and 30 nm accuracy in the lateral and axial position of the dots, respectively. Hence, this tracking routine appears to function for particles within 1 μm of the focal plane.

Recently, Saffarian *et al.* [68] established a new image-based 3-D tracking method that combines widefield and TIRF microscopy, based on the earlier studies of Sarkar *et al.* [69]. These authors determined, experimentally, the distance-dependence of the evanescent field and related the emitted photons of a single particle with its position in this field, achieving an axial accuracy of 10 nm.

1.6 3-D Tracking in Laser Scanning Microscopes

Previously, it was noted that image-based 2-D tracking methods for confocal and TPE microscopy had very low temporal resolution, mainly because it was necessary to scan the laser through the region of interest. Moreover, 3-D tracking required the acquisition of a z-stack of images for each tracking time, which made the procedure even slower.

A completely different approach for tracking particles in three dimensions, using a TPE microscope, was introduced by the present authors' group [70–72], based on the earlier studies of Enderlein [73].

In this approach, the laser is scanned in circles of diameter equal to the PSF radial waist surrounding the particle of interest. The fluorescence intensity is integrated at given points of these orbits as the laser moves around the particle. The scanner performs a given number of orbits, n, which are averaged to improve the SNR. When the laser has finished these orbits, a z-nanopositioner moves the objective to a different z-plane, where the orbits are repeated. The two z-planes are separated from each other by the z-waist of the PSF. The particle position is recovered from the intensity data obtained in a cycle of scanning (e.g., n orbits in the two different z-planes), as described below.

Kis-Petikova and Gratton [70] and Berland *et al.* [74] have demonstrated that the fluorescence intensity (F) during scanning is a periodic function of time (t):

$$F(t)=\frac{2F_0/\pi}{1+\frac{\lambda^2(z_p-z_s(t))^2}{w_0^4\pi^2}}\exp\left[-\frac{2\left[(x_p-x_s(t))^2+(y_p-y_s(t))^2\right]}{w_0^2+\frac{\lambda^2(z_p-z_s(t))^2}{w_0^2\pi^2}}\right]+B \qquad (1.2)$$

where w_0 is the beam waist, λ is the wavelength, B is the background intensity, and F_0 is a constant related to the peak fluorescence intensity. The subscripts p and s refer to the particle and the scanner coordinates, respectively. According to the description above, the scanner coordinates vary as a function of time as follows:

$$\begin{aligned} x_s(t)&=r_{xy}\cos(2\pi f_{\text{orbit}}t)\\ y_s(t)&=r_{xy}\sin(2\pi f_{\text{orbit}}t)\\ z_s(t)&=\begin{cases} r_z & 0<\frac{f_{\text{orbit}}t}{n}<1,2<\frac{f_{\text{orbit}}t}{n}<3\ldots.\\ -r_z & 1<\frac{f_{\text{orbit}}t}{n}t<2,3<\frac{f_{\text{orbit}}t}{n}<4\ldots. \end{cases} \end{aligned} \qquad (1.3)$$

where r_{xy} is the xy-circular orbit radius, r_z is half the amplitude of the z-square wave, f_{orbit} is the frequency of the circular orbit, and n is the number of circular orbits made before changing the z position.

According to these equations, the absolute position of the particle (x_p, y_p, z_p) is encoded in the fluorescence intensity registered during the tracking cycle. Figure 1.5a shows intensity profiles expected for particles situated at different positions

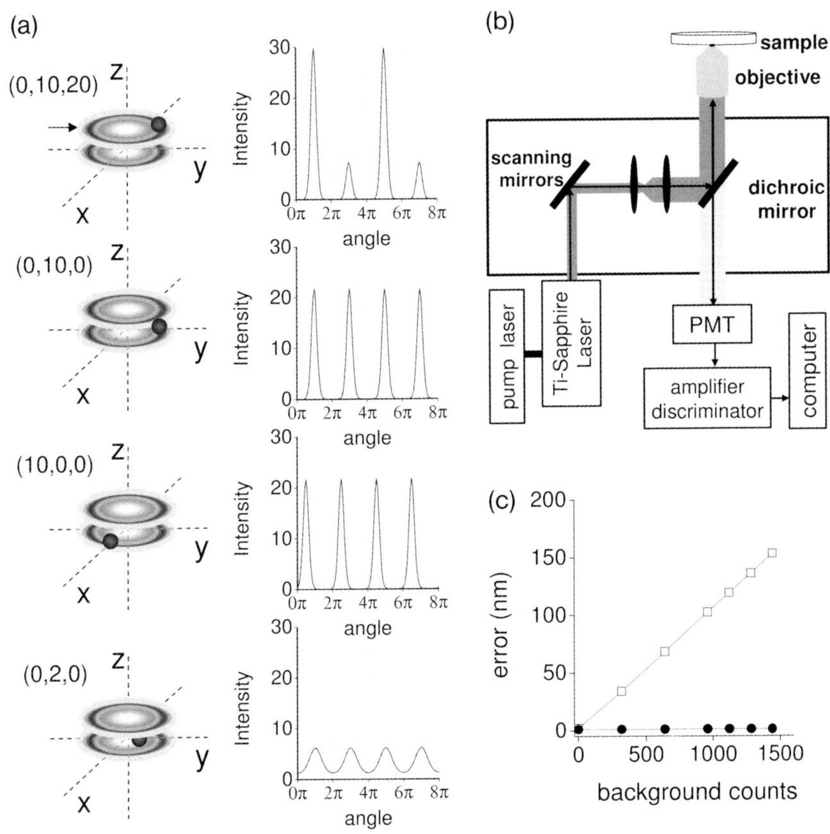

Figure 1.5 Tracking in a two-photon excitation (TPE) microscope. (a) The intensity profile determined along two cycles of the tracking routine is represented as function of the angle of rotation of the laser, for different relative positions of the particle with respect to the center of scanning (right panels). In the examples, each cycle of the tracking routine consisted in two orbits, each one at a different *z*-plane. The left panels show diagrams of the relative position of the particle (dark circle) with respect to the center of scanning. The laser orbits are represented as ovals. The arrow shows the starting point of the tracking cycle. The right panel shows the measured fluorescence intensity. (b) Schematic of the TPE microscope used for single-particle tracking experiments. PMT = photomultiplier tube. (c) The error for the orbiting-based routine (filled circles) was calculated as the standard deviation of the radial position recovered for a fixed particle after 1000 cycles of tracking in two dimensions. The background counts are the average number of counts measured during each cycle of the tracking routine in the absence of the particle. The background noise was assumed to have a Poisson distribution. For comparison, the error from the Gaussian-fitting routine (open squares) calculated from Equation 1.1 is shown. In both cases, it was assumed that the total number of photons detected from the particle was 12 000, either during one cycle of the circular-scanning routine or in one frame of the Gaussian-fitting routine. The pixel size and waist of the point spread function were 250 nm. (a) and (c) reproduced from Ref. [20] with permission from Springer Science+Business Media.

relative to the scanning center. Kis-Petikova and Gratton [70] demonstrated that the particle coordinates could be calculated from the phase and modulation obtained by fast Fourier transform (FFT) of the intensity signal measured during a tracking cycle. This approach is much faster than fitting Equation 1.2 to the intensity trace, and can be achieved "on-the-fly," which is essential for this tracking routine.

The procedure, as described thus far, does not allow tracking of a moving particle as the intensity will drop to zero when the particle moves away from the center of the scan. In order to continuously track a particle, its position must be determined on-the-fly (as described above); then, using a feedback loop, the center of scanning must be moved to the new position of the particle where a new cycle of the tracking routine starts. In other words, during the tracking routine, the scanner follows the particle by changing its position to that calculated for the particle in the previous cycle. In an ideal tracking experiment, the scanner is always on top of the particle, such that the positions of the scanner and the calculated position for the particle are identical.

By using this procedure it was possible to locate single particles with a precision of 20 nm [71] in approximately 16 ms when tracking in two dimensions, and in 32 ms when tracking in three dimensions. As the laser follows the particle in three dimensions, the x, y and z ranges in which the tracking routine works are only limited by the working range of the (x,y) scanning mirrors (see below) and the working distance of the objective, respectively.

A method related to the above-described orbiting method has been proposed by Lessard *et al.* [75]. Here, the experimental set-up is very similar to a confocal microscope, in that a one-photon excitation laser is focused on the sample, while fluorescence emitting from a particle located within the excitation volume is collected by four optical fibers. These fibers act as confocal pinholes, with each one coupled to an avalanche photodiode detector. The fiber faces are arranged in pairs such that their projection back into the sample space forms a tetrahedron. The 3-D position of the particle is recovered in 5 ms by analyzing the different signals collected by the detectors. Each pair of fibers is aligned to provide the position of the particle in the x and y axes; the two pairs of fibers are then offset along the optical axis, which enables location of the particle in the z direction. Based on the results of simulations, these authors reported an error on the particle position determination of approximately 200 nm. Tracking is achieved by using a feedback algorithm similar to that described before; the position of the particle is determined after a cycle of tracking, and the sample stage is then moved so as to bring the target closer to the center of the laser focus.

1.7
Instrumentation

In the two-photon microscope set-up used for the 3-D orbital tracking routine (see Figure 1.5b), a mode-locked titanium–sapphire laser is used as the excitation source. These lasers are ideal for two-photon microscopy because they provide femtosecond pulses with a repetition rate of ~100 MHz, and can be tuned in the range of 700–1000 nm so as to cover the excitation of common fluorescent probes, including most of the variants of fluorescent proteins. The laser is then directed into the microscope by two galvomotor-driven scanning mirrors controlled by the voltage generated from a computer card. When the mirrors are synchronized to move with sine waves shifted by 90° relative to one another, the laser beam moves in a circular path, with a radius determined by the amplitude of the sine waves. The position of the center of scanning is given by the offset values of the sine waves.

The laser light is reflected in a low-pass dichroic mirror and focused onto the sample by the objective. Fluorescence emission is collected by the objective and passed through the dichroic mirror and a short-pass filter to eliminate any reflected excitation light. The emission beam then exits the microscope to the photomultiplier detector with single-photon counting capability, and the photons are counted using a data acquisition card.

To enable changes of the focal plane, a piezoelectric z-nanopositioner equipped with a linear voltage differential transformer feedback sensor, and operated in a closed-loop configuration, is placed below the objective. During each cycle of the tracking routine the computer card generates a square-wave voltage which drives the motion of the z-nanopositioner between two z planes that are separated by a distance given by the amplitude of the square wave. The position of the center of z-scanning is given by the DC offset.

1.8 Background Noise

It is predicted by Equation 1.1 that the position of a particle obtained by Gaussian fitting depends sensitively on the intensity of the background. In Figure 1.5c, it is shown that the determination accuracy of the particle position decreases abruptly upon adding a background representing a small percentage of the total counts. Thus, extreme care must be taken when using this method, and its application should be restricted to cases in which the brightness of the particle is high. In contrast, the precision of particle position determination when using a tracking routine based on circular scanning is approximately constant with the background intensity in a wide range, due to the fact that the FFT of the intensity signal is not affected by a locally homogeneous background noise. Thus, the orbiting method described here has a major advantage with respect to image-based approaches when dealing with samples that have high background levels.

1.9 Simultaneous Two-Particle Tracking

The main disadvantage of the orbiting method is that it allows the tracking of only one particle at a time, although the routine may be modified to recover the position of two (or more) different particles successively. The tracking routine starts on top of one of the particles and, after one tracking cycle, the laser is moved to the position of the second particle to collect another cycle. When this second cycle is finished, the center of scanning is moved back to the position determined in the previous cycle for the first particle, and so on. In this way the positions of the particles are recovered alternately. This method does not require the particles to be at the same z-plane; rather, they can be followed even if they move several microns away from each other in the axial direction. The only requirement is that the particle is still in the vicinity where it was last seen when the algorithm returns. As the positions of the particles are recovered successively, the overall time resolution is cut in half; in other words, when tracking two particles, the time taken to obtain each particle position will be twice that required for the single-particle tracking mode.

The two-particle tracking option represents an interesting choice when conducting SPT tracking experiments in living cells, because the cells may migrate during the experiment and complicate any interpretation of the results. In some cases this problem can be corrected by using an internal reference – that is, a second particle located close to the studied particle. Then, if both particles are moving according to the same mechanism, the analysis of the temporal evolution of the distance between them will provide information regarding their motion [76]. Moreover, if one of the particles can be considered fixed, then the distance between them will provide information regarding the intrinsic motion of the particle [77].

1.10 Application: Chromatin Dynamics in Interphase Cells

The organization of chromatin in interphase nuclei has been the subject of much debate over the past decade [78], with several reports suggesting (e.g., [79–82]) that the chromosomes occupy well-defined volumes within the nucleus, termed "chromosomes territories." In fact, a significant amount of evidence has accumulated recently pointing towards a model in which a specific local reorganization of regions of these territories occurs in order to activate transcription [81, 83–86]. These antecedents show that the spatial organization of chromatin is intimately related to gene expression. However, the most important – and as yet unanswered – questions relate to how the chromatin organization is achieved, and how specific regions change their position within the nucleus.

A major breakthrough in this field was made with the development of a method whereby specific DNA sequences in live cells could be labeled by the insertion of lac operator repeats at specific chromosome locations. This locus could then be detected through binding of the lac repressor protein fused to green fluorescent protein (GFP) and engineered to contain a nuclear localization signal [87, 88]. The fluorescently labeled sequence could be identified as a bright dot in the nucleus, which itself would be dimly fluorescent due to the unbound EGFP-Lac repressor

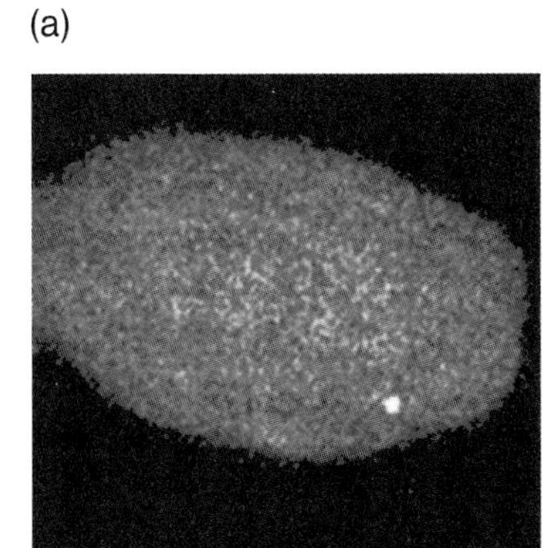

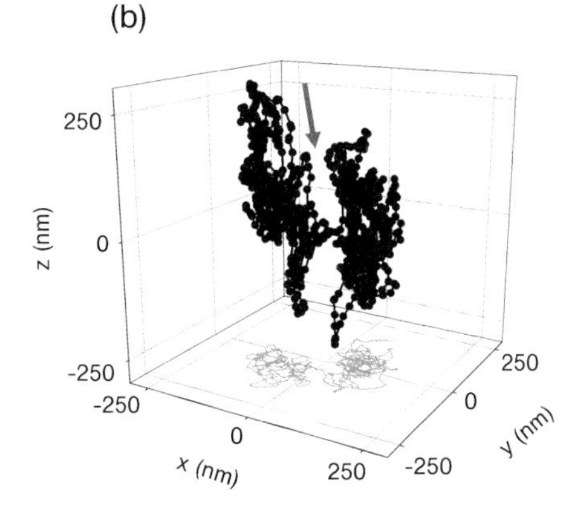

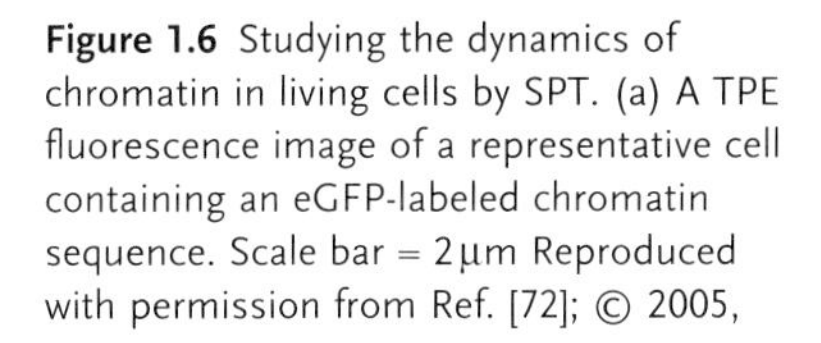

Figure 1.6 Studying the dynamics of chromatin in living cells by SPT. (a) A TPE fluorescence image of a representative cell containing an eGFP-labeled chromatin sequence. Scale bar = 2 μm Reproduced with permission from Ref. [72]; © 2005, Biophysical Society. (b) The trajectory was collected with a time resolution of 32 ms. To facilitate the visualization of a region of curvilinear motion (arrow), the (x,y) projection of the trajectory is also shown (gray).

(Figure 1.6a). The motion of this bright dot could then be followed, and the recovered trajectory analyzed to obtain information concerning the mechanism by which the tagged sequence had moved.

By using image-based tracking approaches, several groups have shown that, during interphase, the GFP-tagged chromatin loci undergo Brownian motion which is limited to a subregion of the nucleus [76, 89, 90]. This motion seemed to be more complex than constrained, passive diffusion, however. For example, Rosa *et al.* [91] showed that two chromosomal sites which exhibited a preferential association with the nuclear membrane were confined to regions of different size, with the site having higher levels of transcription exploring larger regions. Tumbar and Belmont [92] also showed that a specific DNA region in Chinese hamster ovary (CHO) cells could change its position, from the nucleus periphery to the center, in response to VP16 transcriptional activator.

One important factor that limits the temporal resolution and observation time window in studies of chromatin dynamics is that the motion of the labeled sequence is highly sensitive to photodamage [93, 94]. For example, Chuang *et al.* [94] showed that the motion of a tagged chromatin sequence was significantly altered after taking as few as 10 images when using a mercury lamp as an excitation source, under regular imaging conditions.

Subsequently, the dynamics of enhanced GFP (eGFP)-tagged chromatin sequences in interphase cells were reinspected by using the two-photon microscopy tracking technique described above [72]. As noted previously, TPE normally causes less out-of-focus photobleaching and photodamage than does one-photon excitation. In addition, the excitation laser moves within a very small volume of the cell during the tracking routine, and consequently does not introduce any damage in regions far from the tagged sequence. For these measurements, a typical spatial resolution of 20 nm was employed at a temporal resolution of 30 ms.

The trajectories obtained for the eGFP-tagged sequence showed long periods of confinement in regions where the size was similar to the 30 nm fiber of chromatin. This probably reflected local, thermal fluctuations, interspersed by short periods in which the sequence moved approximately 150 nm and followed a curvilinear path. An example of one such trajectory, obtained by tracking the sequence motion in three dimensions, is shown in Figure 1.6b.

The presence of these jumps in position was intriguing, since (to the present authors' knowledge) they had not been observed previously in any study relating to chromatin motion. Moreover, as the jumps occurred within a time range of 0.3–2 s, they could not be observed using a tracking method with a low temporal resolution. The characteristic jump distance of 150 nm was also similar to the spatial resolution of previously used methods.

The possibility of the jumps being due to motion of the nucleus was also eliminated by the use of a second cell line that presented multiple lac operator-repeats insertions, and also expressed the EGFP-lac repressor protein. Each of the insertions was visualized by fluorescence microscopy as a bright dot that was separated from neighboring dots by less than 1–2 μm. Whilst the trajectories of these labeled DNA sequences also presented jumps, they occurred independently from each other; this indicated that the jumps reflected local, short-distance motions of the chromatin sequence.

Consequently, a set of new, statistical tools was designed to obtain insight into the mechanism underlying the motion of the sequence during jumps. These analyses were complemented by experiments which tested the hypothesis derived from the statistical analysis. The statistical analyses were designed to compare the motion of the sequence during jumps with the predictions obtained by considering only a passive diffusion process. Ultimately, the results of all analyses indicated that, on average, the sequence moved fourfold faster than during the periods between jumps, and followed paths that were more rectilinear than might be predicted for random diffusion motion. Overall, these data suggested that an active process was responsible for transport of the sequence during short periods of time. Moreover, the hypothesis was supported by experiments which showed that no jumps occurred in the trajectories following ATP depletion. It appeared, therefore, that the jumps most likely reflected energy-dependent chromatin movements.

1.11 Conclusions

In recent years, SPT techniques have provided unique information with regards to the dynamics of processes in a wide variety of systems, most notably in the field of cell biology. The relevance of these new techniques is reflected by the fact that the number of reports in which SPT has been applied has grown exponentially with time.

The main reasons for such growth has been the development of new microscopy techniques and their application to SPT studies, as well as improvements in the technology required (e.g., CCD cameras) and the development of new tracking routines. Together, these contributions have led to significant advances in both the speed and spatial resolution of SPT.

Over the past few years, much effort has been expended in the design of SPT techniques capable of recovering the trajectories of particles in three dimensions. In this chapter, we have described several methods designed to track in three dimensions, and also introduced a new technique which is based on a two-photon microscopy set-up and offers several advantages with respect to the classical, image-based methods. The new method has high spatial and temporal resolution and, in contrast to image-based approaches, is able to follow the particle even when it moves several microns from its original position. The level of photodamage to the sample, which is introduced during tracking with the orbiting method, is substantially low as the laser is consistently on top of the tracked particle such that the remainder of the sample is not exposed to the excitation light. The tracking is also relatively insensitive to the background noise, which makes it suitable for applications that involve high levels of background intensity, for example, in living cells.

One important characteristic of the orbiting tracking method is that the trajectory is measured on-the-fly, in contrast to other approaches where the trajectory of the particle is recovered *a posteriori*. This major difference introduces a significant advantage, namely that as the laser focus is retained on top of the tracked particle, other spectroscopic parameters (e.g., fluorescence lifetime, emission spectrum, polarization, intensity dynamics) can be measured simultaneously with tracking of the particle [95]. For example, by measuring the fluorescence lifetime parallel to the tracking, the Förster resonance energy transfer (FRET) – and thus association events – can be precisely detected so as to provide information regarding the exact moment and position where the particle interacted with a given target labeled with a FRET acceptor fluorophore.

In a recent study, Hellriegel and Gratton [95] demonstrated the possibility of recording the fluorescence emission spectrum of different-colored fluorescent beads as they were followed in three dimensions using the orbiting-based method. This new capability was also applied to track small, fluorescently labeled protein assemblies in living cells.

This simultaneous multiparameter spectroscopy and 3-D tracking has opened a new window to exciting, new applications for SPT. In the future, tracking experiments will provide not only information with regards to the motion properties of the particle, but also information concerning the molecular parameters that may be used to construct a more complete view of dynamics in complex systems, such as the living cell.

Acknowledgments

These studies were supported in part by grants NIH-P41-RRO3155 and P50-GM076516 (to E.G.). V.L. is a member of CONICET, Argentina.

References

1 Weiss, S. (1999) Fluorescence spectroscopy of single biomolecules. *Science*, **283**, 1676–1683.
2 Axelrod, D., Koppel, D.E., Schlessinger, J., Elson, E., and Webb, W.W. (1976) Mobility measurement by analysis of fluorescence photobleaching recovery kinetics. *Biophys. J.*, **16**, 1055–1069.
3 Magde, D., Elson, E., and Webb, W.W. (1972) Thermodynamics fluctuations in a reactin system: measurement by fluorescence correlation spectroscopy. *Phys. Rev. Lett.*, **29**, 705–708.
4 Reits, E.A. and Neefjes, J.J. (2001) From fixed to FRAP: measuring protein mobility and activity in living cells. *Nat. Cell Biol.*, **3**, E145–E147.
5 Ruan, Q., Cheng, M.A., Levi, M., Gratton, E., and Mantulin, W.W. (2004) Spatial-temporal studies of membrane dynamics: scanning fluorescence correlation spectroscopy (SFCS). *Biophys. J.*, **87**, 1260–1267.
6 Digman, M.A., Brown, C.M., Sengupta, P., Wiseman, P.W., Horwitz, A.R., and Gratton, E. (2005) Measuring fast dynamics in solutions and cells with a laser scanning microscope. *Biophys. J.*, **89**, 1317–1327.
7 Hebert, B., Costantino, S., and Wiseman, P.W. (2005) Spatiotemporal image correlation spectroscopy (STICS) theory, verification, and application to protein velocity mapping in living CHO cells. *Biophys. J.*, **88**, 3601–3614.
8 Saxton, M.J. and Jacobson, K. (1997) Single-particle tracking: applications to membrane dynamics. *Annu. Rev. Biophys. Biomol. Struct.*, **26**, 373–399.
9 Saxton, M.J. (1994) Anomalous diffusion due to obstacles: a Monte Carlo study. *Biophys. J.*, **66**, 394–401.
10 Saxton, M.J. (1994) Single-particle tracking: models of directed transport. *Biophys. J.*, **67**, 2110–2119.
11 Saxton, M.J. (1997) Single-particle tracking: the distribution of diffusion coefficients. *Biophys. J.*, **72**, 1744–1753.
12 Saxton, M.J. (1996) Anomalous diffusion due to binding: a Monte Carlo study. *Biophys. J.*, **70**, 1250–1262.
13 Saxton, M.J. (1993) Lateral diffusion in an archipelago. Single-particle diffusion. *Biophys. J.*, **64**, 1766–1780.
14 Saxton, M.J. (1995) Single-particle tracking: effects of corrals. *Biophys. J.*, **69**, 389–398.
15 Qian, H., Sheetz, M.P., and Elson, E.L. (1991) Single particle tracking. Analysis of diffusion and flow in two-dimensional systems. *Biophys. J.*, **60**, 910–921.
16 Gelles, J., Schnapp, B.J., and Sheetz, M.P. (1988) Tracking kinesin-driven movements with nanometre-scale precision. *Nature*, **331**, 450–453.
17 Lee, G.M., Ishihara, A., and Jacobson, K. (1991) Direct observation of brownian motion of lipids in a membrane. *Proc. Natl Acad. Sci. USA*, **88**, 6274–6278.
18 Geerts, H., de Brabander, M., and Nuydens, R. (1991) Nanovid microscopy. *Nature*, **351**, 765–766.
19 Geerts, H., De Brabander, M., Nuydens, R., Geuens, S., Moeremans, M., De Mey, J., and Hollenbeck, P.J. (1987) Nanovid tracking: a new automatic method for the study of mobility in living cells based on colloidal gold and video microscopy. *Biophys. J.*, **52**, 775–782.
20 Levi, V. and Gratton, E. (2007) Exploring dynamics in living cells by tracking single particles. *Cell Biochem. Biophys.*, **48**, 1–15.
21 Carter, B., Shubeita, G., and Gross, S. (2005) Tracking single particles: a user-friendly quantitative evaluation. *Phys. Biol.*, **2**, 60–72.
22 Cheezum, M.K., Walker, W.F., and Guilford, W.H. (2001) Quantitative comparison of algorithms for tracking

single fluorescent particles. *Biophys. J.*, **81**, 2378–2388.

23 Yildiz, A., Forkey, J.N., McKinney, S.A., Ha, T., Goldman, Y.E., and Selvin, P.R. (2003) Myosin V walks hand-over-hand: single fluorophore imaging with 1.5-nm localization. *Science*, **300**, 2061–2065.

24 Yildiz, A., Tomishige, M., Vale, R.D., and Selvin, P.R. (2004) Kinesin walks hand-over-hand. *Science*, **303**, 676–678.

25 Levi, V., Serpinskaya, A.S., Gratton, E., and Gelfand, V. (2006) Organelle transport along microtubules in Xenopus melanophores: evidence for cooperation between multiple motors. *Biophys. J.*, **90**, 318–327.

26 Fujiwara, T., Ritchie, K., Murakoshi, H., Jacobson, K., and Kusumi, A. (2002) Phospholipids undergo hop diffusion in compartmentalized cell membrane. *J. Cell Biol.*, **157**, 1071–1081.

27 Kusumi, A., Nakada, C., Ritchie, K., Murase, K., Suzuki, K., Murakoshi, H., Kasai, R.S., Kondo, J., and Fujiwara, T. (2005) Paradigm shift of the plasma membrane concept from the two-dimensional continuum fluid to the partitioned fluid: high-speed single-molecule tracking of membrane molecules. *Annu. Rev. Biophys. Biomol. Struct.*, **34**, 351–378.

28 Murase, K., Fujiwara, T., Umemura, Y., Suzuki, K., Iino, R., Yamashita, H., Saito, M., Murakoshi, H., Ritchie, K., and Kusumi, A. (2004) Ultrafine membrane compartments for molecular diffusion as revealed by single molecule techniques. *Biophys. J.*, **86**, 4075–4093.

29 Thompson, R.E., Larson, D.R., and Webb, W.W. (2002) Precise nanometer localization analysis for individual fluorescent probes. *Biophys. J.*, **82**, 2775–2783.

30 Sheppard, C.J.R. and Shotton, D.M. (1997) *Confocal Laser Scanning Microscopy*, Springer-Verlag New York Inc., New York.

31 Levi, V., Gelfand, V.I., Serpinskaya, A.S., and Gratton, E. (2006) Melanosomes transported by myosin-V in Xenopus melanophores perform slow 35 nm steps. *Biophys. J.*, **90**, L7–L9.

32 Spring, K.R. (2003) Cameras for digital microscopy, in *Digital Microscopy: A Second Edition of Video Microscopy*, vol. 72 (eds G. Sluder and D.E. Wolf), Elsevier Academic Press, Amsterdam, pp. 87–132.

33 Oshiro, M. and Moomaw, B. (2003) Cooled vs. intensified vs. electron bombardment CCD cameras – applications and relative advantages, in *Digital Microscopy: A Second Edition of Video Microscopy*, vol. 72 (eds G. Sluder and D.E. Wolf), Elsevier Academic Press, Amsterdam, pp. 133–156.

34 Berland, K.M., Jacobson, K.A., Frenche, T., and Rajfur, Z. (2003) Electronic cameras for low-light microscopy, in *Digital Microscopy: A Second Edition of Video Microscopy*, vol. 72 (eds G. Sluder and D.E. Wolf), Elsevier Academic Press, Amsterdam, pp. 103–132.

35 Kural, C., Serpinskaya, A.S., Chou, Y., Goldman, R.D., Gelfand, V.I., and Selvin, P.R. (2007) Tracking melanosomes inside a cell to study molecular motors and their interaction. *Proc. Natl Acad. Sci. USA*, **104**, 5378–5382.

36 Graf, R., Rietdorf, J., and Zimmermann, T. (2005) Live cell spinning disk microscopy. *Adv. Biochem. Eng. Biotechnol.*, **95**, 57–75.

37 Arhel, N., Genovesio, A., Kim, K.A., Miko, S., Perret, E., Olivo-Marin, J.C., Shorte, S., and Charneau, P. (2006) Quantitative four-dimensional tracking of cytoplasmic and nuclear HIV-1 complexes. *Nat. Methods*, **3**, 817–824.

38 Lange, S., Katayama, Y., Schmid, M., Burkacky, O., Brauchle, C., Lamb, D.C., and Jansen, R. (2008) Simultaneous transport of different localized mRNA species revealed by live-cell imaging. *Traffic*, **9**, 1256–1267.

39 Wazawa, T. and Ueda, M. (2005) Total internal reflection fluorescence microscopy in single molecule nanobioscience. *Adv. Biochem. Eng. Biotechnol.*, **95**, 77–106.

40 Axelrod, D. (2001) Total internal reflection fluorescence microscopy in cell biology. *Traffic*, **2**, 764–774.

41 Sako, Y., Minoghchi, S., and Yanagida, T. (2000) Single-molecule imaging of EGFR signalling on the surface of living cells. *Nat. Cell Biol.*, **2**, 168–172.

42 Vale, R.D., Funatsu, T., Pierce, D.W., Romberg, L., Harada, Y., and Yanagida,

T. (1996) Direct observation of single kinesin molecules moving along microtubules. *Nature*, **380**, 451–453.

43 Hell, S.W. (2007) Far-field optical nanoscopy. *Science*, **316**, 1153–1158.

44 Gustafsson, M.G. (2005) Nonlinear structured-illumination microscopy: wide-field fluorescence imaging with theoretically unlimited resolution. *Proc. Natl Acad. Sci. USA*, **102**, 13081–13086.

45 Rust, M.J., Bates, M., and Zhuang, X. (2006) Sub-diffraction-limit imaging by stochastic optical reconstruction microscopy (STORM). *Nat. Methods*, **3**, 793–795.

46 Betzig, E., Patterson, G.H., Sougrat, R., Lindwasser, O.W., Olenych, S., Bonifacino, J.S., Davidson, M.W., Lippincott-Schwartz, J., and Hess, H.F. (2006) Imaging intracellular fluorescent proteins at nanometer resolution. *Science*, **313**, 1642–1645.

47 Hess, S.T., Girirajan, T.P., and Mason, M.D. (2006) Ultra-high resolution imaging by fluorescence photoactivation localization microscopy. *Biophys. J.*, **91**, 4258–4272.

48 Sharonov, A. and Hochstrasser, R.M. (2006) Wide-field subdiffraction imaging by accumulated binding of diffusing probes. *Proc. Natl Acad. Sci. USA*, **103**, 18911–18916.

49 Egner, A., Geisler, C., von Middendorff, C., Bock, H., Wenzel, D., Medda, R., Andresen, M., Stiel, A.C., Jakobs, S., Eggeling, C., Schonle, A., and Hell, S.W. (2007) Fluorescence nanoscopy in whole cells by asynchronous localization of photoswitching emitters. *Biophys. J.*, **93**, 3285–3290.

50 Huang, B., Wang, W., Bates, M., and Zhuang, X. (2008) Three-dimensional super-resolution imaging by stochastic optical reconstruction microscopy. *Science*, **319**, 810–813.

51 Donnert, G., Keller, J., Medda, R., Andrei, M.A., Rizzoli, S.O., Luhrmann, R., Jahn, R., Eggeling, C., and Hell, S.W. (2006) Macromolecular-scale resolution in biological fluorescence microscopy. *Proc. Natl Acad. Sci. USA*, **103**, 11440–11445.

52 Sieber, J.J., Willig, K.I., Kutzner, C., Gerding-Reimers, C., Harke, B., Donnert, G., Rammner, B., Eggeling, C., Hell, S.W., Grubmuller, H., and Lang, T. (2007) Anatomy and dynamics of a supramolecular membrane protein cluster. *Science*, **317**, 1072–1076.

53 Kellner, R.R., Baier, C.J., Willig, K.I., Hell, S.W., and Barrantes, F.J. (2007) Nanoscale organization of nicotinic acetylcholine receptors revealed by stimulated emission depletion microscopy. *Neuroscience*, **144**, 135–143.

54 Kittel, R.J., Wichmann, C., Rasse, T.M., Fouquet, W., Schmidt, M., Schmid, A., Wagh, D.A., Pawlu, C., Kellner, R.R., Willig, K.I., Hell, S.W., Buchner, E., Heckmann, M., and Sigrist, S.J. (2006) Bruchpilot promotes active zone assembly, Ca^{2+} channel clustering, and vesicle release. *Science*, **312**, 1051–1054.

55 Westphal, V., Rizzoli, S.O., Lauterbach, M.A., Kamin, D., Jahn, R., and Hell, S.W. (2008) Video-rate far-field optical nanoscopy dissects synaptic vesicle movement. *Science*, **320**, 246–249.

56 Göppert-Mayer, M. (1931) Über Elementarakte mit zwei Quantensprüngen. *Ann. Phys.*, **9**, 273–295.

57 Denk, W., Strickler, J.H., and Webb, W.W. (1990) Two-photon laser scanning fluorescence microscopy. *Science*, **248**, 73–76.

58 Diaspro, A., Bianchini, P., Vicidomini, G., Faretta, M., Ramoino, P., and Usai, C. (2006) Multi-photon excitation microscopy. *Biomed. Eng. Online*, **5**, 36.

59 Diaspro, A., Chirico, G., and Collini, M. (2005) Two-photon fluorescence excitation and related techniques in biological microscopy. *Q. Rev. Biophys.*, **38**, 97–166.

60 So, P.T., Dong, C.Y., Masters, B.R., and Berland, K.M. (2000) Two-photon excitation fluorescence microscopy. *Annu. Rev. Biomed. Eng.*, **2**, 399–429.

61 Svoboda, K. and Yasuda, R. (2006) Principles of two-photon excitation microscopy and its applications to neuroscience. *Neuron*, **50**, 823–839.

62 Bornfleth, H., Edelmann, P., Zink, D., Cremer, T., and Cremer, C. (1999) Quantitative motion analysis of subchromosomal foci in living cells

using four-dimensional microscopy. *Biophys. J.*, **77**, 2871–2886.
63 Speidel, M., Jonas, A., and Florin, E.L. (2003) Three-dimensional tracking of fluorescent nanoparticles with subnanometer precision by use of off-focus imaging. *Opt. Lett.*, **28**, 69–71.
64 Prabhat, P., Gan, Z., Chao, J., Ram, S., Vaccaro, C., Gibbons, S., Ober, R.J., and Ward, E.S. (2007) Elucidation of intracellular recycling pathways leading to exocytosis of the Fc receptor, FcRn, by using multifocal plane microscopy. *Proc. Natl Acad. Sci. USA*, **104**, 5889–5894.
65 Toprak, E., Balci, H., Blehm, B.H., and Selvin, P.R. (2007) Three-dimensional particle tracking via bifocal imaging. *Nanoletters*, **7**, 2043–2045.
66 Kao, H.P. and Verkman, A.S. (1994) Tracking of single fluorescent particles in three dimensions: use of cylindrical optics to encode particle position. *Biophys. J.*, **67**, 1291–1300.
67 Holtzer, L., Meckel, T., and Schmidt, T. (2007) Nanometric three-dimensional tracking of individual quantum dots in cells. *Appl. Phys. Lett.*, **90**, 053902, 1–3.
68 Saffarian, S. and Kirchhausen, T. (2008) Differential evanescence nanometry: live-cell fluorescence measurements with 10-nm axial resolution on the plasma membrane. *Biophys. J.*, **94**, 2333–2342.
69 Sarkar, A., Robertson, R.B., and Fernandez, J.M. (2004) Simultaneous atomic force microscope and fluorescence measurements of protein unfolding using a calibrated evanescent wave. *Proc. Natl.Acad. Sci. USA*, **101**, 12882–12886.
70 Kis-Petikova, K. and Gratton, E. (2004) Distance measurement by circular scanning of the excitation beam in the two-photon microscope. *Microsc. Res. Tech.*, **63**, 34–49.
71 Levi, V., Ruan, Q., and Gratton, E. (2005) 3-D particle tracking in a two-photon microscope: application to the study of molecular dynamics in cells. *Biophys. J.*, **88**, 2919–2928.
72 Levi, V., Ruan, Q., Plutz, M., Belmont, A.S., and Gratton, E. (2005) Chromatin dynamics in interphase cells revealed by tracking in a two-photon excitation microscope. *Biophys. J.*, **89**, 4275–4285.
73 Enderlein, J. (2000) Tracking of fluorescent molecules diffusing within membranes. *Appl. Phys. B*, **71**, 773–777.
74 Berland, K.M., So, P.T., and Gratton, E. (1995) Two-photon fluorescence correlation spectroscopy: method and application to the intracellular environment. *Biophys. J.*, **68**, 694–701.
75 Lessard, G., Goodwin, P.M., and Werner, J.H. (2007) Three-dimensional tracking of individual quantum dots. *Appl. Phys. Lett.*, **91**, 224106, 1–3, 6.
76 Vazquez, J., Belmont, A.S., and Sedat, J.W. (2001) Multiple regimes of constrained chromosome motion are regulated in the interphase *Drosophila* nucleus. *Curr. Biol.*, **11**, 1227–1239.
77 Heun, P., Laroche, T., Shimada, K., Furrer, P., and Gasser, S.M. (2001) Chromosome dynamics in the yeast interphase nucleus. *Science*, **294**, 2181–2186.
78 Heard, E. and Bickmore, W. (2007) The ins and outs of gene regulation and chromosome territory organisation. *Curr. Opin. Cell Biol.*, **19**, 311–316.
79 Lichter, P., Cremer, T., Borden, J., Manuelidis, L., and Ward, D.C. (1988) Delineation of individual human chromosomes in metaphase and interphase cells by in situ suppression hybridization using recombinant DNA libraries. *Hum. Genet.*, **80**, 224–234.
80 Cremer, T., Cremer, C., Schneider, T., Baumann, H., Hens, L., and Kirsch-Volders, M. (1982) Analysis of chromosome positions in the interphase nucleus of Chinese hamster cells by laser-UV-microirradiation experiments. *Hum. Genet.*, **62**, 201–209.
81 Branco, M.R. and Pombo, A. (2006) Intermingling of chromosome territories in interphase suggests role in translocations and transcription-dependent associations. *PLoS Biol.*, **4**, e138.
82 Cremer, T. and Cremer, C. (2006) Rise, fall and resurrection of chromosome territories: a historical perspective. Part I. The rise of chromosome territories. *Eur. J. Histochem.*, **50**, 161–176.
83 Branco, M.R. and Pombo, A. (2007) Chromosome organization: new facts,

new models. *Trends Cell Biol.*, **17**, 127–134.

84 Kurz, A., Lampel, S., Nickolenko, J.E., Bradl, J., Benner, A., Zirbel, R.M., Cremer, T., and Lichter, P. (1996) Active and inactive genes localize preferentially in the periphery of chromosome territories. *J. Cell Biol.*, **135**, 1195–1205.

85 Chambeyron, S., Da Silva, N.R., Lawson, K.A., and Bickmore, W.A. (2005) Nuclear re-organisation of the Hoxb complex during mouse embryonic development. *Development*, **132**, 2215–2223.

86 Volpi, E.V., Chevret, E., Jones, T., Vatcheva, R., Williamson, J., Beck, S., Campbell, R.D., Goldsworthy, M., Powis, S.H., Ragoussis, J., Trowsdale, J., and Sheer, D. (2000) Large-scale chromatin organization of the major histocompatibility complex and other regions of human chromosome 6 and its response to interferon in interphase nuclei. *J. Cell Sci.*, **113** (Pt 9), 1565–1576.

87 Robinett, C.C., Straight, A., Li, G., Willhelm, C., Sudlow, G., Murray, A., and Belmont, A.S. (1996) *In vivo* localization of DNA sequences and visualization of large-scale chromatin organization using lac operator/repressor recognition. *J. Cell Biol.*, **135**, 1685–1700.

88 Belmont, A.S., Li, G., Sudlow, G., and Robinett, C. (1999) Visualization of large-scale chromatin structure and dynamics using the lac operator/lac repressor reporter system. *Methods Cell Biol.*, **58**, 203–222.

89 Marshall, W.F., Straight, A., Marko, J.F., Swedlow, J., Dernburg, A., Belmont, A., Murray, A.W., Agard, D.A., and Sedat, J.W. (1997) Interphase chromosomes undergo constrained diffusional motion in living cells. *Curr. Biol.*, **7**, 930–939.

90 Chubb, J.R., Boyle, S., Perry, P., and Bickmore, W.A. (2002) Chromatin motion is constrained by association with nuclear compartments in human cells. *Curr. Biol.*, **12**, 439–445.

91 Rosa, A., Maddocks, J.H., Neumann, F.R., Gasser, S.M., and Stasiak, A. (2006) Measuring limits of telomere movement on nuclear envelope. *Biophys. J.*, **90**, L24–L26.

92 Tumbar, T. and Belmont, A.S. (2001) Interphase movements of a DNA chromosome region modulated by VP16 transcriptional activator. *Nat. Cell Biol.*, **3**, 134–139.

93 Hediger, F., Taddei, A., Neumann, F.R., and Gasser, S.M. (2004) Methods for visualizing chromatin dynamics in living yeast. *Methods Enzymol.*, **375**, 345–365.

94 Chuang, C.H., Carpenter, A.E., Fuchsova, B., Johnson, T., de Lanerolle, P., and Belmont, A.S. (2006) Long-range directional movement of an interphase chromosome site. *Curr. Biol.*, **16**, 825–831.

95 Hellriegel, C. and Gratton, E. (2009) Real-time multi-parameter spectroscopy and localization in three-dimensional single-particle tracking. *J. R. Soc. Interface*, **6** (Suppl. 1), S3–S14.

2
The Tracking of Individual Molecules in Cells and Tissues

Laurent Holtzer and Thomas Schmidt

2.1
Introduction

For many years, microscopy has been the primary technique utilized in biological investigations. In particular, light microscopy, which allows the direct observation of biological processes *in vivo*, is used on an everyday basis in most biology laboratories. The major characteristic of a live system is the constant movement of all its components. The mobility of ions, of small molecules such as ligands, of proteins (whether membrane-bound [1–8] or located in the cytosol [9–11]), and of larger assemblies such as vesicles [12, 13], the nucleus [14] or viruses [15], provides the determining points of how a system will evolve and self-regulate. Hence, in the quest for an understanding of living systems on a microscopic mechanistic basis, the main desire is first to characterize the mobility of the system's components, and then to combine this knowledge with its functional state.

It follows, therefore, that the mobility of a system may be rather complex. The two main classes of mobility–that is, *unrestrictive diffusion* and *linear directed motion*–are rather the exception in the context of the complex environment of cells and tissues [16, 17]. Although initially, proteins might be able to diffuse freely through the cytosol, due to a series of binding events or restrictions in their diffusional space their mobility will become slowed over longer time and length scales. Likewise, a vesicle, which may be immobile during an initial phase of observation, will be actively transported at a later phase due to "molecular motors" that follow a microtubular track [18–21]. The recording and classification of such complex mobility behavior in a statistically significant manner requires a relevant and careful effort in technology development and automated analysis tools in order to achieve success.

In this chapter, we describe the foundations for the technology that has been developed to permit the tracking of individual molecules and small molecular assemblies. Attention will be mainly focused on applications in biomembranes and in cells and tissues, and illustrated by selected examples of how biological information can be extracted by a detailed analysis of molecular mobility.

Single Particle Tracking and Single Molecule Energy Transfer
Edited by Christoph Bräuchle, Don C. Lamb, and Jens Michaelis

ISBN: 978-3-527-32296-1

2.2
Single-Molecule and Single-Particle Localization

The parallel (and hence rapid) data acquisition mode of wide-field fluorescence microscopy (WFFM) is the most appropriate technology for tracking moving molecules and objects. In WFFM, an isotropic emitter which is smaller than the diffraction limit will appear as a diffraction-limited spot in the image plane [22]. The image of the emitter is characterized by a symmetrical signal distribution around the center, with the maximum intensity at the center of the spot. The intensity distribution $I(x, y)$ of such an object on a highly-sensitive charge-coupled device (CCD) camera, as used in these experiments described, is determined by the point spread function (PSF). A good approximation of the PSF is given by a two-dimensional (2-D) Gaussian with a full-width-at-half-maximum (FWHM) equal to $w = 1.03\lambda/\text{NA}$ [1, 23, 24]:

$$I(x, y) = N\frac{4\ln 2}{\pi w^2}\exp\left[-4\ln 2\left(\frac{(x-\mu_x)^2}{w^2}+\frac{(y-\mu_y)^2}{w^2}\right)\right] \tag{2.1}$$

where μ_x and μ_y are the x and y coordinates of the object, and N is the total number of detected photon-counts. At this point it should be noted that the positions are determined with nanometer precision, although the typical size of the generated image, w, is larger than the object [25]. This fact, which is referred to as "super-resolution," will be described in detail later in the chapter.

The identification of individual molecules is complicated by unavoidable background signals in living cells due to out-of-focus objects and autofluorescent particles. Therefore, image pre-processing and reliable background removal is necessary. Subsequently, it was found that such background signals could be satisfactorily removed by applying a spatial low-pass filter to the image with a cut-off frequency of $5/w$, which is far below those generated by the objects of interest. Subtraction of the filtered image from the original image reliably yields an image with a zero background. Likewise, static objects are faithfully removed using a temporal low-pass filter on the movie stack, and are subsequently subtracted from the original image. However, the latter method must be very carefully applied in order not to remove any slowly moving or static objects of interest.

The automatic object identification and position analysis is performed following any appropriate background subtraction. One simple and rapid way of determining the position of the object in the object plane is to calculate the center of mass, or centroid, of its image for each axis:

$$\mu_x = \sum_{i=1}^{M_y}\sum_{j=1}^{M_x}(x_i\cdot\mathbf{I}_{ij})\Big/\sum_{i=1}^{M_y}\sum_{j=1}^{M_x}\mathbf{I}_{ij} \tag{2.2}$$

where $\mathbf{I}_{ij}$ is the signal at a pixel (i, j) [26, 27]. It is important that the image has no offset, as this will bias the position of the particle towards the center. The advantage of this method is that it does not use any prior knowledge concerning the shape of the intensity profile, and can therefore by applied to objects even

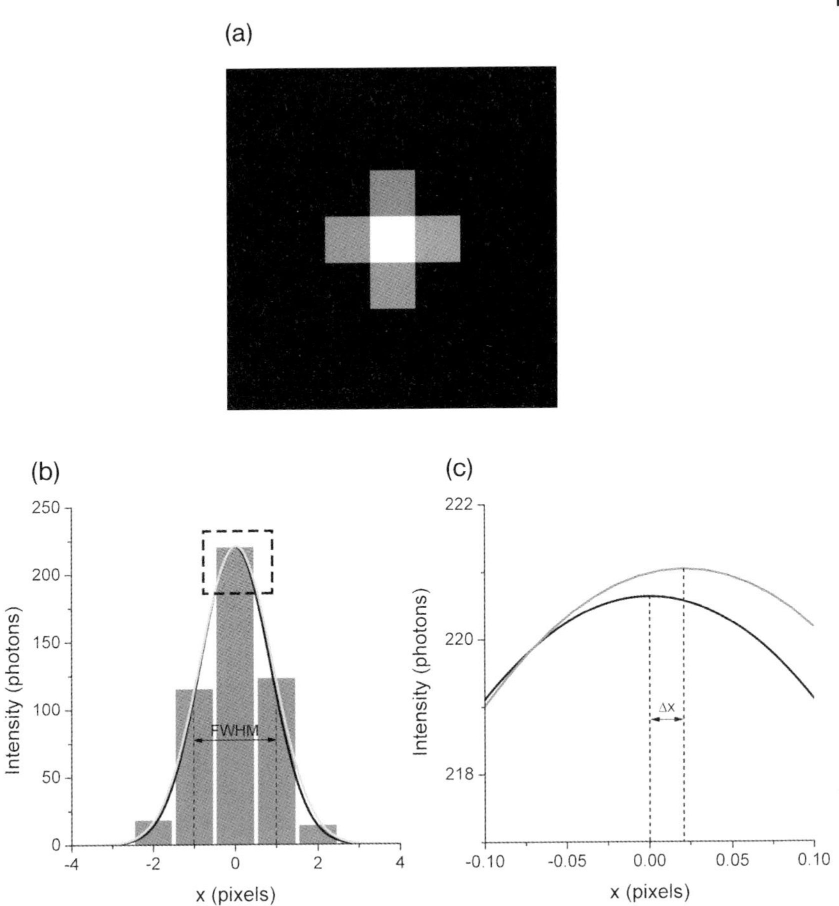

Figure 2.1 (a) Simulated image of a diffraction limited spot approximated by a 2-D Gaussian intensity profile. Poisson noise was added to account for the stochastic nature of photon emission ($w = 2$ pixels, $N = 1000$). (b) Intensity of the image along a slice through the center. In black, a 1-D Gaussian is shown calculated directly from the input parameters. A fit to the data is shown in gray. (c) An enlarged view of panel (b), as indicated by the square. It can be clearly seen that the Gaussian fit determines the position of the particle with high accuracy ($\Delta x = 0.02$ pixels).

in the case of imaging errors, or to objects which are larger than the diffraction limit [28].

The most reliable (but computationally more demanding) method of determining the position of an object is by fitting the image to the 2-D Gaussian intensity profile of the PSF, as presented in Equation 2.1 (see Figure 2.1a and b). A fit of the intensity distribution to Equation 2.1 determines the position of the object with nanometer precision (Figure 2.1c) [7, 25]. The accuracy is thereby inversely proportional to the signal-to-noise ratio (SNR), and is approximated by $w/\sqrt{8N\ln 2}$ [29]. This approximation assumes that additional noise due to background signals is negligible. In typical applications using autofluorescent proteins, a positional accuracy of <30 nm is achieved at video rate [6, 30].

The position determination, as described so far, solely allows the extraction of information with regards to the lateral position of an object. In recent years, several methods for determining the axial position have been described [31–39], the most straightforward and cost-effective of these being to determine the z-position of a single particle by introducing a slight astigmatism into the detection beampath [29, 40]. This can be achieved by placing a weak cylindrical lens ($f \approx 10$ m) into the infinity part of the optical path of the microscope, as shown in Figure 2.2a. The image that is formed becomes elongated in one direction, depending on positive or negative defocus. When the image is circular, the axial position is defined as $z = 0$. By fitting an adapted 2-D Gaussian:

$$I(x, y) = N\frac{4\ln 2}{\pi w_r^2}\exp\left[-4\ln 2\left(\frac{(x-\mu_x)^2}{w_r^2/\varepsilon^2} + \frac{(y-\mu_y)^2}{w_r^2\varepsilon^2}\right)\right] \quad (2.3)$$

to the image, the ellipticity $\varepsilon = \sqrt{\sigma_y/\sigma_x}$ and generalized FWHM $w_r = \sqrt{w_x^2 w_y^2}$ can be determined (Figure 2.2b). The z-position is easily calculated using: [29]

$$z(w_r, \varepsilon) = \begin{cases} \dfrac{z_r}{w_0}\sqrt{\dfrac{w_r^2}{\varepsilon^2} - w_0^2} - \gamma & \varepsilon < 1 \\ -\dfrac{z_r}{w_0}\sqrt{w_r^2\varepsilon^2 - w_0^2} + \gamma & \varepsilon > 1 \end{cases} \quad (2.4)$$

in which γ is the astigmatism parameter, which indicates the distance between the focus in x and the focus in y, and $z_r = 1.51\lambda/\text{NA}^2$ is the focal depth or Rayleigh range. The axial accuracy obtained by using this method is about 2.5-fold that of the lateral positional accuracy. Typically, <75 nm is achieved in live cell experiments.

The one-plane approach as described above allows determination of the axial positions within the Rayleigh-range of ~1 µm. However, for larger-image volumes, the simultaneous imaging of multiple planes onto one CCD chip must be employed. This is achieved by inserting a beamsplitter in front of the CCD, so as to create two light paths with different image distances [41, 42]. Another comparable method was developed in which multiple cameras were used, with each camera focusing on a different plane in the sample [43, 44]. While the latter method has the advantage that a larger volume can be imaged at a faster rate, it is very costly and much more complex software is needed to synchronize all the elements in the set-up.

In cases where image rates are less important, different planes can be imaged consecutively by moving a piezo-mounted objective in an axial direction. The ideal distance between the planes is given by the axial range of the astigmatism method of ~1 µm. Care must be taken that these stacks of images are recorded faster than the typical movement of the particle of interest, in order to avoid any movement of the particle during imaging. If this is not possible, then the difference in time must be taken into account during the data analysis. Whilst it is still possible to

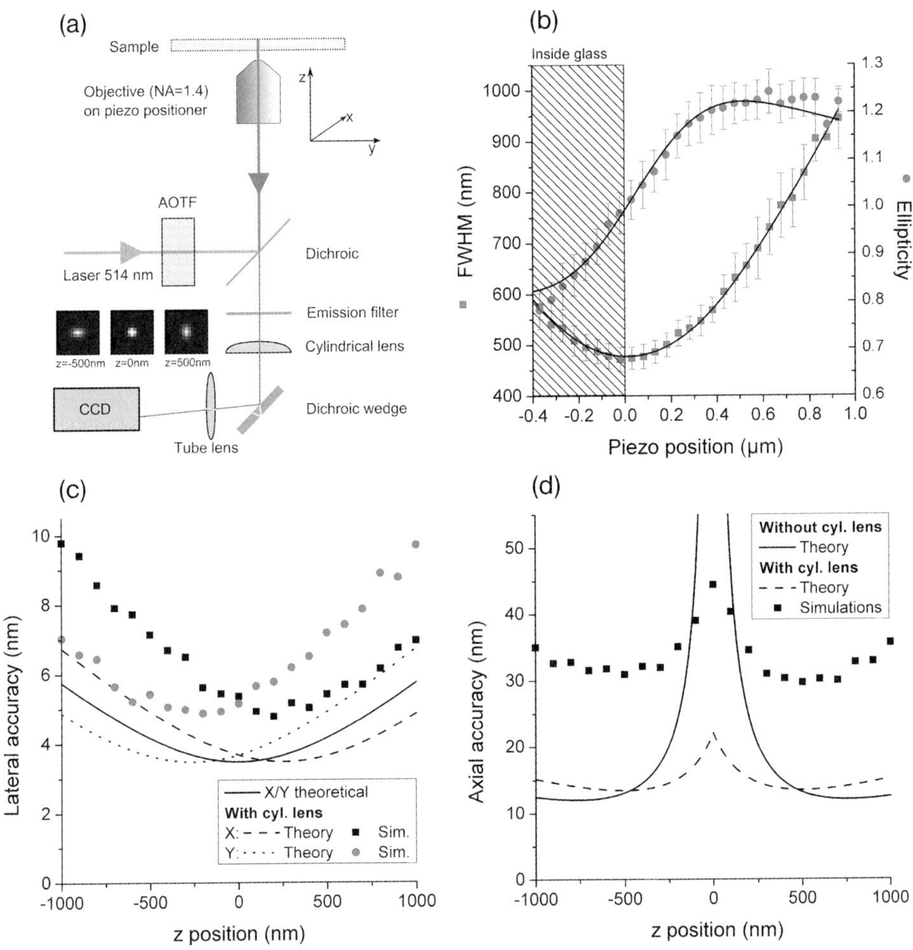

Figure 2.2 (a) A schematic diagram of a 3-D wide-field fluorescence set-up. An acousto-optic tunable filter (AOTF) is used to select the desired excitation wavelength. The dichroic mirror separates the emitted light from the excitation beam, and the emission filter selects for the emission wavelength of the fluorescent molecule. A cylindrical lens ($f = 10\,\text{m}$) is introduced to obtain accurate information about the z position. Finally, a dichroic wedge is installed to separate the two different fluorescent dyes. The three images shown at three different z-positions are of a particle that is imaged using this set-up and demonstrating the effect of the cylindrical lens. (b) w and ε for quantum dots immobilized on a glass substrate. Ten images containing nine QDs on average were taken per point. For $z < 0$, the data deviate from the fit because the focal plane lies inside the coverslip of a different index of refraction. (c, d) The positional accuracy in the lateral (x, y) (c) and in the axial (z) (d) direction for the detection of a fluorescing point object calculated according to the Cramer–Rao bound (lines) and compared to computer simulations (symbols). In the simulation, each point object emitted an average of 4000 photons per frame. Each data point represents an average of 1000 simulations.

fit 2-D Gaussian profiles to each image in a stack, a better alternative is to fit all images in a stack using a global fitting approach. For this, the 2-D Gaussian must be extended to three-dimensional (3-D), bearing in mind that the width of the Gaussian w_r varies with the axial position z (i.e., the inverse of Equation 2.4). While the total intensity of the Gaussian in each plane is constant, it transpires that an allowance should be made for a varying offset per image within the stack, in order to cope with any possible differences in spurious background signals. In focal planes far from the position of the particle, the intensity will be spread widely, effectively increasing the background signal. However, a variable offset can compensate for this effect.

The experimental conditions in single-molecule fluorescence experiments are normally chosen such that the concentration of the fluorescent molecules is low, and that only a few molecules are visible in an image of typical size ($10 \times 10\,\mu m^2$). For low densities, the distance between each molecule is sufficiently large ($>3\,w$) that their intensity profiles are independent. However, if such low densities are not achievable, then a recursive fitting approach must be applied. When using this system, after the initial round of fitting all but one of the fitted molecules must be subtracted from the image; the one molecule that is left is then refitted without the influence of the others. Several of these recursive runs are required to obtain the correct position and intensity of all individual molecules. In this way, densities of up to one molecule per μm^2 can be reliably handled. The details of a similar method were recently reported [45].

The above-described methods may be further developed to allow for the simultaneous imaging of multiple detection channels to separate, for example, different colors or polarizations. Here, a dichroic-wedge in the emission beam path is used to generate two separate images on the CCD coding for two colors, and/or a Wollaston prism is placed in the infinity beam path to generate two images of perpendicular polarization [46]. Such techniques may be used to image two different types of particle simultaneously, by labeling each of the objects with different fluorescent dyes, the emission spectra of which are well separated. It should be noted that any aberrations introduced by placing the dichroic wedge or a Wollaston prism are very small compared to the positional accuracy of the system.

2.3 Positional Accuracy

The emission of photons is a statistical process. Hence, the more photons that are detected, the more accurate the position of the particle that is determined. The positional accuracy of an experimental set-up depends on many factors, including the camera noise, the amount of photons detected per particle, the localization method used, and the magnification of the set-up. A general method used to calculate the error in a position measurement, when applied to single-molecule imaging, has shown that the lateral positional accuracy in typical experiments is equal to 30 nm [25, 30].

One fundamental approach to specify the achievable position accuracy can be calculated from the amount of information obtained in a given dataset. This measure is termed the Cramer–Rao bound (CRB), and is specified by the inverse of the Fisher information matrix I [47, 48]. For X the observed data and θ the unknown parameters are: $I(\theta) = \mathrm{E}\left\{\left[\frac{\partial}{\partial\theta}\ln f(X;\theta)\right]^2 \middle| \theta\right\}$. The CRB yields a lower bound to the variance for any unbiased estimator–that is, in the case of imaging the precision by which the position of a single particle is determined.

As discussed above, the PSF is approximated by a 2-D Gaussian (Equation 2.1). However, if it is assumed that the camera-pixelation and camera read-out-noise are negligible, then the lower limit for the positional error becomes:

$$\sigma_{\mu_x} = \frac{w_r/\varepsilon}{\sqrt{8N\ln 2}}; \quad \sigma_{\mu_y} = \frac{w_r\varepsilon}{\sqrt{8N\ln 2}} \tag{2.5}$$

in the lateral direction, and

$$\sigma_{\mu_z} = \frac{1}{\sqrt{N}}\left(\frac{\sqrt{5}z_r^2}{4(z\pm\gamma)} + \frac{\sqrt{5}}{4}(z\pm\gamma)\right) \quad \varepsilon \lessgtr 1 \tag{2.6}$$

in the vertical direction. Whereas, $\sigma_{\mu_{xy}}$ is independent of the lateral position of the object, σ_{μ_z} varies with z (Figure 2.2c and d), and is lowest in focus. For an experimental set-up without a cylindrical lens $\varepsilon = 1$, and therefore σ_{μ_x} and σ_{μ_y} are equal and σ_{μ_z} is undefined.

In order to calculate limit of the positional accuracy in an actual experiment, account must be taken of all those sources that may influence the image formed on the CCD [47]. This will include the camera pixelation, the position of the object relative to the center of a camera pixel, camera noise, the magnification of the set-up, and any other noise sources present. Furthermore, an Airy function should be used to describe the image formed by the object of interest, in place of the simple Gaussian as used in Equation 2.1. While such extended analytical calculations of the CRB have been performed for some cases [47, 48], the present authors have tested this strategy by means of extensive simulations in which all aspects mentioned were taken into account. The results showed an excellent overlap with the simplified approximation given in Equations 2.5 and 2.6 (Figure 2.2c,d).

The great accuracy with which individual molecules are localized has recently been utilized to provide a tremendous increase in the resolution of light microscopy. In methods now referred to as photo-activated localization microscopy (PALM) [49], fluorescence photo-activated localization microscopy (FPALM) [50], stochastic optical reconstruction microscopy (STORM) [51, 52] and stimulated emission depletion (STED) microscopy [53], the position of individual molecules may be determined and subsequently used to generate an image in which each molecule contributes with a PSF (according to Equation 2.1), but with a width given by the positional accuracy $w/\sqrt{8N\ln 2}$ in place of w. In this way, the "Abbe-limit," which describes the optical resolution (R) of the microscope, has been broken by an order of magnitude. In addition it has been realized recently that, in principle, there is no limit to the resolution of a microscope as this parameter is set solely by the signal obtained from an individual object:

$$R = 1.22\frac{\lambda}{2\mathrm{NA}}\frac{1}{\sqrt{N}} \tag{2.7}$$

2.4 Tracking

Although the process of obtaining the trajectories of sparsely distributed and relatively immobile objects is straightforward [54, 55], larger particle densities per frame and a higher mobility of the particles renders the connection of particles in consecutive images increasingly complex [26]. The computational effort required to solve such connectivity maps is equivalent to the well-known "traveling-salesman" problem in operations research. The novel tracking algorithms described here are based on a numerical approximation developed by Vogel [56] for the field of operations research. First, a translational matrix $p_i(j, k)$ is built up that describes the probabilities that particle j in image i (containing L objects) at position $\vec{r}_{j,i}$ moves to particle k in image $i + 1$ (containing M objects) at position $\vec{r}_{k,i+1}$ by diffusion in a d-dimensional system characterized by a diffusion constant D:

$$p_i(j,k) = \exp\left\{-\frac{(\vec{r}_{j,i} - \vec{r}_{k,i+1})^2}{2dDt}\right\} \tag{2.8}$$

The translational matrix further allows particles to disappear from the observed area by either diffusion or photobleaching, $p(j, k > L) = p_{\mathrm{bleach}}$, and particles are allowed to move into the observed area or to become reactivated, $p(j > M, k) = p_{\mathrm{activation}}$. Probabilities that account for particles that are accidentally not detected in an image are also included. Altogether, this leads to a probability matrix p of size $\{(L + M) \times (L + M)\}$. Trajectories are constructed which optimize for the total probability of all connections between two images, $max(\log(P) = \Sigma_{j,k}\log(p(j, k)))$. Even in the case of a sizable amount of molecules per image, the Vogel algorithm enhances the number of faithfully reconstructed trajectories. More elaborate algorithms have been developed for more complex systems with for example, a high particle density, particle motion heterogeneity, or particle splitting or merging [57–61].

For a reliable analysis of molecular mobility, unavoidable mechanical drift must be corrected for. A simple and efficient way to do this is to calculate the center-of-mass of all objects, given that a sufficient number of continuously tracked objects ($n > 10$) is available per frame. Such bootstrap-type correction algorithms are particularly suited in diffusion-governed systems, since all movements should average to zero and any deviation from zero directly measures the correction needed. In the case that there are not sufficient continuous trajectories available, significantly more molecular positions must be averaged in order to reduce the drift correction below the positional accuracy. For an image of size X, the number of objects must be larger than $(X/\sigma_{\mu_{xy}})^2$. In order to achieve a resolution of $\sigma_{\mu_{xy}} = 10\,\mathrm{nm}$ in a full-view image of $X = 10\,\mu\mathrm{m}$, 10^6 positions must be averaged.

2.5 Trajectory Analysis

A multitude of information is extracted from the trajectories of individual objects, ranging from the diffusion coefficient to the presence of multiple fractions of a certain type [62, 63]. A straightforward method to obtain information regarding the mobility of an object is to calculate its mean squared-displacement (MSD) versus time. The MSD is the average movement of an object within a certain amount of time, and is calculated for each object using

$$\mathrm{MSD} = \left\langle (\Delta r_t)^2 \right\rangle = \frac{\sum_{i=1}^{T-t} (r_i - r_{i+t})^2}{T - t} \tag{2.9}$$

where T is the total length of the trajectory. The type of motion of the object is subsequently extracted from the MSD-versus-time plot. For free diffusion, the MSD has a linear dependence on time

$$\mathrm{MSD} = 2dDt_{\mathrm{lag}} + 2\sum_{i \in S} \sigma_i^2 \quad S = \{x, y, z\} \tag{2.10}$$

where d is the dimension of the movement and σ_i is the positional accuracy in the ith dimension. When a particle is transported, for example, by molecular motors inside a cell [15], the MSD shows a supralinear dependence on time:

$$\mathrm{MSD} = 2dDt_{\mathrm{lag}} + (vt_{\mathrm{lag}})^2 + 2\sum_{i \in S} \sigma_i^2 \quad S = \{x, y, z\} \tag{2.11}$$

in which v is the velocity of the particle. A particle diffusing in a 2-D confined area of side length L will have an associated MSD which levels off for large t_{lag}:

$$\mathrm{MSD} = \frac{L^2}{3} \cdot \left[1 - \exp\left(\frac{-12 D_0 t_{\mathrm{lag}}}{L^2} \right) \right] + 4\sigma_{xy}^2 \tag{2.12}$$

where D_0 denotes the initial diffusion constant [8].

Hence, the MSD-versus-time behavior provides a global characterization of the type of motion. Often, however, this behavior is of transient nature, especially for example, in the transport of vesicles [3, 16, 64] or the motion of receptors in the cell membrane. A standard MSD-analysis will therefore fail to detect short periods of a certain type of motion within a trajectory [63]. The difficulty comes from the fact that the accuracy of a mean value for complex motion scales is inversely proportional to the square root of the number of independent observations, which in this case is the number of independent motion steps within a short part of a trajectory [65]. With a rigorous method, introduced by Huet *et al.* [66], different types of transient motion can be detected and distinguished within a single trajectory at a probability level prior set. For each type of motion a specific parameter is calculated along the trajectory using a rolling analysis window, the width of which is variable.

For a stalled particle the diffusion coefficient D will be close to the detection limit of the set-up. This limit can be experimentally determined using Equation 2.10 by measuring the diffusion coefficient D_{im} for immobilized beads on a coverslip at a SNR similar to the experiment. Particles which diffuse with a diffusion constant D which is 10-fold $D_{\min}$ are classified as mobile with high confidence. If, however, the D-value for a particle, calculated from a rolling window analysis, falls below $D_{\min}$ for a prior set period, it is classified as being stalled during this period. In order to reliably obtain D, it is desirable to calculate the MSD from as many data points as possible – that is, to use a large rolling window size. A linear fit to the first N_{diff} points of the MSD plot then gives a reliable D [65, 67]. On the other hand, in order to detect short periods of immobilization, the number of data points must be small. As a compromise, the minimum number of time points $W_{\min}^{\text{stall}}$ needed to calculate the MSD was set to 51, while keeping $N_{\text{diff}} = 5$.

To detect confined motion, the fact is exploited that the MSD of a confined particle shows a downward curvature in comparison to a particle undergoing free diffusion. For a small value of t, the MSD for a confined particle is very similar to the MSD for free diffusion (see Equation 2.12). Therefore, the first N_{diff} points of a rolling MSD analysis are used to calculate an initial MSD for a particle undergoing simple diffusion. The deviation between the MSD for longer time lag and the initial value is a robust parameter which indicates confined diffusion:

$$Conf = \frac{1}{N_{\text{Conf}}} \sum_{n=1}^{n=N_{\text{Conf}}} \frac{\langle r^2(n\Delta t)\rangle - \langle r^2(n\Delta t)\rangle_{\text{diff}}}{\langle r^2(n\Delta t)\rangle_{\text{diff}}} \tag{2.13}$$

For $Conf < 0$ a confined mobility is likely, whereas for $Conf > 0$ it is unlikely. Hence, to obtain a reliable value for $Conf$ we set $N_{\text{Conf}}/N_{\text{diff}} = 10$. As the error in the MSD becomes increasingly large for high values of t_{lag}, the number of points from the MSD curve used for calculating $Conf$ should not exceed the first $2N/3$ points of this curve.

Whilst it is possible to detect directed motion directly from an MSD curve, it is more efficient to examine the shape of a trajectory, as a directed motion will lead to a highly asymmetric trajectory. For this, the radius of gyration tensor of a trajectory, $\mathbf{R}_g$, is calculated:

$$\mathbf{R}_g(i, j) = \langle r_i r_j\rangle - \langle r_i\rangle\langle r_j\rangle \tag{2.14}$$

where r_i and r_j are the three axes and the averages are defined over all N_{Asym} steps of the analyzed rolling window; typically, $N_{\text{Asym}} = N_{\text{diff}}$. The radii of gyration for each direction are the square roots of the eigenvalues $\mathbf{R}_g$. From those, the asymmetry parameter is calculated:

$$Asym = -\log\left(1 - \frac{\left(R_1^2 - R_2^2\right)^2 + \left(R_1^2 - R_3^2\right)^2 + \left(R_2^2 - R_3^2\right)^2}{2\left(R_1^2 + R_2^2 + R_3^2\right)^2}\right) \tag{2.15}$$

For $Asym > 1$, a directed motion is likely, whereas for $Asym < 1$ a directed motion is unlikely.

For the reliable detection of different types of motion, there is an obvious trade-off between statistical significance and window size. These values depend on the system under study, and thus several typical trajectories are used to optimize the values for a particular sample. In the case of Huet *et al.*, and in the present authors' studies, the minimum window size was $W_{min}^{conf} = 75$ consecutive time points for confined motion, and $W_{min}^{dir} = 11$ consecutive time points for directed motion.

When single-particle trajectories are too short to calculate the MSD on a single trajectory with sufficient accuracy (typically, 100 consecutive time points are required), the displacements of all particles in adjacent frames are analyzed. For the 2-D case, the cumulative distribution function (*cdf*) for the squared displacements r^2 is [1]:

$$P(r^2, t_{lag}) = 1 - \exp\left(-\frac{r^2}{\mathrm{MSD}(t_{lag})}\right) \tag{2.16}$$

where $P(r^2, t_{lag})$ describes the probability that a particle starting at the origin will be found in a circle of radius r after a time t_{lag}. The *cdf* is very useful for a system where there are two fractions of a certain particle, which experimentally are only distinguishable by their different D-values [6, 68, 69]. For two fractions, Equation 2.16 becomes

$$P(r^2, t_{lag}) = 1 - \left[\alpha \cdot \exp\left(-\frac{r^2}{\mathrm{MSD}_1(t_{lag})}\right) + (1-\alpha)\cdot \exp\left(-\frac{r^2}{\mathrm{MSD}_2(t_{lag})}\right)\right] \tag{2.17}$$

where α indicates the fraction size, and $\mathrm{MSD}_1(t_{lag})$ and $\mathrm{MSD}_2(t_{lag})$ the two MSD-values at t_{lag}, respectively. It should be noted that such an ensemble-type analysis does not even require a previous, computationally demanding, trajectory analysis (as outlined in Section 2.4). The position data can likewise be directly analyzed using particle image correlation analysis (PICA), as developed by Semrau *et al.* [70].

2.6 Applications

The examples in Figures 2.3 to 2.5 show that the previously described techniques are powerful tools to obtain information about biological systems. Figure 2.3 provides an example obtained from an experiment where the diffusional behavior of several membrane anchors (K-Ras, H-Ras, and Lck) in live cells were compared, the aim being to study the occurrence and size of lipid rafts in the cytoplasmic leaflet [71]. It was speculated that the association of particular sets of proteins with lipid rafts plays an important role in a variety of signal transduction pathways [72]. Whilst there is evidence for lipid rafts in the exoplasmic leaflet, very little was known regarding the cytoplasmic leaflet. A link between the lipid rafts in the two leaflets has been predicted to be of importance in the transduction of cellular signals from the outside to the inside of the cell. In Figure 2.3a, 3T3-A14 cells are shown expressing eYFP-C14KRas [71]. K-Ras is generally used as a non-raft

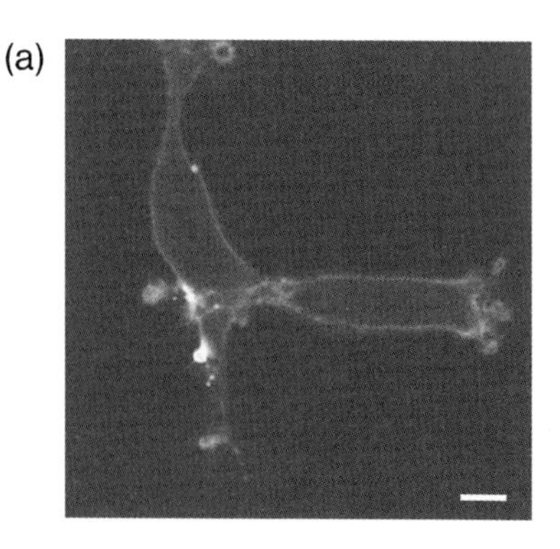

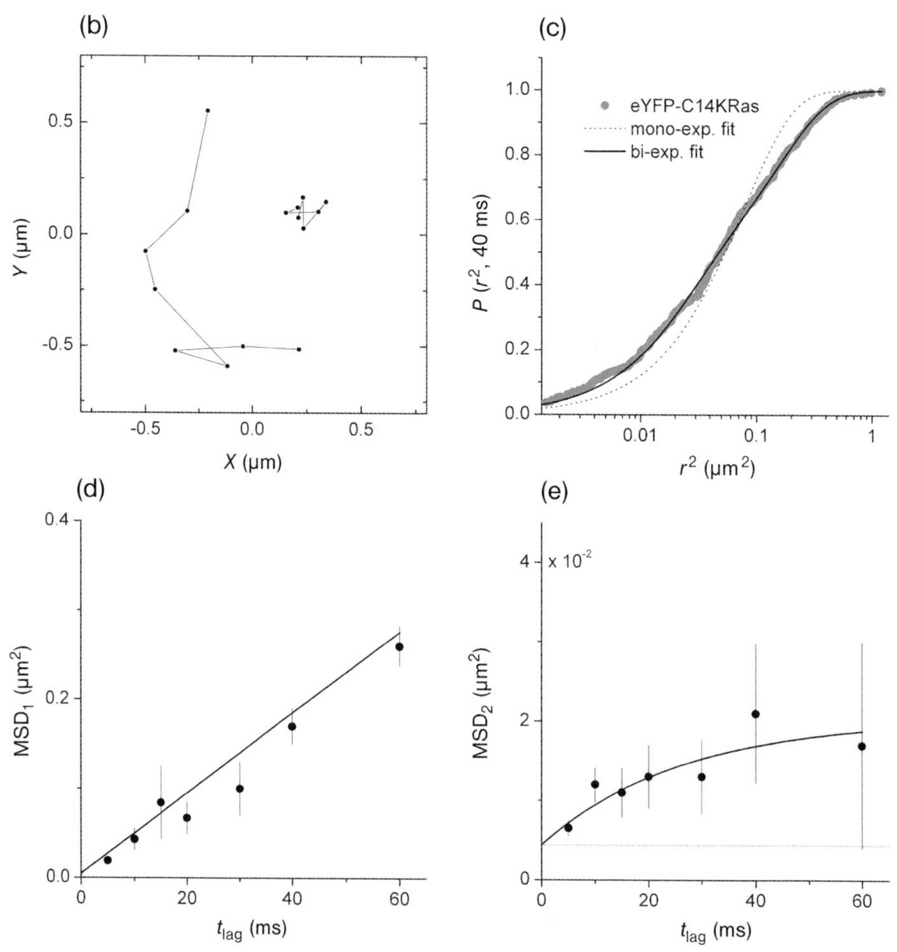

Figure 2.3 (a) Confocal fluorescence microscopy image of 3T3-A14 cells expressing eYFP-C14KRas, at 2 days after transfection. A clear plasma membrane localization was observed. Scale bar = 10 μm. (b) Trajectories of eYFP-C14KRas molecules diffusing in the apical membrane of a 3T3-A14 cell. The time between consecutive points was 20 ms.

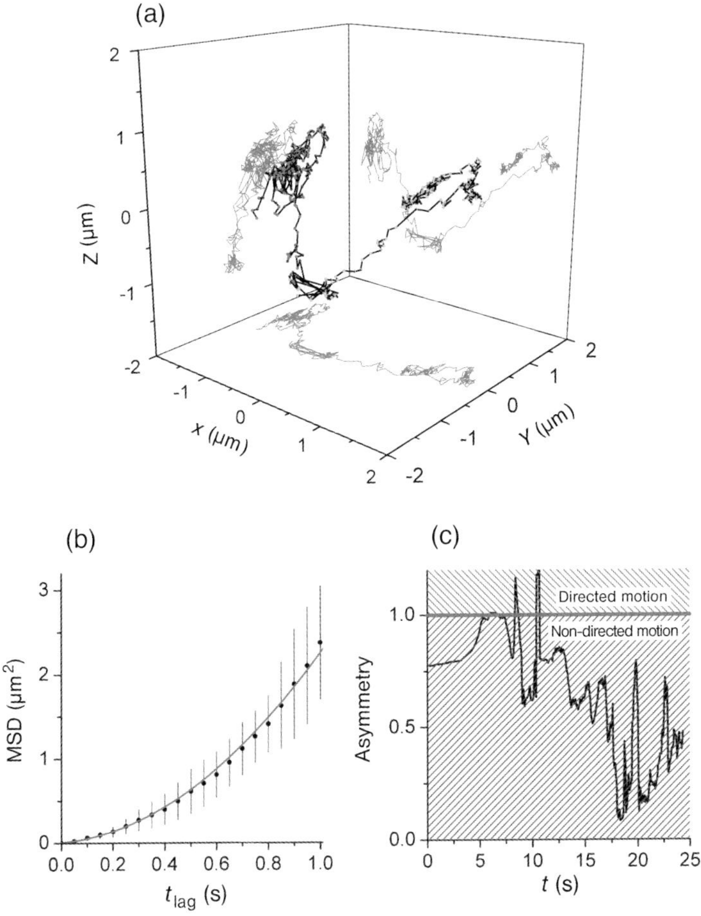

Figure 2.4 (a) Trajectory of a QD loaded into HEK293-cells obtained with a frame rate of 20 Hz for a total time of 25 s. Only one plane was imaged for each time point. In two parts of the trajectory, directed transport can be clearly seen. (b) MSD versus time for the first part of the trajectory where directed motion is observed. The supralinear behavior of the MSD confirms that transport takes place. A fit to the data shows that the QD has a velocity of $1.4 \pm 0.1\,\mu m/s$. (c) Calculation of the asymmetry parameter clearly shows the two parts of the trajectory where directed motion takes place (Asym > 1).

(c) Cumulative probability, $P(r^2, t_{lag})$, versus the square displacements, r^2 with a time lag of 40 ms. Fits to a one-component model (dashed line) and a two-component model (solid line) clearly show that the latter model fits better. (d) The mean squared displacements (MSDs) of the fast fraction plotted versus t_{lag}. The data were fitted according to a free-diffusion model and a diffusion constant $D = 1.00 \pm 0.04\,\mu m^2 s^{-1}$ was obtained. (e) The MSDs of the slow fraction plotted versus t_{lag}. A fit according to a confined diffusion model is shown as a solid line. An average domain size of $L = 219 \pm 71$ nm was found. The dotted line represents the offset due to the limited positional accuracy. The same offset is present in panel (d), although there the dotted line is omitted for clarity.

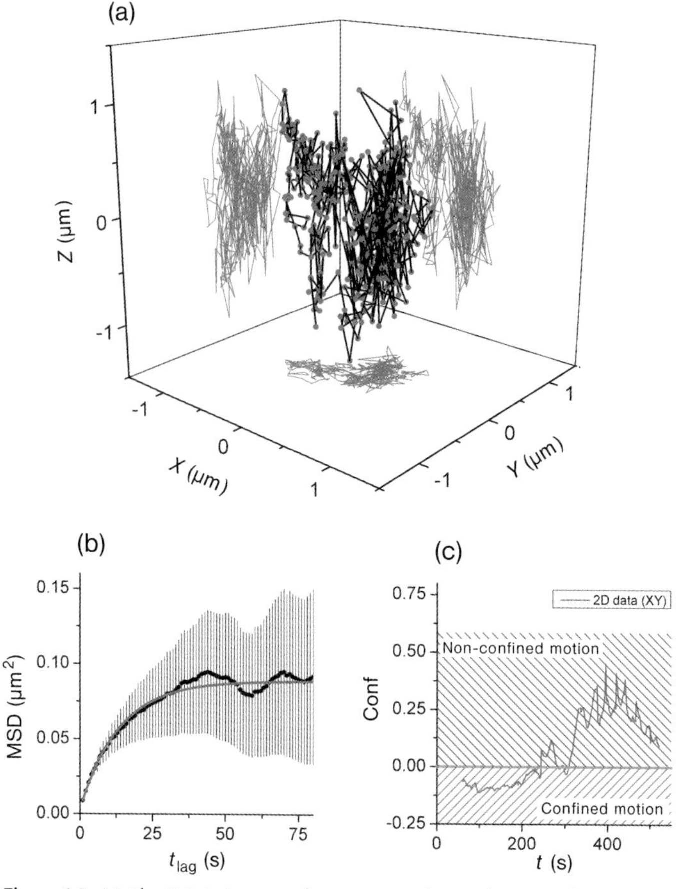

Figure 2.5 (a) The 3-D trajectory of an endosome containing Dpp molecules labeled with Venus YFP [71]. The endosome was followed for almost 500 s at a frame rate of 1 Hz. Each image stack consists of seven image planes. (b) MSD versus time plot for the *xy*-projection of the first 190 s of the trajectory. It can be clearly seen that the movement of the endosome is confined during this period. Fitting Equation 2.12 yielded an initial diffusion constant D_0 of $(1.60 \pm 0.02) \times 10^{-3}\,\mu m^2\,s^{-1}$ and a lateral confinement of side length $L = 508 \pm 2\,nm$. (c) Calculation of the deviation parameter shows that the endosomes shows confined motion in the first part of the trajectory ($Conf < 0$).

marker, and comparison with the raft-marker Lck should provide insights into the presence of lipid rafts. The imaging of large numbers of trajectories (>2500) for eYFP-c14Kras was used for the analysis, of which two examples are shown in Figure 2.3b. It should be noted that the trajectories are relatively short because of the rapid photobleaching of eYFP. From these trajectories, square displacement distributions were constructed (as described in Section 2.5). The cumulative probability distribution versus the square displacement for a time lag of 40 ms is shown in Figure 2.3c. A fit of the data to Equation 2.16 shows clearly that the data cannot be described by a one-component model. However, the fit improves significantly when a two-component model is used (Equation 2.17), while a three-component model does not improve the goodness-of-fit. Fitting the data to Equation 2.17 yielded a fast-diffusing fraction, $\alpha = 0.62 \pm 0.13$ ($MSD_1 = 0.16 \pm 0.04\,\mu m^2 s^{-1}$), and a slow-diffusing fraction ($MSD_2 = 0.021 \pm 0.006\,\mu m^2 s^{-1}$). This analysis was subsequently performed for all time lags from 5 to 60 ms, and the resulting MSDs were plotted versus the time lag. Figure 2.3d,e show the MSD-versus-time lag plots for the fast- and slow-diffusing fraction of the eYFP-C14KRas membrane anchor. A fit to Equation 2.10 yields a diffusion constant $D = 1.00 \pm 0.04\,\mu m^2 s^{-1}$ for the fast-diffusing fraction (Figure 2.3d). However, for the slow-diffusing fraction, the MSD-plot (Figure 2.3e) indicates that the movement of this fraction is confined; a fit of the data to Equation 2.12 yielded an average domain size of 219 ± 71 nm. A study of the diffusional behavior of the Lck anchor in similar manner, showed that the Lck anchor was not significantly slowed-down as compared to the K-Ras anchor. This result does not exclude the presence of rafts in the cytoplasmic leaflet, although the size of these was estimated to be less than 130 nm, which was the detection limit achieved in those experiments.

While the previous example focused on the mobility of proteins in the cell membrane, the processes inside the cell were subsequently studied using 3-D wide-field microscopy (as described above). The use of quantum dots (QDs) as fluorescent markers of biomolecules in cells enables those molecules to be followed for very long time periods that are limited only by the lifespan of the cells. Figure 2.4 shows the results of an experiment where human embryonic kidney (HEK) cells (HEK293) were incubated with a solution containing 0.1 nM QDs [29]. The QDs were internalized within 2 h, after which the HEK cells were imaged using a 3-D wide-field fluorescence set-up via the astigmatism method. Multiple QDs were followed simultaneously in three dimensions with great accuracy (30 nm) and at high frame rates ($f = 20$ Hz), without producing image stacks. One of these trajectories is shown in Figure 2.4a. What was suspected by examining at the trajectory, namely two short periods of directed transport, could be confirmed unambiguously. When the MSD curve for the first period of directed transport was plotted (see Figure 2.4b), the supralinear behavior of the MSD curve confirmed that transport had taken place, while fitting the 3-D MSD yielded a velocity of $v = 1.41 \pm 0.14\,\mu m\, s^{-1}$. The asymmetry parameter was able to reliably identify the two periods of directed transport (see Figure 2.4c). Calculation of the MSD for the initial part of the trajectory confirmed that the QDs had followed a random diffusion during this period ($D = 0.015 \pm 0.001\,\mu m^2 s^{-1}$).

The current techniques may also be applied to more complex systems, an example being the wing imaginal disc of a *Drosophila melanogaster* larva (see Figure 2.5). The wing disc consists of two layers of cells, with the cells in one layer being highly elongated (the study of such topology requires 3-D wide-field microscopy). During the development of the fly wing, morphogens such as decapentaplegic (Dpp) form concentration gradients in these columnar cells, providing positional information for each cell in the tissue [73]. Dpp is produced in the center of the wing imaginal disc at the anterior–posterior compartment boundary [74]. After dissection, the disc was placed onto the microscope and the receiving cells were imaged at a distance of 20 μm from the Dpp source. The fact that Dpp is located mainly in endosomes (up to 100 molecules per endosome) makes it possible to track endosomes for hundreds of frames, even though the Dpp is labeled with a variant of the yellow fluorescent protein. The elongated nature of the cells required the preparation of stacks consisting of seven image planes, each separated by 1 μm. The trajectory of one endosome containing Dpp is shown in Figure 2.5a. From projections onto the 2-D planes, the 3-D trajectory was clearly visible, and the endosome appeared to be confined in a lateral direction. Calculating the MSD curve for the *xy*-projection of the first 190 s (Figure 2.5b) clearly showed that the movement of the endosome was confined during this period. Fitting Equation 2.12 yielded an initial diffusion constant $D_0 = (1.60 \pm 0.02) \cdot 10^{-3} \mu m^2 s^{-1}$ and a lateral confinement of side length $L = 580 \pm 2$ nm. It should be noted here that the size of confinement was significantly less than the lateral size of the cells (~3 μm). Calculation of the *Conf* parameter (Figure 2.5c) confirmed the observed confinement for this endosome in the first part of the trajectory.

2.7 Conclusions

The three examples described clearly demonstrate that single-particle tracking has become an invaluable technique to study processes in live cells and tissues. In recent years, 2-D WFFM has become a widely used technique that has been complemented by several methods so as to provide information concerning the third dimension, with great accuracy. Consequently, by extending such an established methodology the range of biological questions that can be addressed may be significantly broadened. In combination with super-resolution techniques, 2-D WFFM should prove highly valuable, and may help to remove the ambiguities associated with current models of intercellular and intracellular transport. It is not foreseen that, ultimately, single-molecule tracking will allow the identification of intricate signaling pathways in space and time, even within such complex environments as tissues. However, the results of such studies will undoubtedly yield unexpected results and, more importantly, will surely form the solid base for a quantitative mechanistic understanding of cellular processes *in vivo*.

References

1 Anderson, C.M., Georgiou, G.N., Morrison, I.E., Stevenson, G.V., and Cherry, R.J. (1992) *J. Cell Sci.*, **101** (Pt 2), 415–425.

2 Dahan, M., Lévi, S., Luccardini, C., Rostaing, P., Riveau, B., and Triller, A. (2003) *Science*, **302** (5644), 442–445.

3 Dietrich, C., Yang, B., Fujiwara, T., Kusumi, A., and Jacobson, K. (2002) *Biophys. J.*, **82** (1 Pt 1), 274–284.

4 Douglass, A.D. and Vale, R.D. (2005) *Cell*, **121** (6), 937–950.

5 Harms, G., Cognet, L., Lommerse, P., Blab, G., Kahr, H., Gamsjager, R.,

Spaink, H., Soldatov, N., Romanin, C., and Schmidt, T. (2001) *Biophys. J.*, **81** (5), 2639–2646.
6 Lommerse, P., Blab, G., Cognet, L., Harms, G., Snaar-Jagaiska, B., Spaink, H., and Schmidt, T. (2004) *Biophys. J.*, **86** (1), 609–616.
7 Schmidt, T., Schütz, G., Baumgartner, W., Gruber, H., and Schindler, H. (1996) *Proc. Natl Acad. Sci. USA*, **93** (7), 2926–2929.
8 Kusumi, A., Sako, Y., and Yamamoto, M. (1993) *Biophys. J.*, **65** (5), 2021–2040.
9 Kubitscheck, U. (2002) *Single Mol.*, **3** (5-6), 267–274.
10 Goulian, M. and Simon, S. (2000) *Biophys. J.*, **79** (4), 2188–2198.
11 Köhler, R.H., Schwille, P., Webb, W.W., and Hanson, M.R. (2000) *J. Cell Sci.*, **113** (Pt 22), 3921–3930.
12 Schütz, G., Axmann, M., Freudenthaler, S., Schindler, H., Kandror, K., Roder, J., and Jeromin, A. (2004) *Microsc. Res. Tech.*, **63** (3), 159–167.
13 Chen, H., Yang, J., Low, P.S., and Cheng, J.-X. (2008) *Biophys. J.*, **94** (4), 1508–1520.
14 Babcock, H.P., Chen, C., and Zhuang, X. (2004) *Biophys. J.*, **87** (4), 2749–2758.
15 Seisenberger, G., Ried, M., Endress, T., Büning, H., Hallek, M., and Bräuchle, C. (2001) *Science*, **294** (5548), 1929–1932.
16 Simson, R., Sheets, E., and Jacobson, K. (1995) *Biophys. J.*, **69** (3), 989–993.
17 Feder, T., Brust-Mascher, I., Slattery, J., Baird, B., and Webb, W. (1996) *Biophys. J.*, **70** (6), 2767–2773.
18 Gelles, J., Schnapp, B.J., and Sheetz, M.P. (1988) *Nature*, **331** (6155), 450–453.
19 Kural, C., Kim, H., Syed, S., Goshima, G., Gelfand, V., and Selvin, P. (2005) *Science*, **308** (5727), 1469–1472.
20 Nan, X., Sims, P., Chen, P., and Xie, X. (2005) *J. Phys. Chem. B*, **109** (51), 24220–24224.
21 Courty, S., Luccardini, C., Bellaiche, Y., Cappello, G., and Dahan, M. (2006) *Nano Lett.*, **6** (7), 1491–1495.
22 Hecht, E. (1987) *Optics*, Addison-Wesley.
23 Schmidt, T., Schültz, G., Baumgartner, W., Gruber, H., and Schindler, H. (1995) *J. Phys. Chem.*, **99** (49), 17662–17668.
24 Zhang, B., Zerubia, J., and Olivo-Marin, J. (2007) *Appl. Opt.*, **46** (10), 1819–1829.
25 Thompson, R., Larson, D., and Webb, W. (2002) *Biophys. J.*, **82** (5), 2775–2783.
26 Cheezum, M., Walker, W., and Guilford, W. (2001) *Biophys. J.*, **81** (4), 2378–2388.
27 Carter, B., Shubeita, G., and Gross, S. (2005) *Phys. Biol.*, **2** (1), 60–72.
28 Falcón-Pérez, J.M., Nazarian, R., Sabatti, C., and Dell'Angelica, E.C. (2005) *J. Cell Sci.*, **118** (Pt 22), 5243–5255.
29 Holtzer, L., Meckel, T., and Schmidt, T. (2007) *Appl. Phys. Lett.*, **90** (5), 053902.
30 Bobroff, N. (1986) *Rev. Sci. Instrum.*, **57** (6), 1152–1157.
31 Peters, I., de Grooth, B., Schins, J., Figdor, C., and Greve, J. (1998) *Rev. Sci. Instrum.*, **69** (7), 2762–2766.
32 Gustafsson, M., Agard, D., and Sedat, J. (1999) *J. Microsc. (Oxf.)*, **195**, 10–16.
33 McNally, J., Karpova, T., Cooper, J., and Conchello, J.-A. (1999) *Methods*, **19** (3), 373–385.
34 Schütz, G., Axmann, M., and Schindler, H. (2001) *Single Mol.*, **2** (2), 69–73.
35 Speidel, M., Jonas, A., and Florin, E. (2003) *Opt. Lett.*, **28** (2), 69–71.
36 Li, D., Xiong, J., Qu, A., and Xu, T. (2004) *Biophys. J.*, **87** (3), 1991–2001.
37 Levi, V., Ruan, Q., and Gratton, E. (2005) *Biophys. J.*, **88** (4), 2919–2928.
38 Lu, P.J., Sims, P.A., Oki, H., Macarthur, J.B., and Weitz, D.A. (2007) *Opt. Express*, **15** (14), 8702–8712.
39 (a) Cang, H., Wong, C.M., Xu, C.S., Rizivi, A.H., and Yang, H. (2006) *Appl. Phys. Lett.*, **88**, 223901; (b) Lessard, G., Goodwin, P., and Werner, J. (2007) *Appl. Phys. Lett.*, **91** (22), 224106.
40 Kao, H. and Verkman, A. (1994) *Biophys. J.*, **67** (3), 1291–1300.
41 Toprak, E., Balci, H., Blehm, B., and Selvin, P. (2007) *Nano Lett.*, **7** (7), 2043–2045.
42 Juette, M.F., Gould, T.J., Lessard, M.D., Mlodzianoski, M.J., Nagpure, B.S., Bennett, B.T., Hess, S.T., and Bewersdorf, J. (2008) *Nat. Methods*, **5** (6), 527–529.
43 Prabhat, P., Ram, S., Ward, E., and Ober, R. (2004) *IEEE Trans. Nanobioscience*, **3** (4), 237–242.
44 Prabhat, P., Gan, Z., Chao, J., Ram, S., Vaccaro, C., Gibbons, S., Ober, R.J., and Ward, E.S. (2007) *Proc. Natl Acad. Sci. USA*, **104** (14), 5889–5894.

45 Serge, A., Bertaux, N., Rigneault, H., and Marguet, D. (2008) *Nat. Methods*, **5** (8), 687–694.

46 Cognet, L., Harms, G., Blab, G., Lommerse, P., and Schmidt, T. (2000) *Appl. Phys. Lett.*, **77** (24), 4052–4054.

47 Ober, R., Ram, S., and Ward, E. (2004) *Biophys. J.*, **86** (2), 1185–1200.

48 Aguet, F., Ville, D.V.D., and Unser, M. (2005) *Opt. Express*, **13** (26), 10503–10522.

49 Betzig, E., Patterson, G.H., Sougrat, R., Lindwasser, O.W., Olenych, S., Bonifacino, J.S., Davidson, M.W., Lippincott-Schwartz, J., and Hess, H.F. (2006) *Science*, **313** (5793), 1642–1645.

50 Hess, S.T., Girirajan, T.P.K., and Mason, M.D. (2006) *Biophys. J.*, **91** (11), 4258–4272.

51 Rust, M., Bates, M., and Zhuang, X. (2006) *Nat. Methods*, **3** (10), 793–795.

52 Huang, B., Wang, W., Bates, M., and Zhuang, X. (2008) *Science*, **319** (5864), 810–813.

53 Willig, K., Rizzoli, S., Westphal, V., Jahn, R., and Hell, S. (2006) *Nature*, **440** (7086), 935–939.

54 Geerts, H., Brabander, M.D., Nuydens, R., Geuens, S., Moeremans, M., Mey, J.D., and Hollenbeck, P. (1987) *Biophys. J.*, **52** (5), 775–782.

55 Ghosh, R. and Webb, W. (1994) *Biophys. J.*, **66** (5), 1301–1318.

56 Reinfeld, N. (1958) *Mathematical Programming*, Prentice-Hall, Englewood Cliffs, New Jersey.

57 Veenman, C., Reinders, M., and Backer, E. (2001) *IEEE Trans. Pattern Anal.*, **23** (1), 54–72.

58 Bonneau, S., Dahan, M., and Cohen, L. (2005) *IEEE Trans. Image Process.*, **14** (9), 1384–1395.

59 Sbalzarini, I. and Koumoutsakos, P. (2005) *J. Struct. Biol.*, **151** (2), 182–195.

60 Genovesio, A., Liedl, T., Emiliani, V., Parak, W., Coppey-Moisan, M., and Olivo-Marin, J. (2006) *IEEE Trans. Image Process.*, **15** (5), 1062–1070.

61 Jaqaman, K., Loerke, D., Mettlen, M., Kuwata, H., Grinstein, S., Schmid, S.L., and Danuser, G. (2008) *Nat. Methods*, **5** (8), 695–702.

62 Saxton, M. and Jacobson, K. (1992) *Annu. Rev. Biophys. Biomol. Struct.*, **26**, 373–399.

63 Coscoy, S., Huguet, E., and Amblard, F. (2007) *Bull. Math. Biol.*, **69** (8), 2467–2492.

64 Meilhac, N., Le Guyader, L., Salome, L., and Destainville, N. (2006) *Phys. Rev. E*, **73** (1), 011915.

65 Qian, H., Sheetz, M., and Elson, E. (1991) *Biophys. J.*, **60** (4), 910–921.

66 Huet, S., Karatekin, E., Tran, V., Fanget, I., Cribier, S., and Henry, J. (2006) *Biophys. J.*, **91** (9), 3542–3559.

67 Saxton, M. (1997) *Biophys. J.*, **72** (4), 1744–1753.

68 Schütz, G., Schindler, H., and Schmidt, T. (1997) *Biophys. J.*, **73** (2), 1073–1080.

69 Deverall, M., Gindl, E., Sinner, E., Besir, H., Ruehe, J., Saxton, M., and Naumann, C. (2005) *Biophys. J.*, **88** (3), 1875–1886.

70 Semrau, S. and Schmidt, T. (2007) *Biophys. J.*, **92** (2), 613–621.

71 Lommerse, P.H.M., Vastenhoud, K., Pirinen, N.J., Magee, A.I., Spaink, H.P., and Schmidt, T. (2006) *Biophys. J.*, **91** (3), 1090–1097.

72 Simons, K. and Toomre, D. (2000) *Nat. Rev. Mol. Cell. Biol.*, **1**, 31–39.

73 Kicheva, A., Pantazis, P., Bollenbach, T., Kalaidzidis, Y., Bittig, T., Jülicher, F., and González-Gaitán, M. (2007) *Science*, **315** (5811), 521–525.

74 Basler, K. and Struhl, G. (1994) *Nature*, **368** (6468), 208–214.

3
Messenger RNA Trafficking in Living Cells

Ulrich Kubitscheck, Roman Veith, Jörg Ritter, and Jan-Peter Siebrasse

3.1
Intranuclear Structure and Dynamics

The nucleus is the key organelle of eukaryotic cells. Vital processes such as the storage of genetic material, DNA replication and transcription, RNA processing and export occur within this cellular compartment. Despite great efforts, however, the present understanding of the internal structure and dynamics of the cell nucleus remains very limited. The labeling of specific nuclear structures with antibodies, and more recently with autofluorescent proteins, has revealed a complex spatiotemporal nuclear organization, with many distinct subcompartments that are involved in both DNA and RNA metabolism (for reviews, see Refs [1, 2]). In most cases, the specific processes that take place in the distinct compartments are relatively unknown. In contrast to the cytoplasmic organelles, nuclear structures are not enclosed by membranes, which in turn raises fundamental questions such as: (i) Which processes lead to their formation and continuation; and (ii) To what extent do they possibly restrict or enhance the mobility of functional molecules?

Diffusion measurements of inert macromolecules using fluorescence recovery after photobleaching (FRAP), photoactivation (PA), fluorescence correlation spectroscopy (FCS), and time-resolved anisotropy have revealed that diffusion in the cell nucleus occurs four to eight times more slowly than in aqueous solution [3]. A view has emerged of the nuclear interior as being a crowded and highly structured environment rather than a homogeneous viscous gel. The movements and interactions of functional molecules such as transcription factors or messenger RNA (mRNA), respectively particles containing RNA and proteins – the so-called ribonucleoprotein particles (RNPs) – are even less understood. Studies of the dynamics of fluorescently tagged nuclear proteins using FRAP, PA and FCS have become feasible only during the past few years [4]. Such studies have resulted in a highly dynamic view of the nuclear interior, with proteins diffusing rapidly within the nucleus and demonstrating highly dynamic exchanges at their target sites. It is known, however, that it is very difficult to correctly account for inter-

Single Particle Tracking and Single Molecule Energy Transfer
Edited by Christoph Bräuchle, Don C. Lamb, and Jens Michaelis

ISBN: 978-3-527-32296-1

actions when analyzing FRAP data or to correctly measure binding times [5]. Likewise, it is problematic to measure mobility by using FRAP, when small fluorescent ligands undergo frequent exchange with further mobile binding partners, as this leads to complex recovery kinetics [6].

One of the most intricate processes within the cell nucleus is RNA biogenesis [7]. Already at the transcription site, the pre-mRNA molecules become associated with numerous proteins to form messenger ribonucleoprotein particles (pre-mRNPs), which are then subject to a complex process of biochemical modification, in which they are finally prepared for export through the nuclear pore complexes in the nuclear envelope [8, 9].

3.2 FCS and FRAP Studies of Nuclear mRNP Mobility

During the past decade, several groups have focused on acquiring an understanding of the intranuclear mobility of (m)RNA. The analysis of data acquired with FCS and FRAP have demonstrated that mRNA moves inside the nucleoplasm by passive diffusion rather than by active transport. Yet, energy-dependent processes presumably determine such mobility, most likely by impacting upon the particle binding or confinement in chromatin-formed structures. The pioneering FCS measurements of Politz and coworkers provided the first estimation of the diffusion coefficient (*D*) of poly(A) RNAs within the open spaces between chromatin domains – the interchromatin channels – with *D*-values of up to $10\,\mu m^2 s^{-1}$ [10], whereas FRAP and PA studies rather yielded an estimation of the effective, long-range mobility of mRNA with a $D_{LR} \sim 0.7\,\mu m^2 s^{-1}$ [11–15]. In all such experiments, several further mobility fractions were found that indicated a strong tendency for mRNPs either to bind to specific sites inside the nucleus or to be retarded by nonspecific interactions. At least partially, the retardation is presumably due to the vast amount of decondensed chromatin inside the interphase nucleus.

3.3 Single-Particle Tracking of mRNA Molecules

3.3.1 The Aim and Purpose of Intranuclear Single mRNA Tracking

Measurements of mRNA bulk mobility have yielded important insights into the dynamics of mRNA *in vivo*. Both, FCS and photobleaching/PA methods are fast and reliable techniques that provide a rapid overview of the dynamics of the molecules in question. However, direct microscopic observations of functional single molecules (such as mRNA) in living cells can offer a significantly more detailed view of intracellular dynamics. Notably, such direct observations have one major advantage, namely that the mobility of functional molecules can be followed in detail, providing information on the *mobility distribution* of the molecular species and on its possible time dependence. Within the nucleus, mRNA undergoes numerous biochemical modifications that involve a plethora of different interaction partners before its export into the cytoplasm. For a single mRNA, these processes occur in sequential manner, but when considering the total pool of mRNA all of these modifications occur simultaneously at different locations within the cell nucleus. Consequently, most of the RNA molecules will be engaged in different molecular complexes that comprise different protein factors, and hence have different masses and mobilities. Although, in a complex system such as the cell nucleus it is not possible to synchronize these processes, by using single-molecule imaging it becomes possible to analyze them separately. Furthermore, this technique reveals directly any molecular interactions with intracellular structures or binding events. Moreover, it is not only the duration of the binding events that can be evaluated, but also the time sequence of sequential processes, if a greater number of interaction partners is involved. As noted in Section 3.5.2, it is relatively straightforward to discriminate directly between situations where molecules move with different mobilities, and where they alter their type of motion along a pathway, which might well include several molecular modifications. Such a discrimination is not possible by using bulk techniques. In addition, molecular mobility in functionally or structurally different spatial domains within the cell nucleus, or the binding to small distinct regions or intracellular structures such as fibers or filaments, can be directly visualized and analyzed. Finally, single-molecule imaging within cells can reveal rare events that might be hidden among bulk measurements. This is especially valuable when examining mRNA dynamics, since it is well known that pre-mRNA processing involves a number of steps that include poly-adenylation, 3′ end capping, and the splicing of introns. Whilst it has been recognized that all of these steps require the action of multienzyme complexes, it is not quite known where–and at what time–they occur. It can be assumed, however, that all such events comprise only a tiny fraction of the mRNA's lifetime within the nucleus, and so would not be observed using bulk measurements, where pure average values of the dynamics are determined. It is quite likely that short binding and processing events would not be observable by using any technique other than single-particle tracking. It should be noted that all of the above-mentioned processes of mRNP maturation and trafficking can be regulated, thus governing the outcome of gene expression. This process of regulation is, however, very poorly understood.

3.3.2
Single-Particle Tracking of "Designer" mRNPs

In order to gain a deeper insight into intracellular mRNA trafficking, artificial mRNAs have been designed that can specifically bind multiple fluorophores. This allows them to be labeled so strongly that they can be observed for long time periods, on a single-particle basis.

To date, two different labeling approaches of this type have been used successfully. The first approach employs the high-affinity interaction ($K_d \approx 1\,nM$) between the viral capsid protein MS2 and sequence-specific RNA stem-loops [16]. The results showed that 24 MS2-GFP (green fluorescent protein) fusion proteins, when bound to the corresponding number of stem-loop repeats, formed a fluorescence signal that was bright enough to be detected with sensitive video microscopy. Clearly, this system would require the simultaneous expression of MS2-GFP and the viral stem-loop RNA. Since this reporter pair was encoded by plasmid vectors they could be inserted by transfection, thus avoiding any mechanical stress to the cell (as would be exerted, for example, by the microinjection of probes). The labeled mRNA produced in this way had a length of 2.8 kilobases (kb) and, together with the bound MS2-GFP fusion proteins, an estimated molecular mass of 2.7 MDa. The Stokes diameter of the final mRNP complex was estimated at approximately 1 μm. Clearly, such a bulky probe might encounter difficulties when maneuvering itself through the DNA-filled mammalian cell nucleus!

A second, rather elegant, approach for labeling mRNPs was to use "molecular beacons" [17]; these are modified oligonucleotides that contain a fluorescent dye at one end, a quencher at the other end, and have several bases located close to both dyes that form a short stem-loop structure. The stem-loop causes a close approximation between the fluorophore and the quencher for the unbound molecular beacon, which is almost nonfluorescent in this state. Between the two complementary strands of the stem-loop structure is located a specific oligonucleotide sequence, which is designed complementary to certain strands on a specific RNA. Upon binding to this target RNA, the stem-loop is unfolded and the quenching effect of the fluorophore is abrogated. This system has one great advantage, in that the reporter molecules become visible only when they are bound to the target RNA; consequently, the final signal-to-noise ratio (SNR) of mRNA labeled by these probes is excellent. Recently, Vargas *et al.* [17] have designed a special RNA, which comprised 96 repeats of a 50 nucleotide (nt)-long target sequence.

By using either approach, highly fluorescent artificial mRNPs could be constructed that were easily visualized and tracked one at a time. It was confirmed that, although mRNPs moved by free diffusion, such diffusion was often spatially constrained. Rather, the emerging picture was that mRNPs traveled predominantly in the extranucleolar interchromatin space; however, when they diffused into domains containing dense chromatin their motion was significantly hindered, or even stalled. In the single-particle tracking experiments, ATP depletion had no effect on the individual, still-mobile mRNPs, but further constrained the movement of the particles, most likely due to a more condensed organization of chromatin [16, 17].

In both experimental approaches, quite low diffusion constants were obtained. Shav-Tal *et al.* reported a *D*-value of $0.04\,\mu m^2 s^{-1}$ for the 2.8 kb RNA labeled by numerous MS2-GFP molecules [16]. The mRNA engineered by Tyagi and coworkers (2005) had a size of 4.8 kb (respectively, 9.6 kb including the fluorescent oligonucleotides), and displayed an intranuclear *D*-value of $0.03\,\mu m^2 s^{-1}$ [17]. The retardation was most likely due to a temporary corralling in the complex chromatin network, which interfered with the mobility of the mRNPs. It was supposed that the constraining effect of chromatin was substantial, or that the mRNPs in mammalian nuclei would interact continuously with major nonchromosomal structures, such as spliceosomes or supraspliceosomes, for extended time periods.

However, in this context, it is very important to note that the chosen method of data analysis may have an immense impact on the *D*-value determined. For example, the mean square displacement (MSD) analysis used by Shav-Tal *et al.* clearly had a bias for low diffusion constants, because only mRNPs, which remained for a minimum time of 3 s in focus, were considered [16]. This led to the selection of slow particles, because fast particles rapidly left the focal plane of a high numerical aperture objective lens. For example, particles with $D = 4\,\mu m^2 s^{-1}$ were seen to reside for maximal 0.5 s in the focal region of such an objective lens. Likewise, Vargas *et al.* evaluated the trajectories only of those mRNPs that remained in focus for extended periods of time [17]. Hence, the true discrepancy between *D*-values determined by single mRNP tracking and FRAP or FCS might be smaller than indicated by the published values.

3.4 Single-Particle Tracking of Specific, Native mRNPs

The results obtained so far have suggested that the chromatin network might represent a major hindrance to mRNP trafficking. However, to address this question directly, a cellular system with large nucleoplasmic regions devoid of chromatin would be required. The nuclei in the salivary gland cells of *Chironomus tentans* larvae represent such a system, and allowed the movement of individual, specific and native mRNPs to be followed, without the complex influence of chromatin [18]. Not surprisingly, a markedly enhanced intranuclear mobility of mRNPs was observed compared to the results of previous mRNP tracking studies. In addition, the fact that the mRNPs were seen to move in a discontinuous fashion suggested that the particles might interact transiently with supramolecular entities other than chromatin.

3.4.1 Properties of *Chironomus tentans* Salivary Gland Cells

The adult female of *Chironomus tentans*, a member of the insect order Diptera [19, 20], lays between 300 and 3000 eggs at a time, into water. After leaving the egg mass, the larvae live and grow in tubes that are built up from a silk-like material produced by the salivary glands. When pupation is finished, the adult midge is released at the water surface. The larvae have two salivary glands, each of which contain 30–40 very large cells with nuclei that are approximately 75 μm in diameter. These cell nuclei contain very compact so-called "polytene" chromosomes that comprise 10 000–30 000 individual double-stranded DNA (dsDNA) molecules with their packing proteins, and are perfectly aligned one with another. The polytene chromosomes show a special banding pattern due to different unfolding states of the DNA during transcription. The chromosomes are very large, having a diameter of 10 μm and a length of between 45 and 100 μm. The four polytene chromosomes are compacted in distinct areas inside the nucleus, and are surrounded by nucleoli and a seemingly unstructured nucleoplasm (Figure 3.1).

Some of the silk-like proteins which comprise the larval housing tube may be up to 1.4 MDa in size, while the corresponding genes contain 27–40 kb. At chromosome IV, the transcription sites of three of the genes form two large puffs, the so-called Balbiani rings (BRs), referred to as BR1, BR2.1, and BR2.2. The BR genes contain four short introns, three of which are located close to the 5′ end, and one to the 3′ end [20, 21]. The BR2.1 mRNA is large and has a repetitive structure of short units present in 130–150 copies of 200 bases, all of which are virtually identical at the nucleotide level [20]. The single units themselves contain a repetitive region; this offers the possibility of labeling the mRNPs with oligonucleotides which are complementary to parts of the internal repetitive sequence. The spliced mRNA is packed into very bulky mRNPs which have a diameter of approximately 50 nm and contain numerous heterogeneous nuclear proteins (hnRNPs) [21–23]. The mRNA is coiled around a filamentous core of proteins, and at least parts of the mRNA sequence are still accessible [23]. Several of the estimated 400–500 hnRNP proteins, which are added cotranscriptionally to the BR mRNP, have been characterized [19, 24–27].

In a sophisticated immunoelectron microscopic study, Singh *et al.* showed that the BR mRNPs could travel through the nucleoplasm by random diffusion, rather than be actively transported [28]. The *D*-value of the particle was estimated to be $\geq 0.12\,\mu m^2 s^{-1}$. Based on the results of electron tomography studies, it is known that the processed mRNPs can attach to nucleoplasmic nonchromatin structures [29]. Miralles *et al.* showed that one-third of the nucleoplasmic BR particles appeared to be associated with larger complexes, and one-third was attached to fibers with a diameter of ~7 nm and a length of between 15 and 50 nm. Occasionally, these so-called *connecting fibers* (CFs) merged into larger structures that were designated

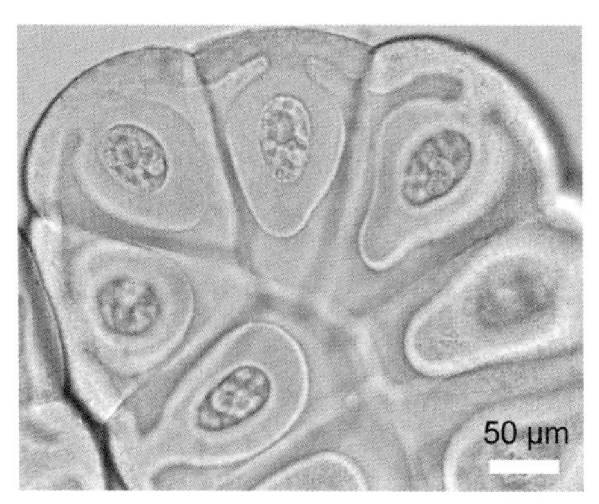

Figure 3.1 Bright field image of a *C. tentans* salivary gland, showing several salivary gland cells with nuclei.

as *fibrogranular clusters* (FGCs). The nature and function of both the CFs and FGCs was unclear. The remaining fraction of the BR mRNPs did not show any sign of interactions, and these particles were most likely mobile. One major component of the CFs is the protein hrp65, a 65 kDa RNA binding protein [30]. Miralles and coworkers speculated how the CFs might influence the mobility of the particles, and suggested that a CF-mediated mRNP interaction with larger structures (such as the FGCs) could be either transient or stable. In the latter case, this would suggest that mRNP movement toward the nuclear pore was promoted by the FGCs, possibly in a directional manner.

The *C. tentans* experimental system is an established, valuable tool which can be used to study the processing and trafficking of the native, endogenous mRNPs that displays all features of mRNA biogenesis in normal mammalian cells. The unique nuclear architecture permits study of the mRNP movement in nucleoplasmic regions devoid of chromatin, and therefore offers the potential to discover interactions, which are not mediated by chromatin.

3.4.2
Measurement of Nuclear Viscosity

Studies on the intranuclear diffusion benefit from a knowledge of the intranuclear viscosity, η_{nuc}, which has been measured in a range of mammalian cell systems using FRAP and FCS. In general, η_{nuc} was found to be four- to eightfold greater than that of an aqueous solution [3, 31]. As the salivary gland cell nuclei contain polytene chromosomes and not extended chromatin, it may well be that the viscosity of the remaining nucleoplasmic space is quite different from that of normal mammalian cells.

In order to determine η_{nuc} experimentally, 500 kDa fluorescein isothiocyanate (FITC)-labeled dextrans with an approximate Stokes diameter of ~50 nm, red fluorescent polystyrol microspheres with a diameter of 210 nm, and streptavidin-conjugated quantum dots (QDs) (655-Sav) with a diameter of 26 nm, were microinjected into the gland cell nuclei. The mobility of beads and quantum dots was studied using single-particle tracking (SPT), while the dextran mobility was monitored using FRAP. A *D*-value was determined for all probes using the Stokes–Einstein relationship, $D = kT/6\pi\eta R_s$. With the additional knowledge of the absolute temperature, T, and the hydrodynamic radius, R_s, the viscosity (η) was deduced to be approximately 3 cP (this is discussed below in greater detail).

3.4.3
Mobility Measurements by Photobleaching

All FRAP experiments with FITC-labeled, 500 kDa dextrans were performed using a point-scanning laser scanning microscope, after careful microinjection into a salivary gland cell nucleus. Following the acquisition of pre-bleach images, a circular region with a radius of 6.5 µm was bleached with a laser power of 0.55 mW (measured at the sample) for 130 ms. For bleaching and imaging, a 20×/0.5

numerical aperture (NA) objective was used, as its low numerical aperture created an almost cylindrical bleaching profile. During the subsequent image acquisition, the laser power was reduced to 0.1% to prevent further photobleaching. The fluorescence recovery of the circular area was monitored and normalized according to Refs [3, 31]:

$$I_{\text{norm}}(t) = \frac{[B(t) - Bg] \cdot [N_0 - Bg]}{[N(t) - Bg] \cdot [B_0 - Bg]} \tag{3.1}$$

Here, $B(t)$ is the time-dependent, mean intensity of the bleached region of interest, B_0 is the mean pre-bleach intensity, $N(t)$ is the time-dependent, mean intensity of the whole nucleus, and N_0 is mean pre-bleach intensity of the whole nucleus. The mean intensity of the background, Bg, was measured with zero illumination intensity. The total measurement duration was 123 s (350 images). For a cylindrical bleaching profile, the recovery process can be simplified to a two-dimensional (2-D) problem according to [32]:

$$I_{\text{norm}}(t) = e^{-K} + k \cdot \left[1 + \left(e^{-K} - 1\right) \cdot \left[1 - e^{-2\frac{\tau_r}{t}} \left(\mathrm{I}_0\left(2\frac{\tau_r}{t}\right) + \mathrm{I}_1\left(2\frac{\tau_r}{t}\right) \right) \right] - e^{-K} \right] \tag{3.2}$$

Here, K designates a fitting parameter quantifying the extent of bleaching, k takes an immobile component into account and τ_r is the characteristic time of diffusion. Typical parameters for a data set were $\tau_r = 5.2 \pm 0.04\,\text{s}$, $K = 0.6 \pm 0.003$, and $k = 1.0 \pm 0.001$. As expected, no immobile component was detected. The characteristic time of diffusion was related to the diffusion coefficient D according to:

$$D = \frac{w^2}{4\tau_r} \tag{3.3}$$

Here, w corresponded to the radius of the bleached circle. For the 500 kDa FITC-labeled dextran molecules, a D-value of $2.4 \pm 0.2\,\mu\text{m}^2\,\text{s}^{-1}$ was determined in 12 nuclei from different salivary glands. Reference measurements of FITC-labeled 500 kDa dextran (0.05%, w/v) in buffer were carried out in a small sample chamber, and yielded a mean D-value of $7.7 \pm 0.4\,\mu\text{m}^2\,\text{s}^{-1}$. This value corresponded to a Stokes diameter of approximately 57 nm for the 500 kDa dextran molecule; a value was quite close to that for the BR2 mRNPs. As D is proportional to the reciprocal of the viscosity, the ratio of the diffusion coefficients was $D_{\text{buffer}}/D_{\text{nuc}} = \eta_{\text{nuc}}/\eta_{\text{buffer}}$. Since $\eta_{\text{buffer}} \approx 1\,\text{cP}$ for the aqueous transport buffer the viscosity of the nucleoplasm for the 500 kDa FITC-dextrans corresponded to $\eta_{\text{nuc}} \approx 3\,\text{cP}$.

3.4.4 Analysis of Mobility by Single-Particle Tracking

The major prerequisites for intracellular single-molecule imaging are a high sensitivity, a reduction of the background light, and the use of picomolar concentrations of the probe of interest. Laser light sources provide the required irradiance of $\geq$0.1 kW cm^{-2} for efficient fluorescence excitation. The light transmission in the detection pathway of the microscope was optimized, and fast camera systems of high sensitivity were used for fluorescence detection. Due to the nature of light, the resolution of a microscope is limited by diffraction to approximately 200 nm. Therefore, the image of a single molecule reproduces the point-spread function of the optical system, which corresponds to a blurred spot with an almost Gaussian intensity profile. The position of the molecule, however, can be determined with very high precision by fitting a 2-D Gaussian to the blurred spot. The localization precision is only dependent on the SNR, and is usually in the range of 20 to 40 nm, and therefore subresolution particle movements can easily be detected. The image processing involved the identification, fitting, and tracking of the single-particle signals from recorded movies, which was accomplished using Diatrack 3.02 (Semasopht, Chavannes, Switzerland).

The observation of single molecules by high-speed cameras allows the tracing of single-molecule trajectories. The single-frame integration must be short enough that their movement during image acquisition is negligible. Recently, it was demonstrated that frame rates of 300–400 Hz were suitable for following the trajectories of molecules with molecular weights as low as 40 kDa in aqueous buffer [33]. Moreover, the motions of single fluorescent QDs with such frame rates could also easily be followed.

Each trajectory is defined by coordinates (x_i, y_i) with $1 \leq i \leq N$, where N denotes the trajectory length. In the case of Brownian motion, the MSDs, $\langle r^2(t)\rangle$, were linearly related to time, t, and to D, according to:

$$\langle r^2(t)\rangle = 4Dt \tag{3.4}$$

The analysis of particle trajectories according to Equation 3.4 is not adequate when different mobility fractions exist, or when the particles change their diffusion constant along trajectories, for example, in case of binding events. Such heterogeneous populations must be analyzed by more complex methods, for example, by a jump distance analysis [34, 35] or an analysis of the cumulated jump distances [36]. A jump distance analysis is straightforward and is based on the following considerations. When a particle starts at the origin, the probability that it will be encountered within a shell of radius $r + dr$ at time t is given as:

$$p(r,t)dr = \frac{1}{4\pi Dt} e^{-r^2/4Dt} 2\pi r dr \tag{3.5}$$

This probability distribution function can be approximated by a frequency distribution by quantifying the number of measured jump distances within respective intervals [r, $r + \Delta r$] traveled by single particles in a given time, for example, the cycle time of movie acquisition or multiplies thereof. Thus, a frequency histogram of all measured jumps can be constructed. When particles change their diffusion speed during their trajectory or different particle populations exist, the jump distance distributions cannot satisfactorily be fitted by Equation 3.5. More than one mobility fraction can be accounted for by summing up several terms according to

Equation 3.5. For example, a jump distance distribution containing three species with different diffusion constants, is given by

$$p'(r,t)dr = \sum_{j=1}^{3} \frac{Mf_j}{2D_j t} e^{-r^2/4D_j t} r dr \qquad (3.6)$$

In this expression, f_1, f_2 and f_3 denote the relative fractions with diffusion constants D_1, D_2 and D_3, respectively.

3.4.5
Tracer Particles in Cell Nuclei

In order to confirm and complement the FRAP results with dextran molecules, highly diluted QDs (Stokes diameter 26 nm) and microspheres (210 nm diameter) were microinjected into the nucleoplasm of salivary gland cells, and their mobility analyzed by using SPT. Single QDs could be imaged with excellent SNRs, yielding very long trajectories at a frame rate of ~100 Hz. The microspheres could even be observed using line-scanning confocal microscopy, at an acquisition rate of 15 Hz. For both particle species, two different mobility fractions were found. The data of the slower fractions, which were presumably due to particle aggregates and/or particles being entrapped in some intranuclear structures, were disregarded. For example, those particles which reached the polytene chromosomes usually became stalled. The fast particles were identified as monomers. For the fast QD population, a jump distance analysis yielded $D_{1,QD} = 4.2 \pm 0.4\,\mu m\,s^{-2}$, and for the microspheres $D_{1,sphere} = 0.43 \pm 0.05\,\mu m\,s^{-2}$. As noted above, the apparent viscosity of the nucleoplasm (η_{nuc}) could be calculated from comparative SPT measurements in buffer solution, and was determined to be approximately 4–5 cP. Thus, in view of the complexity of the medium "nucleoplasm" and the involved analytical approaches, the FRAP and SPT experiments were in very good agreement. In summary, the effective intranuclear viscosity was found to be in the range of 3–5 cP.

3.5
In Vivo Labeling of Native BR2 mRNPs

Within the salivary gland cell nuclei there exist between 10 to 50 BR particles per femtoliter (corresponding to 1 μm^3); when imaged using electron microscopy these were seen to be particles with a diameter of approximately 50 nm [29, 37]. As discussed above, the BR2 mRNPs contain a highly repetitive sequence, which can be used to specifically label them using complementary DNA- or RNA-oligonucleotides. Following the nuclear injection of small amounts of fluorescent DNA-oligonucleotides, the BR2 transcription site on chromosome IV became clearly visible; this showed that the DNA-oligonucleotides had bound almost instantaneously to the nascent mRNA transcripts. However, not only the nascent mRNPs at the BR were labeled, but also mRNPs in the nucleoplasmic space.

Notably, no labeled particles were found within the polytene chromosomes and within the nucleoli, which appeared as dark structures in the nucleus. The assumption of specific BR2 mRNP labeling could be verified by two control experiments. First, the transcription inhibitor DRB was applied; this specifically hinders RNA polymerase II to leave the promoter, but does not inhibit later transcription steps. The specific labeling of the BR transcription site disappeared after 10 min, when DRB was coinjected with the oligonucleotides. DRB inhibited the start of new RNA transcription processes, and thus no new mRNA transcripts were formed. Hence, after finishing the initiated transcription processes, the fluorescence labeling of the Balbiani ring faded away. Second, experiments using labeled control DNA- or RNA oligonucleotides of the same length but of reverse sequence did not yield any labeling of the BR transcription site. Neither could any distinct particles be detected in the nucleoplasm.

3.5.1 Mobility Analysis of BR2 mRNPs

In these studies, the visualization and tracking of single mRNP particles in the nucleoplasm was optimized by using high-speed microscopy. A line-scanning confocal laser scanning microscope (LSM 5 Live; Zeiss, Jena, Germany) yielded images with a high SNR at high frame rates. However, due to the high density of the labeled mRNPs, it was necessary to carry out a short photobleaching phase in order to reduce the density of fluorescent particles for single-particle observation and tracking. To this end, a rectangular region was selected which comprised approximately one-fifth of the nucleoplasm, which was rapidly bleached with high laser power. Any new, unbleached particles moving back into that region could then be clearly distinguished one from another, and recorded by imaging with a frame rate of 20 Hz. Alternatively, a video microscope was employed for mRNP observation. When using this instrument, a circular region within the nucleus with a diameter of ~15 μm was bleached with maximum laser intensity, before movies of unbleached mRNPs moving within the bleached volume were acquired at reduced irradiance. These movies comprised several hundred images, and lasted for up to 10 s. Example images from such movies of individual BR2 mRNPs are shown in Figure 3.2a. Five consecutive frames from a single LSM movie are displayed, in which several single BR2 mRNPs can be perceived. At the bottom of the figure, tracks of several particles determined in the first 16 frames of the complete image series were shown. Figure 3.2b shows a filtered version of the image series. During longer time periods the particle trajectories covered the nucleoplasm quite homogeneously, but did not enter those areas that were occupied by polytene chromosomes or nucleoli, as could be judged in respective bright-field images.

In order to quantify the mobility of the BR2 particles, their jump distance histograms at various lag times were analyzed. The complete analysis was based on more than 11 000 trajectories from four independent experiments comprising almost 50 000 single jumps between consecutive frames. Analyses were conducted

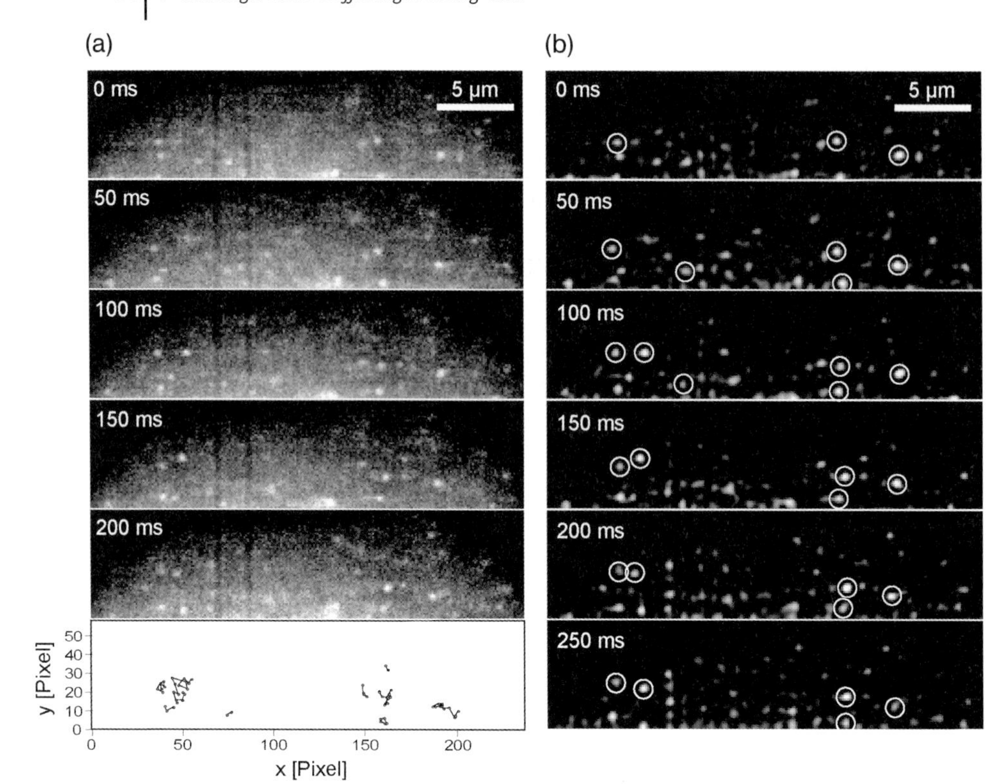

Figure 3.2 Imaging single BR2 mRNPs in *C. tentans* salivary gland cell nuclei. (a) Example frames from a movie showing individual BR2 mRNPs. At the bottom of panel (a) can be seen several mRNP tracks, which were determined when the first 16 frames of the complete movie sequence were plotted. (b) Filtered version of the image series. Several BR mRNPs, the tracks of which were plotted in panel (a), are marked by white circles.

of the histograms for the distances covered after 1, 10, 20, and 30 frames; these jumps were designated as jd1, jd10, jd20, and jd30, and corresponded the distances covered by the particles within times of 0.05, 0.5, 1, and 1.5 s, respectively. The jd1 histogram is shown in Figure 3.3a. Here, a total of four mobility fractions could be identified, each characterized by a specific diffusion coefficient. Three of these (I–III) were determined in the jumps from frame to frame (jd1 histogram). Not unexpectedly, fraction I (14%) had a surprisingly high diffusion coefficient, namely $D_{\mathrm{I}} = 4.0 \pm 1.3\,\mu\mathrm{m}^2\,\mathrm{s}^{-1}$. This fast mRNP fraction could no longer be detected after ≥0.5 s; clearly, the fast mRNPs had diffused out of the focal plane within 0.5 s. This was consistent with a similarly high *D*-value obtained for mRNPs as previously observed by Politz *et al.*, who used FCS to detect the diffusion of fluorescent poly(dT) [10]. These authors speculated that such rapid motion occurred within the interchromatin channels. The salivary gland cell nucleoplasm corresponded

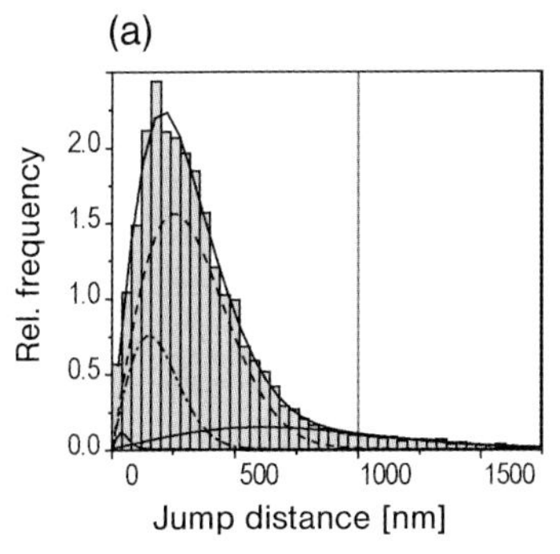

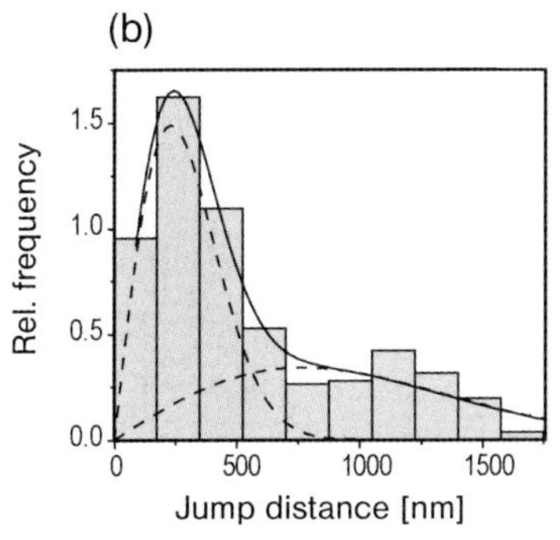

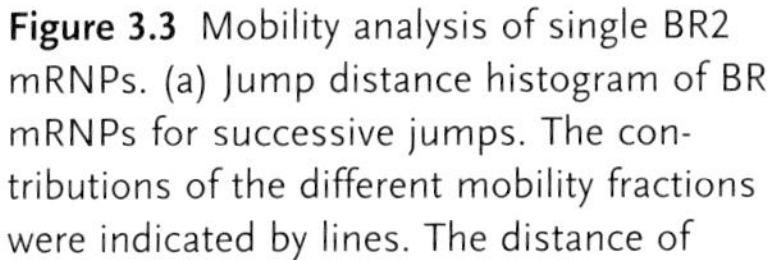

Figure 3.3 Mobility analysis of single BR2 mRNPs. (a) Jump distance histogram of BR mRNPs for successive jumps. The contributions of the different mobility fractions were indicated by lines. The distance of 1000 nm, which was only covered by the fraction of fast particles, is marked. (b) Histogram of those trajectories, which contained at least on jump beyond 1000 nm and four steps (see text for details).

almost everywhere to this nuclear domain, because it contained the chromatin in the very dense form of polytene chromosomes, leaving a major volume void of chromatin. In contrast, the interchromatin domains in mammalian cell nuclei were quite small. Interestingly, taking into account the effective viscosity of the nuclei of 3–5 cP (see Sections 3.4.3–3.4.5), particles of 50 nm diameter would move with a *D*-value of approximately 2.5 $\mu m^2 s^{-1}$ – which was almost equal to the value determined experimentally for the fast BR2 mRNPs, which were about this size. Hence, the BR2 mRNPs can indeed move as fast as would be expected for particles of their physical dimension in a medium with the viscosity of 3 cP. Consequently, the existence of the fast diffusion component fully supported the hypothesis of Politz and Pederson, that the slow-down of the mRNPs – which was always observed by FRAP and PA measuring mobility on a large distance scale – is caused by hindrances imposed on the particles by decondensed chromatin when they travel extended distances. However, on a local scale, mRNP mobility may be fast, since it is governed only by particle size and viscosity.

Besides the fast-diffusing species, decelerated mRNPs were also observed. Fraction II (66%) showed $D_{II} = 0.7 \pm 0.1\,\mu m^2 s^{-1}$, while fraction III (22%) was characterized by $D_{III} = 0.24 \pm 0.05\,\mu m^2 s^{-1}$. In addition to these, a fourth – very slow – fraction (IV) with $D \sim 0.015\,\mu m^2 s^{-1}$ became detectable after 0.5 s, and was more clearly observed after 1 s. As the MSD of this fraction increased linearly with time, it was clearly mobile, although quite slow. The locations where such slow mRNPs were observed were spread over the complete nuclei, and not limited to specific intranuclear domains. In summary, movement of the BR mRNPs in the nucleoplasm was found to be complex, with four BR mRNP fractions each having a different mobility; however, no evidence was found for completely immobilized particles.

3.5.2
BR mRNPs Diffuse in a Discontinuous Fashion

The four BR2 mRNP fractions with different mobilities could correspond either to four classes of different BR2 mRNPs, or to a single type of BR2 mRNPs changing mobility with time, for example, by transient interactions with further large intranuclear components. In order to discriminate between these alternative explanations, the histograms were analyzed in the following manner.

First, all trajectories were selected which fulfilled two conditions: the length was at least three jumps, and at least one jump between successive frames was >1000 nm. In this way, practically exclusively mRNPs of fraction I with $D = 4\,\mu m^2 s^{-1}$ were extracted. The particles of fractions II to IV virtually never jumped beyond 1000 nm between successive frames. Only fraction I particles performed such long jumps with a probability of 0.28. On the one hand, if different particle populations with distinct mobilities were to exist, only the fast fraction would be extracted by this trajectory filtering approach. On the other hand, in case particles would alter their mobility (or, more precisely, their *D*-value) along their track, then trajectories containing long jumps also would contain phases of slow motion, respectively short jumps.

The experimentally determined BR mRNP trajectories were filtered using this approach, and from the thus selected trajectories the jd1 histograms were again constructed. This histogram was clearly bimodal (see Figure 3.3b), while the analysis indeed retrieved two mobility fractions: a slow fraction (46 ± 3%) with $D = 0.58 \pm 0.04\,\mu m^2 s^{-1}$, and a fast fraction (54 ± 3%) with $D = 5.64 \pm 0.62\,\mu m^2 s^{-1}$. The slow fraction that was identified clearly represented a mixture of the fractions II to IV with $D = 0.015$, 0.24, and $0.7\,\mu m^2 s^{-1}$, although an explicit fit of the data with more than two fractions was not meaningful due to the limited number of jumps left after filtering. Altogether, this approach clearly demonstrated that more than one diffusion constant would be required to describe the motion of single BR mRNPs along their trajectories.

A Monte Carlo simulation supported the above analysis. First, two distinct particle populations diffusing with two different D-values ($3.74\,\mu m^2 s^{-1}$ and $0.64\,\mu m^2 s^{-1}$) were simulated in a nucleus-like geometry. The two mobilities were clearly distinguishable in the jump distance histogram of the simulated trajectories by curve fitting. However, when the above-used trajectory filter was applied to these data, only one population was found in the jump distance histogram of the retrieved trajectories featuring $D = 4.56 \pm 0.44\,\mu m^2 s^{-1}$. The thus-determined *D*-value was higher than the input *D*-value because the filter selected only the fastest trajectories. Next, the jump distance histogram of simulated particles, which changed their mobility between $D = 3.74\,\mu m^2 s^{-1}$ and $D = 0.64\,\mu m^2 s^{-1}$, with a probability of 0.5 after each diffusion step, was analyzed. Following application of the discussed trajectory filter, a jump distance histogram was again determined for the retrieved trajectories. As expected, a fit of the histogram yielded two mobility fractions: 60% with $D = 0.66 \pm 0.01\,\mu m^2 s^{-1}$, and 40% with $D = 3.61 \pm 0.2\,\mu m^2 s^{-1}$. In this case, two mobility fractions were obtained, as in the case of the experimental trajectories.

In summary, the BR mRNPs were found to vary their mobility along their tracks – a situation which can be designated as "discontinuous mobility."

3.5.3
Verification of Particle Identity and High mRNP Mobility

Following the specific labeling of BR2.1 mRNPs using a complementary fluorescent DNA oligonucleotide (as described above), it was possible to demonstrate, using a statistical analysis, that the observed particles moved discontinuously, notably with *D*-values in the range of 0.7 to $4\,\mu m^2 s^{-1}$. In order to eliminate the possibility that the fastest-moving component was caused by an RNase H-mediated degradation of the DNA–RNA hybrid molecules to produce faster-diffusing subfragments, extensive experiments were also performed using RNA oligonucleotides. For this, 2′-*O*-methyl-RNA oligonucleotides were 5′ end-labeled with the photostable dye ATTO647N to further improve the SNR. By using this labeling approach, it was possible to verify the previously determined *D*-values for the DNA-oligonucleotide-labeled mRNPs.

As outlined in Section 3.4.1, the heterogeneous nuclear ribonucleoprotein hrp36 is involved in the packing of the mRNPs [26]. This protein can be expressed bacterially, and then purified and fluorescently labeled. As with the experiments using DNA- and RNA-oligonucleotides, when the fluorescent protein was microinjected into living *C. tentans* salivary gland cell nuclei, the BR transcription site on chromosome IV became clearly labeled, thus demonstrating an efficient incorporation of the labeled hrp36 into the BR2 mRNPs (Figure 3.4a and b, arrows); this was in contrast to the coinjected fluorescently labeled bovine serum albumin (BSA). Since hrp36 is not associated exclusively with BR mRNAs, it was also found in several other transcription sites, for example, on chromosome IV or the other polytene chromosomes (as indicated by the asterisks in Figure 3.4b), although neither exhibited the intense hrp36 staining of the BR2 puff. This subnuclear distribution of the microinjected hrp36 in living gland cells corresponded very well with immunofluorescence labeling of the endogenous hrp36 in intact cells [26] or in isolated chromosomes [25].

Upon the injection of very small amounts of this protein into the nuclei, distinct particles were detected in the single-molecule microscope after a minimum incubation time of 10 min (Figure 3.5). Thus, it was assumed that a 10 min period was probably the minimum required for mRNP formation and hrp36 incorporation. As with the oligonucleotide-labeled BR particles, these particles could easily be imaged and tracked on a single-particle basis.

The hrp36-labeled particles had a jump distance distribution that was very comparable to the mobility pattern of the oligonucleotide-labeled BR2 mRNPs (see Table 3.1). Again, a fast component was detected that had a *D*-value which was significantly higher than was previously observed in other studies, and two less-mobile fractions with D-values of 0.73 and $0.1\,\mu m^2 s^{-1}$, respectively. As with the RNA-oligonucleotide labeled particles, those particles which were bound to the fluorescent hnRNP were not prone to digestion by RNAse H.

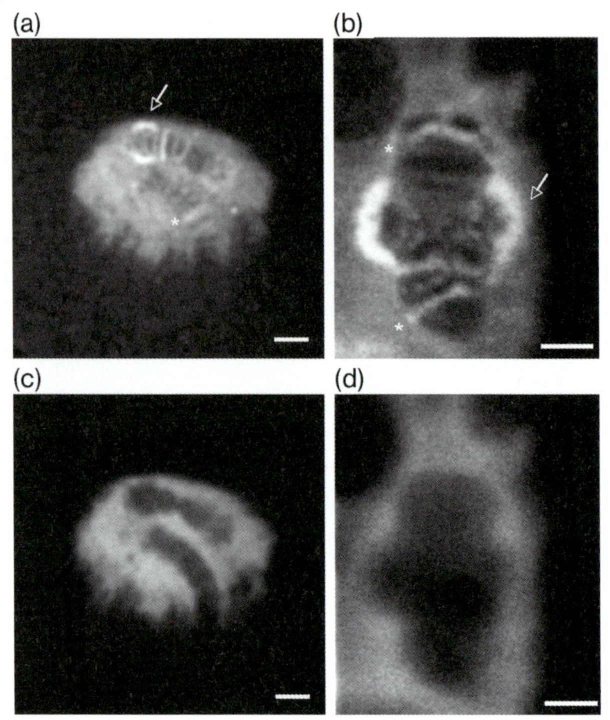

Figure 3.4 Subnuclear distribution of fluorescently labeled hrp36 and bovine serum albumin (BSA) within living gland cells. (a, b) The microinjected hrp36 is mainly found in BR2 puffs (arrows), and to a lesser extent at other transcriptions sites (asterisks). (c, d) Confocal image of coinjected BSA-AlexaFluor633, showing no distinct labeling of transcription sites. Scale bars: 10 μm in panels (a) and (c); 2 μm in panels (b) and (d).

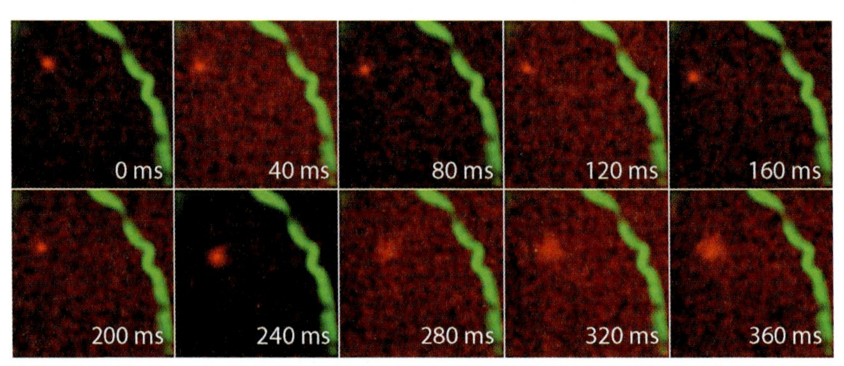

Figure 3.5 A single particle detected upon hrp36-microinjection. The green color indicates p10-AlexaFluor488 labeling the nuclear envelope; the red color indicates hrp36-Atto647N. The movie was acquired 30 min after microinjection. Field of view = $12.2 \times 12.2\,\mu m^2$.

Table 3.1 Comparison of mRNP mobility using the three different labeling approaches.

	A_{III} (%)	D_{III} ($\mu m^2 s^{-1}$)	A_{II} (%)	D_{II} ($\mu m^2 s^{-1}$)	A_I (%)	D_I ($\mu m^2 s^{-1}$)	N
DNA-labeled BR mRNP	22 ± 9	0.24 ± 0.05	65 ± 7	0.7 ± 0.1	14 ± 4	4.0 ± 1.28	49547
RNA-labeled BR mRNP	29 ± 2	0.11 ± 0.04	66 ± 1	0.44 ± 0.01	5 ± 0.8	3.51 ± 1	33048
hrp36-labeled BR mRNP	19 ± 2	0.1 ± 0.02	58 ± 3	0.73 ± 0.06	23 ± 3.5	4.4 ± 1	2546

It was concluded from these experiments that, with each of the labeling approaches, predominantly the very same intact mRNPs were observed. The fact that these different labeling methods produced particles with very comparable mobilities suggested that the labeling *per se* did not alter the particles' dynamic behavior.

3.5.4
A Discussion of mRNP Mobility

Numerous recently conducted studies have demonstrated the potential of single-molecule or single-particle tracking in living cells [38–40]. Single-molecule observations within the cell nuclei have allowed the analysis of nucleocytoplasmic transport and the characterization of intranuclear transport pathways, mobility restrictions, and intranuclear binding processes. The tracking of molecules and particles is especially feasible within the cell nuclei, as green and red autofluorescence in the nuclei is very low. The findings of the past ten years have revealed that the nuclear architecture is far more complex than was previously imagined. Indeed, the emerging view is that the nucleus is a highly organized cellular organelle, in which life processes such as DNA transcription and replication, as well as mRNA processing, take place in complex, spatiotemporal patterns. Even inert mobility probes do not perform with free Brownian motion, but rather suffer from complex restrictions to mobility [41]. This suggests the existence of structural barriers and the occurrence of binding–unbinding events for immobile or slowly moving supramolecular structures, or their enclosure into supramolecular networks.

The intranuclear trafficking of native mRNPs can be studied relatively easily in live cells, in real time, by using high-performance microscopy in a well-established model system for RNA biogenesis [24]. Here, it was found that a specific native mRNP particle, the BR2 mRNP, could be brightly labeled *in vivo* by using fluorescent DNA- or RNA-oligonucleotides. A further labeling strategy employing hrp36 protein marked predominantly the very same particles, mostly BR mRNPs. Hence, labeling by these two distinct approaches created fluorescent particles with very comparable mobilities, and suggested that neither strategy altered the dynamic behavior of the particles; therefore, a mobility analysis using either approach should yield valid results. Initially, a very fast mRNP mobility was observed which occurred in the chromatin-free nucleosol of the salivary gland cells. Hence, in regions devoid of chromatin, the mRNPs were able to move as predicted by the Stokes–Einstein law. Such high mobility was not due to a low viscosity in the salivary gland cell nucleoplasm; rather, an effective intranuclear viscosity of 3–5 cP was determined, this value being only slightly less than that in mammalian cell nuclei (4–8 cP). The Stokes radius of the BR2 mRNPs was determined as 18 ± 6 nm, which agreed well with previous electron microscopically derived values [28]; this verified that complete, single mRNPs, and not aggregates or fragments, had indeed been visualized. We clearly detected that the hrp36-labeled particles were exported out of the nucleus, whereas the oligonucleotide-tagged particles could not be detected at the nuclear envelope. As the hrp36 visualizes the same particles as the oligonucleotides it can be assumed that, prior to export, the oligonucleotides are "stripped off" from the mRNA, possibly by a RNA helicase activity or as the result of a special chemical environment within the nuclear pore complexes, whilst at least some packing proteins (e.g., hrp36) remain associated.

It is reasonable to conclude that most – if not all – BR particles in the nucleoplasm are destined to be transported to the cytoplasm during the conditions used in these studies. Since the analysis of labeled BR particles shows that a single BR particle characteristically displays both fast and slow motion, it is probable that the BR particles to be transported also behave in this manner. However, according to the study results, a firm conclusion will require the direct observation of a single BR particle when it moves to and through the nuclear pore into cytoplasm. Such a demanding experiment remains to be conducted.

Today, high-performance microscopy has enabled the intranuclear trajectories of individual BR2 mRNPs to be traced, and confirmed earlier suggestions that mRNPs move randomly within the nucleus, most likely by Brownian motion. Yet, it has been shown that BR2 mRNPs do not move unhindered, even within the chromatin-free nucleosol. Rather, four different mobility fractions have been identified, with the fastest BR2 mRNPs moving as might be expected by their size and effective nuclear viscosity, while other BR2 mRNP fractions moved considerably more slowly. However, no particles were found that were completely immobilized.

It was shown, unambiguously, that the mRNPs with a high mobility also exhibited extended decelerated phases along their pathways, and this type of motion was designated as "discontinuous." Such very slow diffusive motion occurred all over the nucleoplasm, and was not limited to specific domains, such as near the polytene chromosomes or the nucleoli. Thus, BR particles are likely to be associated transiently with submicroscopic *nonchromatin* structures of large size, corresponding to a low mobility in the nucleoplasm. As outlined above, it has been shown, by using electron tomography, that a large fraction of the BR mRNPs contained connecting fibers that could also establish connections to nonchromatin structures; these were referred to as fibrogranular clusters [29]. The results of the present authors corresponded well with these observed interactions of BR2 particles, because the phases of slow diffusive motion might indicate binding to large molecular structures. These might correspond also to the fibrogranular clusters because, whilst the presumably bound mRNPs were not fixed they still showed a low mobility, as might be expected for large macromolecular aggregates. Taken together, these morphology and mobility data suggest that BR particles bind transiently via the CFs to the FGCs within the nucleoplasm, and this results in a drastic slowing of particle movement. It might also be expected that specific biochemical processes could occur at the FGCs, although at present this is only speculation.

3.6 Outlook: Light Sheet-Based Single-Molecule Microscopy

During the investigations into BR mRNP trafficking, a fast confocal line-scanning microscopy set-up was used for many measurements, on the basis that the oligonucleotide-labeled BR mRNPs were extraordinarily bright. Moreover, this optical sectioning technique yielded a very good SNR. For these studies in general, however, single-particle or single-molecule microscopy is employed, using epifluorescence microscopes equipped with video cameras as the recording devices. One disadvantage of this approach is that molecules within out-of-focus regions become excited and drastically reduce the SNR of those molecules in focus, which in turn reduces the achievable precision of localization.

An elegant solution to this problem has been provided, however, by the use of an older technique, namely *ultramicroscopy* [42], which has recently undergone something of a revival [43–47]. The sample is illuminated using *a light sheet* in a perpendicular direction to the observation axis, so as to create an intrinsic optical sectioning effect. As a consequence, no out-of-focus fluorescence is generated, which in turn causes an efficient reduction in the background signal and greatly improves the SNR. Furthermore, photobleaching is minimized, because only the thin focal plane of interest in the specimen is illuminated, and all illuminated molecules contribute to the image signal generation. This smart principle can easily be incorporated into a high-speed video microscope with single-molecule sensitivity, thus providing a further application of the technique [48].

Light sheet-based microscopy (LSBM) represents an especially promising approach for single-molecule tracking in living cells and tissues. The study of adherent monolayer culture cells using LSBM is feasible but demanding, due mainly to a possible distortion of the strongly focused illumination light sheet by the glass coverslip. Light sheet-based single-molecule (respectively nanoparticle) observations are, however, relatively straightforward for larger specimens such as polymers, tissue slices, 3-D cell cultures, complete organs such as the *C. tentans*

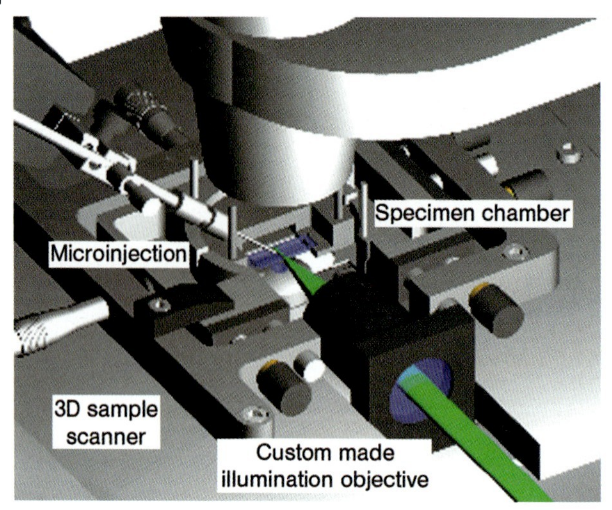

Figure 3.6 Schematic representation of the inverted light sheet-based microscope, using a commercial inverse microscope. This geometry allows microinjection of the sample.

salivary glands, and even whole organisms such as embryos or zebrafish. Often it is required that the utilized specimen in the sample chamber are accessible for microinjection and -manipulation. Clearly, the sample chamber must also allow perpendicular illumination using a cylindrical high-NA optics, and imaging by a high-NA detection objective lens. Fortunately, both of these requirements can be met by using a small, open specimen chamber made from polished glass with a glass bottom and featuring a standard coverslip thickness (170 μm). The chamber is mounted in a special holder on the stage of a commercial inverse microscope (Figures 3.6 and 3.7).

Notably, this technique has further advantages. For example, on occasion it is necessary that a fraction of the probe molecules is bleached, if the concentration is above the level appropriate for individual particle discrimination. This can be achieved with LSBM in an ideal manner, as the focal plane can be photobleached selectively without compromising fluorescence labels in other sample regions. The completely unaffected, fluorescent molecules or particles will then diffuse back into the focal region. Finally, single-particle tracking using LSBM can easily be extended to the real-time tracking of particles in three spatial dimensions. The use of a weak cylindrical lens in the optical imaging path will produce single-particle signals that are circular in focus, but ellipsoidal above and below the focal plane, with the major axis of the ellipsoid shifted by 90° in passing through the focus position [49, 50]. The axial particle coordinate can then be determined in straightforward manner from the signal shape and orientation. In addition, the information relating to the axial particle position can be used to maintain the particle in focus in real time, using a piezo element to control the axial

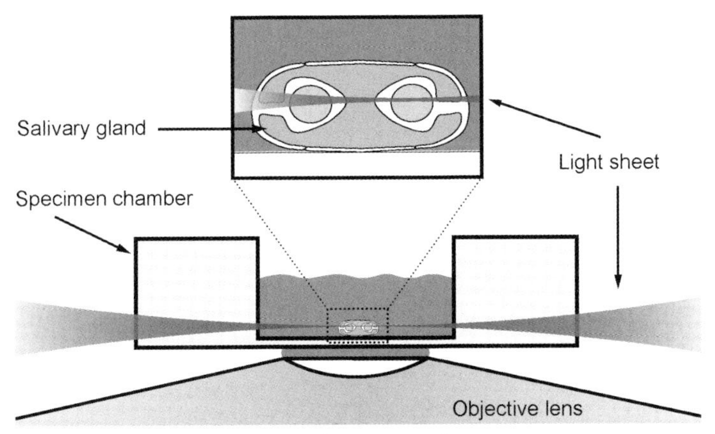

Figure 3.7 Schematic showing the illumination of a *C. tentans* salivary gland in a light sheet-based microscope.

position of the specimen. The full 3-D trajectory of the particle can then be determined.

By employing the superior features of LSBM, the hope is to obtain new insights into the intranuclear dynamics of mRNP particles. Certainly, future investigations into mRNP biogenesis in *C. tentans* salivary gland cells at a single-particle level will profit hugely from the LSBM approach, notably because the cells are large and the distance between the nuclei and the support is ≥100 μm. In particular, it is expected that this approach will allow an examination of the nuclear export of hrp36-labeled mRNPs, with unprecedented temporal and spatial resolution.

Acknowledgments

U.K. and J.-P.S. are indebted to Reiner Peters for successful and longstanding cooperation. U.K. would also like to thank Thorsten Kues and David Grünwald for their great help during his first steps into single-molecule imaging. Finally, the authors thank Werner Wendler for preparing Figure 3.6.

References

1 Handwerger, K.E. and Gall, J.G. (2006) *Trends Cell Biol.*, **16**, 19.
2 Misteli, T. (2007) *Cell*, **128**, 787.
3 Verkman, A.S. (2002) *Trends Biochem. Sci.*, **27**, 27.
4 Gorski, S.A., Dundr, M., and Misteli, T. (2006) *Curr. Opin. Cell Biol.*, **18**, 284.
5 Mueller, F., Wach, P., and McNally, J.G. (2008) *Biophys. J.*, **94**, 3323.
6 Braga, J., McNally, J.G., and Carmo-Fonseca, M. (2007) *Biophys. J.*, **92**, 2694.
7 Stutz, F. and Izaurralde, E. (2003) *Trends Cell Biol.*, **13**, 319.

8 Stewart, M. (2007) *Mol. Cell*, **25**, 327.
9 Kohler, A. and Hurt, E. (2007) *Nat. Rev. Mol. Cell. Biol.*, **8**, 761.
10 Politz, J.C., Browne, E.S., Wolf, D.E., and Pederson, T. (1998) *Proc. Natl Acad. Sci. USA*, **95**, 6043.
11 Politz, J.C., Tuft, R.A., Pederson, T., and Singer, R.H. (1999) *Curr. Biol.*, **9**, 285.
12 Cremer, T., Kurz, A., Zirbel, R., Dietzel, S., Rinke, B., Schrock, E., Speicher, M.R., Mathieu, U., Jauch, A., Emmerich, P., Scherthan, H., Ried, T., Cremer, C., and Lichter, P. (1993) *Cold Spring Harbor Symp. Quant. Biol.*, **LVIII**, 777.
13 Calapez, A., Pereira, H., Calado, A., Braga, J., Rino, J., Carvalho, C., Tavanez, J., Wahle, E., Rosa, A., and Carmo-Fonseca, M. (2002) *J. Cell Biol.*, **159**, 795.
14 Molenaar, C., Abdulle, A., Gena, A., Tanke, H., and Dirks, R. (2004) *J. Cell Biol.*, **165**, 191.
15 Politz, J.C., Tuft, R.A., Prasanth, K.V., Baudendistel, N., Fogarty, K.E., Lifshitz, L.M., Langowski, J., Spector, D.L., and Pederson, T. (2006) *Mol. Biol. Cell*, **17**, 1239.
16 Shav-Tal, Y., Darzacq, X., Shenoy, S.M., Fusco, D., Janicki, S.M., Spector, D.L., and Singer, R.H. (2004) *Science*, **304**, 1797.
17 Vargas, D.Y., Raj, A., Marras, S.A., Kramer, F.R., and Tyagi, S. (2005) *Proc. Natl Acad. Sci. USA*, **102**, 17008.
18 Siebrasse, J.P., Veith, R., Dobay, A., Leonhardt, H., Daneholt, B., and Kubitscheck, U. (2008) *Proc. Natl Acad. Sci. USA*, **105**, 20291.
19 Daneholt, B. (2001) *Chromosoma*, **110**, 173.
20 Wieslander, L. (1994) *Prog. Nucleic Acid Res. Mol. Biol.*, **48**, 275.
21 Kiseleva, E., Wurtz, T., Visa, N., and Daneholt, B. (1994) *EMBO J.*, **13**, 6052.
22 Skoglund, U., Andersson, K., Strandberg, B., and Daneholt, B. (1986) *Nature*, **319**, 560.
23 Dreyfuss, G., Matunis, M.J., Pinol-Roma, S., and Burd, C.G. (1993) *Annu. Rev. Biochem.*, **62**, 289.
24 Kiesler, E. and Visa, N. (2004) *Prog. Mol. Subcell. Biol.*, **35**, 99.
25 Singh, O.P., Visa, N., Wieslander, L., and Daneholt, B. (2006) *Chromosoma*, **115**, 449.
26 Visa, N., Alzhanova-Ericsson, A.T., Sun, X., Kiseleva, E., Bjorkroth, B., Wurtz, T., and Daneholt, B. (1996) *Cell*, **84**, 253.
27 Percipalle, P., Fomproix, N., Kylberg, K., Miralles, F., Bjorkroth, B., Daneholt, B., and Visa, N. (2003) *Proc. Natl Acad. Sci. USA*, **100**, 6475.
28 Singh, O.P., Bjorkroth, B., Masich, S., Wieslander, L., and Daneholt, B. (1999) *Exp. Cell Res.*, **251**, 135.
29 Miralles, F., Overstedt, L., Sabri, N., Aissouni, Y., Hellman, U., Skoglund, U., and Visa, N. (2000) *J. Cell Biol.*, **148**, 271.
30 Miralles, F. and Visa, N. (2001) *Exp. Cell Res.*, **264**, 284.
31 Lang, I., Scholz, M., and Peters, R. (1986) *J. Cell Biol.*, **102**, 1183.
32 Braeckmans, K., Peeters, L., Sanders, N.N., De Smedt, S.C., and Demeester, J. (2003) *Biophys. J.*, **85**, 2240.
33 Grunwald, D., Hoekstra, A., Dange, T., Buschmann, V., and Kubitscheck, U. (2006) *ChemPhysChem*, **7**, 812.
34 Anderson, C.M., Georgiou, G.N., Morrison, I.E.G., Stevenson, G.V.W., and Cherry, R.J. (1992) *J. Cell Sci.*, **101**, 415.
35 Kues, T., Dickmanns, A., Lührmann, R., Peters, R., and Kubitscheck, U. (2001) *Proc. Natl Acad. Sci. USA*, **98**, 12021.
36 Schutz, G.J., Schindler, H., and Schmidt, T. (1997) *Biophys. J.*, **73**, 1073.
37 Skoglund, U., Andersson, K., Bjorkroth, B., Lamb, M.M., and Daneholt, B. (1983) *Cell*, **34**, 847.
38 Peters, R. (2007) *Annu. Rev. Biophys. Biomol. Struct.*, **36**, 371.
39 Joo, C., Balci, H., Ishitsuka, Y., Buranachai, C., and Ha, T. (2008) *Annu. Rev. Biochem.*, **77**, 51.
40 Siebrasse, J.P. and Kubitscheck, U. (2009) *Methods Mol. Biol.*, **464**, 343.
41 Grunwald, D., Martin, R.M., Buschmann, V., Bazett-Jones, D.P., Leonhardt, H., Kubitscheck, U., and Cardoso, M.C. (2008) *Biophys. J.*, **94**, 2847.
42 Siedentopf, H. and Zsigmondy, R. (1903) *Ann. Physik.*, **4**, 1.
43 Voie, A.H., Burns, D.H., and Spelman, F.A. (1993) *J. Microsc.*, **170**, 229.

44 Huisken, J., Swoger, J., Del Bene, F., Wittbrodt, J., and Stelzer, E.H. (2004) *Science*, **305**, 1007.

45 Engelbrecht, C.J. and Stelzer, E.H. (2006) *Opt. Lett.*, **31**, 1477.

46 Dodt, H.U., Leischner, U., Schierloh, A., Jahrling, N., Mauch, C.P., Deininger, K., Deussing, J.M., Eder, M., Zieglgansberger, W., and Becker, K. (2007) *Nat. Methods*, **4**, 331.

47 Huisken, J. and Stainier, D.Y. (2007) *Opt. Lett.*, **32**, 2608.

48 Ritter, J.G., Veith, R., Siebrasse, J.P., and Kubitscheck, U. (2008) *Opt. Express*, **16**, 7142.

49 Kao, H.P. and Verkman, A.S. (1994) *Biophys. J.*, **67**, 1291.

50 Holtzer, L., Meckel, T., and Schmidt, T. (2007) *Appl. Phys. Lett.*, **90**, 053902.

4
Quantum Dots: Inorganic Fluorescent Probes for Single-Molecule Tracking Experiments in Live Cells

Maxime Dahan, Paul Alivisatos, and Wolfgang J. Parak

4.1
Introduction

Light microscopy is central to the development of modern cell biology. Because it is a sensitive and noninvasive approach, optical imaging provides an invaluable tool to decipher complex cellular processes. Over the past decade, advances in optical instrumentation, chemical probes, cell biology techniques and data processing, have enabled ever-more sensitive measurements in live cells; indeed, recent novel approaches have pushed the detection limit down to the ultimate level of single molecules. It has thus become possible to follow, in real time, the motion of individual proteins or nucleic acids in their natural cellular environment. Since they provide direct access to molecular properties in a cellular context, single-molecule experiments promise to elucidate many aspects of cell organization and dynamics. In fact, single-molecule measurements have already been used fruitfully to address important questions related to molecular diffusion [1–4] and transport [5, 6], membrane compartmentation [7], gene expression [8, 9], protein–protein interactions [10], or cell signaling [11].

On the basis of its high sensitivity, *fluorescence microscopy* is a key technology for optical detection and single-molecule experiments [12]. In general, the molecules of interest must be tagged with a fluorescent marker in order to be optically detected, which means it is essential that the properties of the tags used in the experiments must also be known. An "ideal" fluorescence marker must fulfill several requirements. From an optical point of view, it should be very bright (i.e., have a large absorption cross-section and a high quantum yield), be photostable (i.e., no photobleaching), have a low amount of fluorescence intermittency (also known as "blinking"), and have a narrow emission spectrum (to provide an easily identifiable optical signature). From a physico-chemical point of view, the marker should neither affect the molecule to which it is attached, nor the molecule's environment. Furthermore, the marker requires suitable biochemical binding sites for its attachment to the molecule that is to be labeled. At present, no such "ideal" fluorescence marker, with optical, chemical, and biochemical properties

Single Particle Tracking and Single Molecule Energy Transfer
Edited by Christoph Bräuchle, Don C. Lamb, and Jens Michaelis

ISBN: 978-3-527-32296-1

that satisfy all of the above conditions, is available. However, the brightness and photostability of semiconductor quantum dots (QDs) make them very suitable makers for high-sensitivity measurements in general, and for single-molecule tracking experiments in particular.

These advantages will be discussed in this chapter, where several fluorescent labels, both organic and inorganic, that are used for single-molecule tracking, are first introduced, and their respective merits and limitations briefly discussed. Attention is then focused on the optical properties of QDs, and details of their chemical synthesis, solubilization and functionalization are presented. Finally, some recent applications of QDs for single-molecule imaging in live cells are reviewed, and the future prospects of these techniques in cell biology discussed.

4.2 Fluorescent Labels for Single-Molecule Tracking in Cells

4.2.1 Organic Fluorophores

Although, traditionally, the fluorescence markers used for tracing molecules have been organic molecules, today a wide range of dyes is available commercially that spans virtually the whole visible spectrum. Moreover, these dyes can be attached, using well-established conjugation protocols, to a large variety of biomolecules, including proteins, nucleic acids, and sugars [13]. Arguably the biggest advantage of organic fluorophores is their small size; fluorescein, for example, has a molecular weight of only 376 Da. Thus, the attachment of such small labels to bigger molecules (e.g., proteins or oligonucleotides) would be expected to have a negligible effect on the biofunctionality of the molecule. Unfortunately, organic fluorophores exhibit limited photostability and, upon repetitive excitation–emission cycles, tend to be rapidly photo-destroyed. This process, which often is referred to as "photobleaching," imposes a severe limitation for tracing applications as it restricts the time interval over which the target molecule can be tracked optically. When investigating a single organic fluorophore, the observation time is usually limited to a couple of seconds (Table 4.1), which often is too short to fully investigate complex cellular processes.

4.2.2 Fluorescent Proteins

Rather than being labeled with organic fluorophores via bioconjugation chemistry, proteins can be genetically modified to become fluorescent. For this, nucleotide sequences encoding for fluorescent proteins (FPs) can be added to the genome so that, upon protein biosynthesis, a fusion product of the target protein and a fluo-

Table 4.1 Summary of the different markers used for single-molecule imaging.

Markers	Size (nm)	Extinction coefficient ($\times 10^4\ M^{-1}\,cm^{-1}$)	Photostability (s)	Structure
Organic fluorophores	~1	1–20	1–10	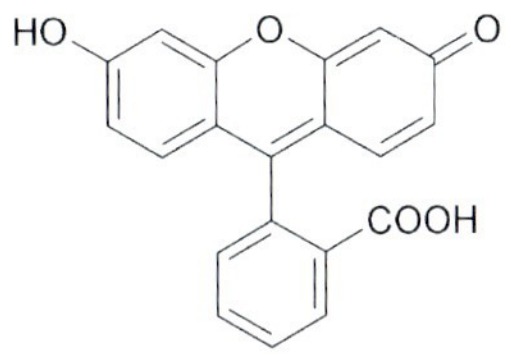
Fluorescent proteins	2–4	1–5	<1	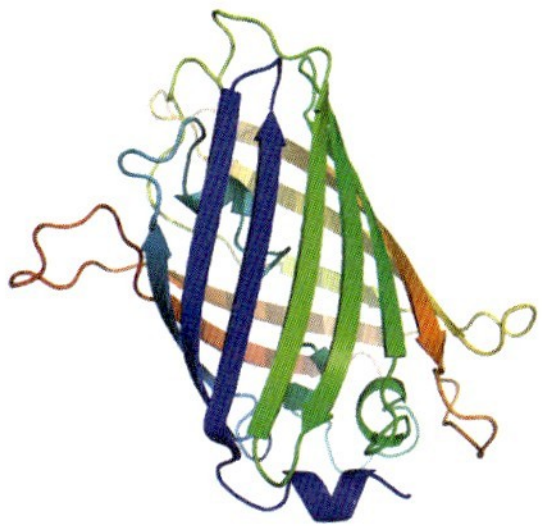
Colloidal QDs	2–10 (size of inorganic core) 10–30 (hydrodynamic diameter)	10–1000	>1000	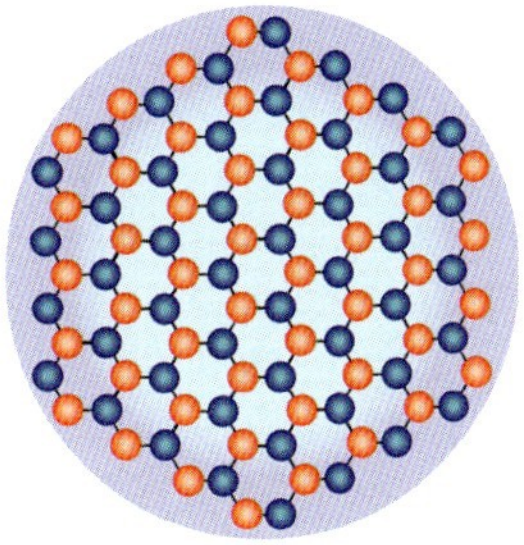

rescent protein will be expressed by the cell. In many cases, the resultant construct conserves the characteristics of the target protein, while being also fluorescent due to the genetically added domain. Unquestionably, the advent of genetically encoded fluorescent markers (for which the Nobel Prize for Chemistry was awarded in 2008) has revolutionized the biochemical methods used to selectively visualize molecules or cellular compartments, and to record their properties in living systems [12, 14]. A multitude of fluorescent reporters and sensors, such as green fluorescent protein (GFP) and its many variants, can now be prepared and inserted into the genome of various organisms, offering a wide flexibility to approach complex biological questions. Genetically modified proteins are particularly useful for the investigation of intracellular processes, as the proteins are directly generated through the natural protein expression pathways. In contrast, labeling a target biomolecule with a fluorophore requires the organic dye to be coupled to a purified version of the biomolecule, outside the cellular environment. As a result, for intracellular applications the labeled proteins must be reintroduced into the cytosol, creating many experimental difficulties.

Nevertheless, fluorescent proteins also present their own limitations, notably for single-molecule experiments. First, they are by no means small labels; typically, their molecular weight is close to 30 kDa, which corresponds to a physical dimension of between 2 and 4 nm (Table 4.1). More problematic is the fact that, compared to organic fluorophores, fluorescent proteins tend to absorb light less efficiently. In fact, their extinction coefficients usually vary between 10^4 and 10^5, a factor of 2 to 10 lower than the best organic dyes. Finally, they are even less photostable than organic fluorophores and, when imaged at the single-molecule level, usually photobleach in less than 1 s.

4.2.3 Fluorescent Microspheres

Strategies have been developed to circumvent some of the problems (notably the limited photostability) encountered with single organic fluorophores or fluorescent proteins. One approach consists of grouping several fluorophores within a bead to form a new single particle. For example, polymeric spheres the size of a few tens of nanometers up to a few micrometers, can be impregnated with organic fluorophores. As each sphere comprises many fluorophores, its photophysical properties correspond to those of an ensemble of dyes. As a result of this averaging, problems associated with individual emitters, such as blinking, can be bypassed. Due to the large number (tens to thousands) of fluorophores within one particle, lower excitation intensities are required in order to achieve a good detection efficiency, which in turn means that photobleaching and other intensity-related problems are also reduced. Naturally, fluorescent spheres are bigger than single organic fluorophores, and therefore might potentially interfere with the molecular properties of the target molecule. Consequently, the improved photophysical properties are acquired at the cost of a size increase or reduced colloidal stability. In fact, practical problems encountered with fluorescent microspheres are such that they are seldom employed for single-molecule experiments in cells.

4.2.4 Colloidal Quantum Dots

The emergence of functional inorganic nanomaterials has recently added new elements to the toolbox with which biological systems can be investigated [15–17]. These nanomaterials possess physical properties (optical, electrical, magnetic, etc.)

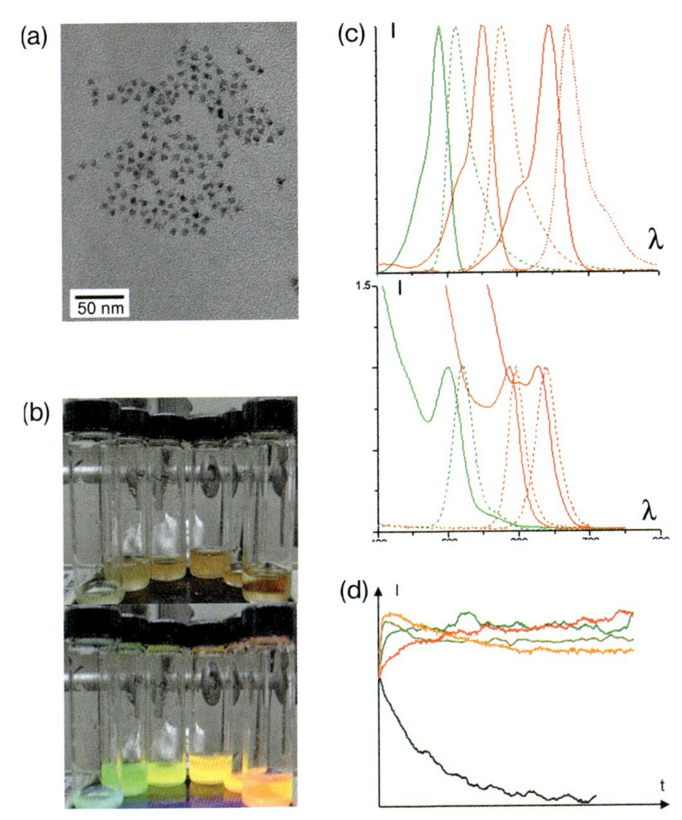

Figure 4.1 (a) Transmission electron microscopy (TEM) image of polymer-coated CdSe/ZnS quantum dots (QDs). Adapted from Ref. [18]. (b) Solutions of QDs of different size (left image). Under illumination with a hand-held UV lamp, the size-dependent fluorescence of the CdSe QDs can be seen. Adopted from Ref. [19]. (c) Comparison of absorption (solid curve) and fluorescence (dashed curve) spectra of organic fluorophores and colloidal QDs. Upper image: Normalized absorption and fluorescence spectra in water of three typical organic fluorophores (fluorescein, tetramethylrhodamine (TAMRA), and Cy5). Lower image: Normalized absorption and fluorescence spectra in water of polymer-coated CdSe/ZnS QDs of different sizes. Adopted from Ref. [16]. (d) Time-dependence of the fluorescence intensity of silanized QDs and rhodamine 6G. The QDs exhibit a stable emission for at least 4 h, while the organic fluorophores bleach after 10 min. The data correspond to nanocrystals of four different colors of emission and to rhodamine 6G. Adopted from Ref. [20].

which are often superior to those of their organic counterparts, opening up fascinating prospects for advanced sensing and detection schemes in fundamental and applied biomedical research.

Colloidal QDs are probably the most prominent example of nanomaterials used in a biological context [17]. Quantum dots are semiconductor nanocrystals in which the diameter of the optically active inorganic core varies between 2 and 10 nm (Table 4.1; Figure 4.1a), intermediate between organic molecules and polymeric spheres. Dependent on the material and the size of the particle, QDs emit fluorescence at different wavelengths in the visible or infrared (IR) region (see Section 4.3; Figure 4.1b). Once solubilized and functionalized, QDs can be used as biological fluorescent probes [21, 22] although, due to their crystalline nature, they have distinct optical properties compared to conventional organic fluorophores or fluorescent proteins. In particular, their extinction coefficient often exceeds 10^6 in the visible spectrum, which means that they are bright emitters that can be detected individually, with a high signal-to-noise ratio (SNR). Furthermore, QDs are much more photostable than organic dyes, most likely due to their crystalline structure, which is more robust against photoinduced damage than is that of organic chains. This combination of brightness and photostability makes QDs especially appealing for single-molecule experiments. During the past few years, a rapidly growing number of experiments have shown that the ability to track individual biomolecules over extended periods of time in live cells, opens many prospects for advanced biological assays. However, the development of novel QD-based assays requires not only a good understanding of their optical properties but also a control of their colloidal and biochemical properties. Hence, details of the optical properties, synthesis, solubilization and functionalization of QDs are provided in the following sections.

4.3 Optical Properties of Colloidal Quantum Dots

4.3.1 Absorption and Emission Properties

As the inorganic core of QDs is only a few nanometers in diameter (typically 2–10 nm; see Figure 4.1a), their energy levels are determined by quantum mechanical effects. Charge carriers (electrons and holes) are confined to the nanometer-dimension of the colloidal particles, which results in a confinement energy [23] that adds to the energy levels of the bulk material. Therefore, the energy gap in a semiconductor nanoparticle, which is the energy difference between the valence band (equivalent to the highest occupied level) and the conduction band (equivalent to the lowest unoccupied level), is larger than that for bulk material, and can be finely tuned with the size of the QD (see Figure 4.1b) [23]. Importantly, it is not only the average size of QDs but also their size distribution that can be controlled at the nanoscale, with a relative dispersion in radius equal or inferior to 5%.

The relaxation of light-generated electron hole pairs may take place not only through fluorescence emission but also through nonradiative pathways. For example, if one of the charge carriers reacts with trap states at the QD surface, then no light will be emitted. In order to raise fluorescence quantum yields as high as possible, the electron-hole pairs must be kept away from the QD surface, and a general strategy in this direction is to overcoat the QDs with a thin layer of a semiconductor material with a higher band gap. For this, one of the most common systems is CdSe/ZnS, in which CdSe cores are overcoated with a shell of ZnS [24]. As the band gap of ZnS is higher than that of CdSe, light-generated electrons and holes do not possess sufficient energy to completely enter from the CdSe core into the ZnS shell, and are thus confined to the CdSe core. As ZnS can be grown quasiepitaxially on CdSe, there are only few surface states at the CdSe/ZnS interface, and thus the fluorescence yield is enhanced. This effect is important both for the stability and intensity of the fluorescence signal. For the best samples, the fluorescence quantum yield can exceed 50%, comparable to good organic fluorophores.

4.3.2
QDs as Fluorescent Biological Probes

The interest in QDs for biological imaging stems from a combination of unique photophysical properties [17, 21, 22]. For example, QDs possess a narrow emission spectrum (<30 nm, without red tail; see Figure 4.1c), with a peak position determined by the energy gap. As indicated above, the gap depends on the size of the semiconductor core, and can be precisely adjusted during their synthesis by inorganic chemistry. Quantum dots also have a large absorption spectrum, due to the fact that virtually any photons with energy higher than the gap can be absorbed by the nanoparticles (see Figure 4.1c). As a consequence, QD samples with a distinguishable emission wavelength can be excited with a single laser line [25], provided that the laser photons have a wavelength lower than the shortest of all emission wavelengths. Because their emission spectra is narrow, there is little crosstalk between emission channels. As QDs are also much less prone to photodestruction, they will fluoresce over much longer durations than organic emitters (Figure 4.1d). For these reasons, QDs appear ideal for multicolor detection, which was the primary motivation when they were introduced into biology [21, 22]. The spectral properties of QDs have also been shown as advantageous for energy-transfer experiments when used as donors [26, 27].

Of note, QDs have other – sometimes overlooked – interesting optical properties. In particular, their radiative lifetime is ~10–20 ns, which is longer than the radiative lifetime of the autofluorescence signal (~1–5 ns) that limits the detection sensitivity in cells or tissues. As a result, efficient time-gating can performed, leading to enhanced SNRs in biological imaging [28]. As they are composed of semiconductor materials, QDs are electron-dense and can be visualized using transmission electron microscopy (TEM) in biological environments [3, 29] (Figure 4.1a). This offers many interesting possibilities for so-called "correlative microscopy," where both optical and electronic imaging are employed to access the structure and dynamics of biological specimens. Finally, the synthesis and engineering of QDs with different semiconductor materials and structures have expanded the range of possible emission wavelengths, from the visible to the red and IR regions, which are more appropriate for imaging in tissues and animals [30].

4.3.3
Single Quantum Dot Detection

An attractive feature of QDs is the sensitivity with which individual nanoparticles can be detected [31] (Figure 4.2a). The high SNR primarily derives from the extinction coefficient ε of the QDs, which is on the order of $10^6\,M^{-1}cm^{-1}$ in the visible spectrum, some 10- to 100-fold higher than that of standard organic fluorophores (dyes or GFP) (Table 4.1). Under continuous excitation with intensity I, the fluorescence photon detection rate S varies as $\alpha Q \varepsilon I$ where Q is the fluorescence quantum yield (typically around 0.5) and α is the detection efficiency (up to 3–5% for good optical microscopes). For a given value of I, this means that it is possible to detects 10- to 100-fold more fluorescence photons with a QD than with a fluorophore. Under typical experimental conditions, S can reach $10^5\,s^{-1}$, such that the SNR (limited by the shot-noise) $\sqrt{S\tau}$ is on the order of 30 for an acquisition time $\tau \sim 10\,ms$. The photostability of QDs means that their fluorescence emission can be recorded over long periods of time. For single organic emitters, a useful parameter is the average number N_p of detected photons before photobleaching; N_p is on the order of 10^5 for a fluorescent protein, and increases up to 10^6 for a good

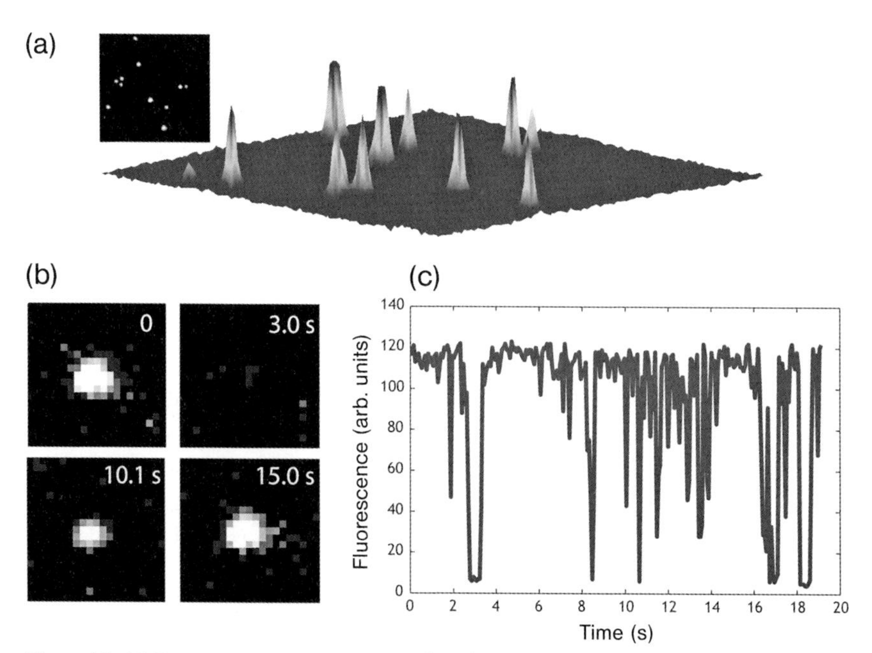

Figure 4.2 (a) Fluorescence intensity image of single QDs. The inset shows a 2-D image of the same field of view. The position of each peak can be determined with an accuracy in the 10–50 nm range, depending on the SNR. (b) Blinking of a QD under continuous excitation. The images are extracted from an image sequence. (c) Fluorescence intensity of the same QD during the full sequence.

organic fluorophore. In comparison, 10^8–10^9 photons can be routinely detected for a single QD. This number does not necessarily reflect an upper bound imposed by photodestruction but, rather, results from other physical or biological factors limiting the total duration of the experiment.

4.3.4
Fluorescence Intermittency of Individual Quantum Dots

During the past decade, single-molecule optical experiments have amply shown that, when observed individually, fluorescence emitters exhibit properties that remain hidden at the ensemble level. With regards to the detection of individual QDs, the most striking–and possibly unwelcome–effect is arguably that of blinking when, under continuous illumination, the emission randomly alternates between bright ("on") and dark ("off") periods (Figure 4.2) [31]. Of note, similar intermittency effects have been observed with virtually all fluorescent emitters (dyes, fluorescent proteins), but such effects are probably more spectacular with QDs due to the long observation times made possible by the photostability of the nanoparticles. Although it is not completely understood, the physical origin of the blinking phenomenon in QDs is commonly attributed to a transient ionization of the nanoparticle [32]. Upon light excitation, an electron-hole pair is created within the QD that can relax through different competing decay channels. Most frequently, the pair recombines radiatively by emitting a fluorescence photon. However, with a low probability, the electron (or the hole) can tunnel out of the QD toward surface traps. While the QD is in this ionized state, the nonradiative recombination rate is enhanced and the fluorescence emission dramatically quenched. At a later time, the trapped charge tunnels back into the inorganic core, and the nanoparticle resumes its regular fluorescence cycles. Both the "on" and "off" times are distributed according to a power-law [32, 33], which means that the mechanisms behind the blinking cannot be described simply in terms of Poisson processes. Blinking is certainly a limitation for the use of QDs as a nanoscopic light source, and much effort has been expended in attempts to reduce it. Recently, progress towards nonblinking nanoparticles has been reported, although not yet for water-soluble nanoparticles, and at the cost of an increase in the particle size [34, 35]. In practice, fluorescence blinking is, nevertheless, a convenient criterion to evaluate whether individual nanoparticles are observed, since it is normally assumed that an intermittent fluorescent spot is indicative of a single QD emission.

4.4
Synthesis of Colloidal Fluorescent Quantum Dots

Colloidal QDs are fluorescent crystalline nanoparticles made from semiconductor materials, which are dispersed in a solvent [16, 17, 36]. Although alternative methods exist, most common colloidal QDs are grown using a wet-chemistry approach that requires at least two different ingredients. First, the molecular precursors must contain the atomistic or molecular building blocks from which the QD will be constructed. Second, surfactant molecules are needed to stabilize the QDs (see Figure 4.3). In a typical synthesis, the precursors are injected into a solvent mixture that includes the surfactant molecules. The reaction of the precursors to the designated material is initiated, for example, by heating, such that a molecular decomposition of the precursors results in reactive atoms/molecules, or by the addition of a reducing agent [37]. Hereby, the kinetics of the reaction is controlled by the surfactant molecules [38], which may coordinate to the precursors. It is possible to imagine two extreme cases: (i) when the surfactants bind very strongly to the precursors, no reaction will occur, as the precursor is caged by surfactant; and (ii) when the surfactants do *not* bind to the precursors, the highly reactive precursors will rapidly agglomerate in an uncontrolled way to a stoichiometrically undefined material driven by van der Waals' interaction. The trick in controlling particle growth is, therefore, based on the choice of surfactants with moderate binding affinity. Wherever a surfactant molecule binds to the surface of a growing particle, this position is temporarily blocked (passivated) for further growth. Due to their moderate binding affinity, surfactant molecules may dynamically bind and unbind, while the particle may grow at parts which are not passivated by surfactants by the attachment of new precursor molecules. The binding affinity of surfactant molecules is also determined by temperature; consequently, the growth rate of QDs at a high temperature is faster, as their surface is less passivated by surfactants. On the other hand, growth can be suppressed by lowering the temperature so that passivation of the particle surface by surfactants prevents any further particle growth.

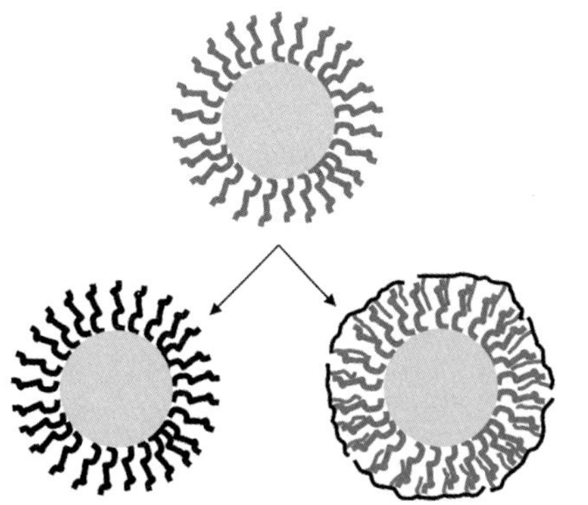

Figure 4.3 Hydrophobic QDs are stabilized by hydrophobic stabilizing molecules, which coordinate at one end to the particle surface. The easiest strategy to transfer such QDs into polar solvent is to exchange hydrophobic for hydrophilic stabilizing molecules. Alternatively, an amphiphilic polymer can be wrapped around the QD, the hydrophobic chains of which intercalate the hydrophobic stabilizing molecules bound to the QD, whereas the hydrophilic parts of the polymer endow water-solubility. Adapted from Ref. [19].

By applying these general principles, QDs can be grown with controlled size from a variety of semiconducting materials, such as CdSe, CdTe, CdS, InP, and InAs. The role of the surfactant allows even for a shape-controlled growth. As each QD is a crystal, and thus possesses different facets, a surfactant can be selected which binds more weakly to one facet than to another. Then, as this particular facet will be less passivated by surfactant molecules, the particle will grow preferentially in this direction. This capability would enable, for example, the growth of rod-shaped rather than spherical particles [39]. Different facets of particles allow even for the growth of defined hybrid structures, in which one particle comprises two oriented domains made from different materials. Here, the particle composed of the first material will act as a seed for nucleation of the precursors used for growth of the second material. As surfactant molecules may stick with less affinity to the tips of rod-shaped semiconductor particles than to the surface parallel to the axis, the second material will grow preferentially on the tips rather than along the walls of the semiconductor rods [40]. It could be said that, today, colloidal chemistry allows for the growth of particles with a defined size, shape, and composition for a large variety of materials [39, 41–45].

Traditionally, colloidal semiconductor nanoparticles are grown from materials in Groups II-VI of the Periodic Table, such as CdSe, CdS, and CdTe, due mainly to the fact that these particles are the most easy to synthesize. However, because of their more ionic structure, they grow more slowly during their synthesis than do nanoparticles from Groups IV-IV or III-V, which have a more covalent structure. At present, the most widespread QDs are the CdSe/ZnS nanoparticles, where the ZnS shell around the CdSe core provides improved optical properties, notably a quantum yield exceeding 50% [24]. Also important for biological applications is the fact the fluorescence of CdSe/ZnS QDs is in the visible spectrum. As their emission wavelengths vary between 500 and 700 nm, these QDs are perfectly compatible with the optical microscopy techniques used in biological laboratories. Unfortunately, working with Cd-based materials raises several problems due to the metal's toxicity [46, 47]; in particular, cadmium cannot be used in the application of nanoparticles to diagnostics or therapeutics in humans. Hence, the future trend in QD synthesis will unquestionably move towards more biocompatible QDs, composed of different materials.

4.5
Surface Chemistry for the Water-Solubilization of Quantum Dots

Whilst the fluorescence of a QD is determined primarily by the material and the size of the semiconductor nanoparticle, its interaction with the environment is mediated by its surface. First, the surface chemistry of the QDs will determine the solubility, or more precisely the colloidal stability, of the particles. Second, the surface chemistry is crucial when linking molecules to the particle, and thus for the design of specific interfaces with the biological world, to which the particles are to be connected.

4.5.1 General Strategies for Water-Solubilization

With regards to the solubility of QDs, two different lines must be considered. As noted above, the particles are synthesized in a surfactant solution that may be based on either aqueous or organic solvents. In the former case, the particles are capped with hydrophilic surfactant molecules, and thus are automatically soluble in aqueous solution. In the latter case, the particles are capped with hydrophobic surfactant molecules, and thus are soluble only in organic solvents and not in aqueous media (Figure 4.3). Before being used for biological tracking applications, the particles must first be rendered hydrophilic, and thus water-soluble. Although, initially, QD syntheses originated via water-based reactions, and these particles are directly soluble in water, there remain certain disadvantages associated with QDs. In contrast to synthesis in organic solvents, a water-based synthesis is limited to a temperature of up to only 100 °C, and to the use of water-soluble precursors and surfactant molecules. The use of a high-temperature synthesis, coupled with a wide variety of metallo-organic precursors and a huge choice of surfactants, permits QDs to be grown which (in general) have an improved crystallinity, a better size distribution, and more complex architectures. Whilst this statement does not apply to metal particles (e.g., gold) and metal-oxides (e.g., iron oxide), it generally holds true for many semiconductor nanoparticles. Nowadays, virtually all colloidal QDs produced by the major manufacturers are based on CdSe/ZnS grown in organic solvents.

Obviously, these particles must be rendered hydrophilic before they can be used in biological applications, and for this two general strategies are available. First, the original hydrophobic surfactant molecules can be replaced in a ligand-exchange procedure with hydrophilic surfactant molecules (Figure 4.3). Second, the originally hydrophobic nanoparticles can be embedded in an amphiphilic shell in the hydrophobic cavity of which is embedded the QD, the hydrophilic outside of which warrants water-solubility. Generally speaking, a ligand-exchange will lead to smaller nanoparticles, whereas an amphiphilic coating will provide more colloidally stable particles. Today, both methods are used in parallel. It should also be pointed out that, although the concepts described here appear to be straightforward, they must be regarded only as model systems. Most often, the exact geometry and molecular order on the surface of the QDs is not known exactly, since its experimental access may be very complicated. Even the hydrodynamic diameters of QDs, as determined by different methods and different groups, will vary to a significant degree [48].

As noted above, the presence of surfactant molecules warrants colloidal stability, and in aqueous solution this can be achieved in either of two ways. Quantum dots with similarly-charged surfaces will repel each other electrostatically; in fact, most QDs are stabilized by negatively charged surfactant molecules bound to their surface. Although charge stabilization is very efficient, it depends critically on the nature of the solvent. Indeed, the charge of the molecules can change with the pH. For example, COO^- groups will lose their charge at a low pH-value by protonation to COOH; therefore, negatively charged QDs will be stable at a high pH, but unstable at a low pH. Another problem emerges due to the presence of salt; ions with a charge opposite to that present on the particle surface will be electrostatically attracted and form a cloud of counterions around the QDs. These counterions will then screen the charge of the particles and thus reduce the electrostatic repulsion between the QDs. As a consequence, the charged particles typically become colloidally unstable in electrolytic solutions with a high salt content. Hydrophilic polymers may also be used for particle stabilization in aqueous solution, as their presence on the particle surface will cause steric repulsion between particles because the surface molecules on adjacent particles cannot penetrate each other. Steric repulsion of the first order depends neither on the pH nor the salt content of the solution, and thus is almost universal. Examples of materials which cause particle-stabilization by steric repulsion include poly(ethylene glycol) (PEG) [49–51] and sugars [52]. Stabilization by charge and stabilization by steric hindrance can also be achieved with ligand exchange, by embedding particles into an additional shell.

4.5.2 Solubilization with Ligand Exchange

Ligand exchange represents a very straightforward approach to solubilization. In this process, QDs capped with the original surfactant are incubated in a solution containing an excess of the new surfactant molecule. As noted above, most surfactants bind with only finite strength to the QD surface, and heating will accelerate the continuous process of binding and unbinding. However, by dispersing QDs (which have been capped by surfactant molecules of type A) in a solution that contains a huge excess of surfactant molecules of type B, this will (statistically) eventually lead to QDs becoming capped by surfactant molecules of type B, even in the situation where the type A surfactant molecules would bind more strongly to the QD surface than would type B. In a final step, the surfactant molecules remaining in solution must be removed.

Ligand exchange can be used to exchange hydrophobic surfactant molecules for hydrophilic molecules, thus transferring QDs from an organic solution to the aqueous phase. In aqueous phase, ligand exchange can also be used to exchange one type of hydrophilic surfactant for another hydrophilic type; this may also be applied to QDs that originally were synthesized in aqueous solution.

Although ligand exchange is a straightforward process, it has several disadvantages. First, all ligand exchange procedures carry an inherent risk of a reduced colloidal stability, especially if a phase transfer from nonpolar to polar solvents is involved. During ligand exchange, at a point when part of the original molecules have been exchanged and some original molecules are still attached to the particle surface, the QDs will possess a mixed surface chemistry, at which point they are most vulnerable to agglomeration. Occasionally, it is impossible to perform a complete ligand exchange [53]. The exposure of QDs to several subsequent ligand exchange procedures inevitably leads to a growing degree of particle aggregation.

As noted above, the ability to perform ligand exchange is based on the fact that the surfactant molecules continuously bind to, and unbind from, the QD surface; such a situation endangers the colloidal stability of particles when a solution is bare of excess surfactants. Hence, when an increasing number of surfactant molecules leave the QD surface with time, and there is not any excess of surfactant molecules available to rebind, then the particles eventually will agglomerate as they lose their colloidal stability. The problem here is that, especially for the surface of the most commonly used CdSe/ZnS QDs, there is no reactive group available that can bind to the surface with high affinity. Most often, surfactant molecules which contain thiol groups as anchors for the attachment to the QD surface are used in this role. However, as the bond of thiols to ZnS is not very strong (notably with regards to photo-oxidation [54]), the ligands will typically leave the particle surface within a few days, leading to the formation of particle agglomerates. In order to increase the affinity of the surfactant molecules to the QD surface, several groups have used multidentate ligands [55–57] or ligand molecules which can be crosslinked [20, 58, 59]. Crosslinking, in particular, can lead to colloidally very stable QDs, although such stability tends to occur via a multilayer of surfactant molecules than by a monolayer. On the positive side, monolayer-protected QDs – by definition – lead to the smallest possible hydrodynamic particle diameters. The choice of different classes of ligand on the particle surface, such as carboxylic acids [22], sugars [52], peptides [60], PEG [49], and polymers [61], allows for an adjustment to be made of the particle surface within the biological environment where the QDs are to be used.

4.5.3 Surface Coating with Amphiphilic Molecules

The overcoating of hydrophobically capped QDs with amphiphilic molecules differs fundamentally from ligand exchange. Here, the original hydrophobic surfactant resulting from the particle synthesis is not removed; rather, an additional layer of amphiphilic molecules is added around the QDs. Whilst the hydrophobic part of the amphiphilic molecules points towards the hydrophobic QD surface, the hydrophilic part points towards the solution and renders the coated particle water-soluble. Whereas, ligand exchange is based on the chemical binding of reactive groups to the particle surface, overcoating with amphiphilic molecules is mediated only by hydrophobic interactions. Moreover, this fundamental interaction is very general and does not depend on the exact chemical nature of the QD surface, nor of the amphiphilic molecule. The amphiphilic shell around the QD can be seen as micelle with a hydrophobic cavity and a hydrophilic surface. Many reports exist of mainly polymeric [18, 25, 62–64] and lipid micelles [65–67]. In this way, virtually any hydrophobically capped QD can be embedded within a variety of amphiphilic molecules, although chemical rather than geometric arguments come into play here. In the case of too-large micelles, several QDs would become embedded in one micelle, such that the single QD character would be lost. On the other hand, the amphiphilic polymer molecules should not be so small that a significant fraction of the entire QD surface would become coated.

Amphiphilic micelles represent an almost universal means of solubilizing the originally hydrophobically capped QDs into aqueous solution. Naturally, the micelle adds to the total particle diameter (which can reach 25–30 nm in the case of CdSe/ZnS QDs, compared to 2–10 nm for the inorganic particle core); however, this loss in performance is often compensated by the excellent colloidal properties provided by this technique. As the stability of the outer shell is not based on the bonds between the individual surfactant molecules and the QD surface, but rather on a collective attachment via hydrophobic interactions, the QDs are colloidally very stable. This technique also allows for different types of QDs to be embedded into shells with the same surface chemistry, a finding which may prove to be a major advantage when seeking a unified strategy for bioconjugation [18].

4.6 Interfacing Quantum Dots with Biology

4.6.1 Conjugation of QDs to Biomolecules

Whilst today, techniques allow for the synthesis of highly colloidally stable QDs, the bioconjugation of these particles remains very much an open subject that is under continuous development. As a huge variety of different functionalization strategies have been reported, this section will provide an overview of the different concepts available rather than detail all of the approaches. At present, two fundamentally different strategies may be applied for bioconjugation: (i) biological molecules can be attached directly to the QD surface; or (ii) they can be bound to the surfactant shell and its subsequent layers.

The direct attachment of biological molecules to the QD surface is based on ligand exchange, in which a portion of the surfactant molecules are replaced with biological molecules. If the original surfactant molecules bind only very weakly to the QD surface, they can be partly replaced with biological molecules which adsorb nonspecifically to the QD surface; however, such constructs are very labile and will be of very limited practical use. This method is more suited to biological molecules bearing a group that binds to the particle surface. In the case of CdSe/ZnS QDs, this is possible for molecules with thiol groups. For example, the amino acid cysteine has one thiol group; hence, peptides with cysteine in their sequence can be bound to the CdSe/ZnS surface simply by replacing the original ligands. Another example of this approach is the attachment of oligonucleotides that have been modified at one end with a thiol group. Unfortunately, as all of the problems relating to limited colloidal stability with regards to ligand exchange also hold true in this case, the direct attachment of biological molecules to QD surfaces would not be the method of choice for preparing stable conjugates.

On most occasions, the biological molecules are attached to the shell of the surfactant molecules and/or to the additional layers around it. The stability of this conjugation is limited, therefore, by the stability of the surfactant layer around the QDs. In the most primitive scenario, biological molecules are attached by nonspecific adsorption; however, a greater stability can be achieved by electrostatically mediated adsorption–that is, by the attraction of oppositely charged molecules to the charged particle surface. One well-established protocol of this type is based on fusing a positively charged zipper domain to proteins, which in turn become stably bound when the zipper domain binds to the surface of negatively charged QDs [57, 68]. The most prominent approach, however is the use of bioconjugate chemistry, in which functional groups on the QD surface (in particular COOH and NH_2) are crosslinked with the functional groups of the biological molecule (notably NH_2, SH, and COOH). Many crosslinker molecules are available for this purpose, and many different protocols have been developed [69]. Although covalent attachment is, arguably, the most stable method of bioconjugation, it should be noted that QD bioconjugation is by far not yet fully established. This relates to the fact that most bioconjugation protocols have been classically developed for the modification of proteins. Due to their different natures, QDs in general do not behave as proteins, which means that established protocols must often be modified. This is especially true for the limited range of QD concentration and for salt stability, as QDs cannot be reasonably concentrated above several tens of micromolar concentration. The high salt concentrations used to link similarly charged proteins so as to reduce their electrostatic repulsion would also screen the charge of the QDs, causing them to agglomerate. Whilst no salt would warrant for QD stability, proteins with the same sign of charge as the QDs would be repelled, but a high salt content would cause QD agglomeration. In practice, whilst identifying the "window of opportunity" is often complicated, the covalent attachment of many different types of biological molecule to QDs has been reported. Of particular interest here is the attachment of proteins, whether of antibodies for molecular recognition or enzymes for functional assays.

To some extent, the conjugation process can be controlled to a point where conjugates with a defined stoichiometry may be prepared [51, 70]. By using electrophoresis in agarose gels, constructs with a QD attached to an exactly defined number of molecules (zero, one, two, or more) can be separated and identified [51, 71], provided that the molecules have a sufficient size and/or charge. This aspect of QD functionalization, although requiring further development, has been shown to be essential for single-molecule experiments which, ideally, require QD–biomolecule constructs in a 1:1 ratio.

Unfortunately, bioconjugation is only as good as the subsequent purification procedure. In order to obtain pure conjugates, any unbound biological molecules remaining in solution must be removed as they would severely interfere with the labeling process. Whilst the free molecules may bind to exactly the same receptors as those molecules conjugated to the QD label, they cannot be visualized because they are unlabelled. Moreover, in the presence of a large excess of unbound molecules, the extent of labeling would be greatly suppressed due to competition between the unbound and conjugated molecules for the designated binding sites. *Purification* is also crucial for the removal of excess surfactant molecules, which might interfere with cells. The purification of QDs is not straightforward, but due to their size (which is comparable for example with proteins), any purification methods based on sorting by size (e.g., size-exclusion chromatography) will have only a limited application. Worse still, the analysis of bioconjugation and purification is also complicated, and consequently a number of reports on applied conjugation and purification strategies have been based on the trust of the authors rather than on any experimental characterization of the final product. To date, no standard protocols have been developed for evaluating the bioconjugation of QDs, and on most occasions sketches regarding the geometry of particle–biomolecule conjugates are accepted as idealized model. These are in agreement with experimental data and appear plausible, although reality they might involve more complex structures. As noted above, even the absolute hydrodynamic diameters can be determined only with an accuracy of a few nanometers [48]. Clearly, much improvement is required in this area before any industrial applications of QDs in bioconjugation processes can be considered.

4.6.2
Cytotoxicity of Semiconductor QDs

One problem associated with CdSe/ZnS QDs is their potential cytotoxicity since, when the QDs corrode, they release toxic cadmium ions [46, 47]. This may cause major difficulties in tracking, tracing, and homing applications with whole cells, which must be labeled with as many QDs as possible in order to remain sufficiently bright over a period of days, also taking into account any dilution of the QD label as the cells divide [72]. Nonetheless, the toxicity of CdSe/ZnS-based QDs can essentially be neglected for single-molecule tracing applications (as described in this chapter), for two reasons. First, the nature of optical single-molecule observations decrees that the QD labels must be very dilute, such that the effective QD concentrations are low. Second, single-molecule tracing is, at best, extended over several hours, which means that the incubation times are relatively short. Nevertheless, future improvements in particle synthesis soon will result in the development of Cd-free QDs [73, 74].

4.7
Single Quantum Dot Tracking Experiments in Live Cells

4.7.1
Why Experiments at the Single Molecule Level?

During the past 15 years, a variety of experiments has shown that the ability to image a single emitter represents a powerful approach to probe complex environments at the nanoscale [75, 76]. Although seen initially as a topic of interest in its own right, the detection of individual molecules has evolved to become a major tool for monitoring a multitude of physical or chemical processes. In fact, the field has grown so much that it has become almost impossible to review all of the research directions and possible applications in physics, chemistry, or biology.

Here, the aim is not to provide an exhaustive view but rather to highlight the concepts that are important for single-molecule experiments in live cells.

From a general point of view, the fundamental advantage of measurements at the single-molecule level is that they provide access to the full distribution of parameter values, rather than to the mere average values as in conventional ensemble measurements [75–77]. In other words, fluctuations around the mean value can be determined. This is important in two distinct cases. First, when a sample is inhomogeneous, molecular subpopulations can be identified by looking at molecules one at a time; these subpopulations can be considered as "static" fluctuations. Second, single-molecule experiments are also useful to look at "dynamic fluctuations"; indeed, for a system driven by biochemical reactions, the temporal evolution is not deterministic. Rather, molecules undergo stochastic processes. Unless molecules can be all synchronized at a given time, it is not possible to determine the kinetic parameters that control their dynamics. In most cases, only equilibrium constants are measured when averaging over an ensemble of molecules. In contrast, single-molecule experiments allow for a full description of the stochastic temporal evolution of a molecule, giving access to the transition rates underlying the biochemical reactions.

The two main motivations for single-molecule experiments – namely probing static and dynamic fluctuations – are of particular interest in live cells. As cells are physical and biochemical systems with a high degree of spatial heterogeneity, it is essential to probe the behavior of molecules as a function of their local environment. Moreover, biochemical reactions can rarely be synchronized in live cells, which means that single-molecule measurements might provide a unique approach to determine *in situ* the kinetics of these reactions. Importantly, in many situations the cell activity is mediated by a low number of proteins, and should preferably be investigated at the single-molecule level. Indeed, studies that involve the over-expression of a fluorescently labeled mutant have a risk to largely perturb the biochemical equilibria within the cell, and to significantly shift the kinetics and amplitude of the cellular response.

An important technical aspect of single molecule experiments is the accuracy with which the position of individual emitters can be determined. As the emitter size is much smaller than λ (the wavelength of light; 400–800 nm in the visible spectrum), its fluorescence spot corresponds to the point spread function (PSF) of the microscope (Figure 4.3). Although this spot has itself a width on the order of λ, its center can be localized with much higher precision. In fact, in the common case of a shot-noise-limited signal, the center position can be determined with an accuracy σ on the order of $\lambda/\sqrt{N}$ (where N is the number of photons detected during the acquisition time) [78, 79]. For a reasonable value of $N \sim 10^4$, σ is as low as 5–10 nm, which means that a single molecule can inform on a complex environment (such as a cell membrane) truly at the nanoscale.

In single-molecule tracking measurements, sequences of consecutive images are acquired and individual trajectories extracted from these sequences. When QDs are used as fluorescent markers, one difficulty derives from the fluorescence intermittency of the nanoparticles. As the markers can transiently disappear from the fluorescence images, specific algorithms are required to track individual QD-tagged molecules [80–82]. In brief, the algorithm can be separated into two parts. First, all of the visible spots within a sequence of images are detected using a conventional cross-correlation with a Gaussian model for the PSF of the microscope. Second, these spots are associated to reconstruct spatial trajectories of the individual proteins. For that purpose, a model of the spatial dynamics (diffusive motion, directed transport, etc.) is used to provide an estimate of the possible location of the molecules, either in consecutive images or after an "off" period.

In general, the output of single QD tracking experiments is a set of trajectories, with a time resolution in the range of 1 to 100 ms, and a spatial resolution between 5 and 50 nm (Figure 4.4a,b). Several methods have been devised to extract statistically meaningful properties from these molecular trajectories. The most common method is based on the computation of the mean square displacement (MSD), which provides an estimate of the parameters controlling the diffusive or directed transport of the molecules (Figure 4.4f) [84, 85]. More advanced techniques have been later introduced to analyze situations in which a molecule does not exhibit a unique mode of motion throughout its trajectory but, rather, dynamically alternates between multiple types of movement [86–89]. This is, for instance, the case when a membrane molecule, which initially is diffusing, becomes temporarily confined in a microdomain, or when a protein is transported due to transient attachment to a molecular motor.

4.7.2 Tracking Single Membrane Receptor Molecules

The use of QDs for single-molecule tracking experiments in cells was introduced for membrane receptors. The first biological question to be investigated using single QD imaging was the diffusion dynamics of glycine receptors in live cultured neurons (Figure 4.4) [3]. Since then, a similar experimental approach has been employed to study a large number of membrane molecules, including (without being exhaustive) other receptors for neurotransmitters [γ-aminobutyric acid (GABA), α-amino-3-hydroxy-5-methyl-4-isoxazolepropionic acid (AMPA), *N*-methyl-D-aspartate (NMDA)] [90–92], potassium channels [93], cystic fibrosis transmembrane conductance regulators [94], aquaporin-4 water channels [95], epidermal growth factor (EGF) receptors [81, 96], lipids, glycosylphosphatidylinositol (GPI)-anchored proteins [97], and immunoglobulin E (IgE) receptors [98].

There are many reasons, both technical and conceptual, why membrane molecules have been the subject of such attention. From a technical point of view, it is relatively simple to label membrane molecules by the use of antibodies against an extracellular epitope (Figure 4.4a–c). In many cases, there exist antibodies with high affinity (<1 nM) that allow binding over many hours. Once combined with a secondary antibody coupled to a QD, this type of affinity labeling permits a rather

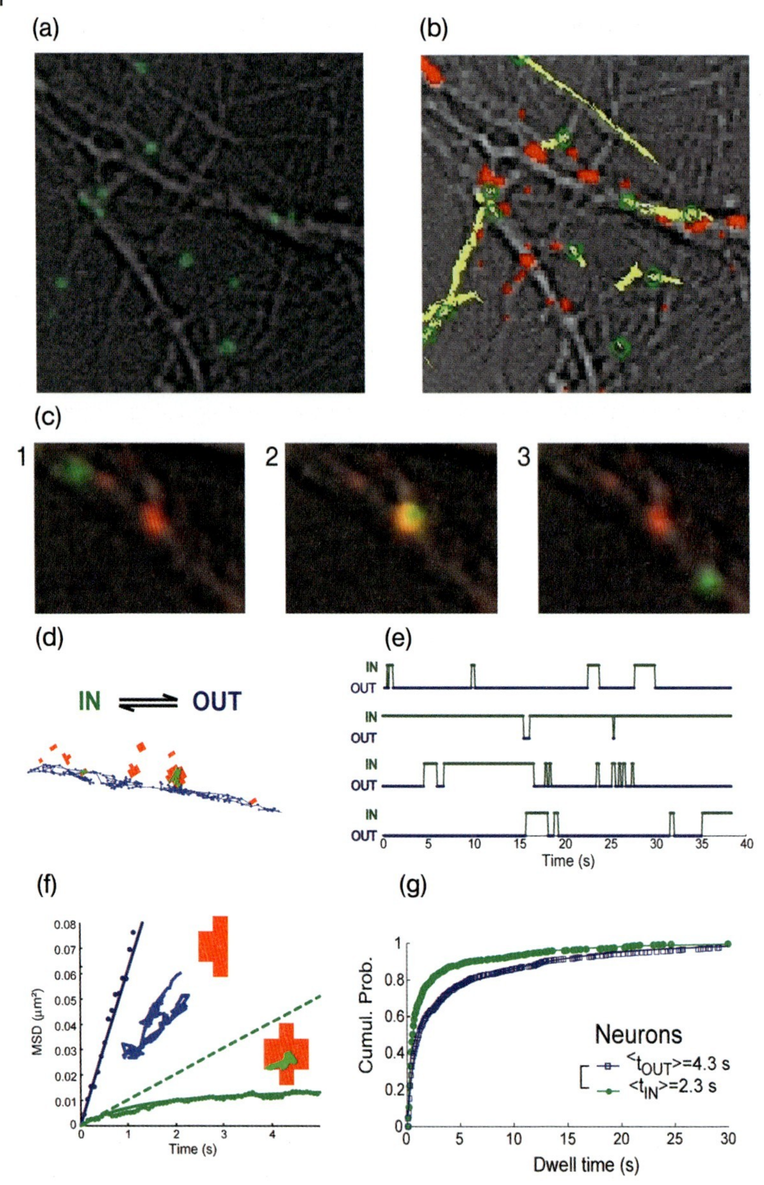

Figure 4.4 (a) Superposition of the fluorescence of individual QDs (green) coupled to glycine receptors and the bright-field image of live neurons. (b) Trajectories (yellow) of the QDs during the image sequence. The red spots correspond to synaptic sites identified using FM-64 markers. (c1–3) Motion of a single glycine receptor (green) diffusing in the membrane of live neuron and transiently interacting with a cluster of GFP-tagged scaffolding proteins gephyrin (red). (d) Trajectory of the QD in the membrane, alternating between IN and OUT states according to the QD position with respect to gephyrin clusters. The green points correspond to part of the trajectory when it colocalizes with the gephyrin cluster (IN state). (e) Dynamic equilibrium of individual receptors entering and exiting synaptic sites. (f) Example of MSD curves obtained for a freely diffusing extrasynaptic receptor (blue) or for a receptor having a confined diffusion at a synapse (green). The red spots indicate the position of the synaptic site. (g) Distribution of residence times in and out of the gephryin clusters in live neurons. Adapted from Ref. [83].

straightforward access to single-molecule dynamics, with detection possible over extended periods of time (tens of minutes, or more). The main requirement is to use labeling conditions such that the membrane density of tagged molecules is low enough, and individual QDs can be identified. Thanks to the brightness of QDs (see Section 4.3), a simple epifluorescence microscope with a UV lamp is often sufficient for these experiments, although a laser-based total internal reflection fluorescence (TIRF) microscope can be useful to achieve a high SNR in the detection of biomolecules in the basal membrane.

From a conceptual point of view, the organization of the plasma membrane raises many fascinating questions. Far from being a homogeneous viscous fluid, as described in the fluid-mosaic model [99], the membrane is a dynamically compartmentalized two-dimensional (2-D) system, and this compartmentalization in microdomains seems to play a key role in cellular activity and signaling. Moreover, the domains that coexist in the membrane exhibit a large degree of heterogeneity, with variability in both composition and size. For all of these reasons, experiments at the single-molecule level have been highly desirable in order to understand the molecular and physical mechanisms that lead to the dynamic and plastic organization of the membrane.

In fact, the field of single-particle tracking predates the advent of QDs. Since the late 1980s, latex beads or gold nanoparticles, detected using differential interference contrast (DIC), have been used to follow the motion of membrane molecules, often with great success. Nevertheless, the emergence of single-molecule fluorescence imaging techniques (using organic fluorophores, FPs, or QDs) has significantly expanded the range of applications of single-particle tracking. One reason for this is that fluorescence probes are much smaller than either the beads (~500 nm) or the gold nanoparticles (~40 nm) initially used. Also, fluorescence measurements are often more sensitive than DIC, and allow the multicolor detection of different species. Lastly, single-particle tracking experiments do not permit access to intracellular targets.

The case of receptors for neurotransmitters provides a clear example of the interest for single-molecule experiments [3, 4]. In neurons, synapses – either inhibitory or excitatory – can be viewed as specialized membrane subdomains (with a typical size of 1 μm) where signal transmission occurs between pre- and postsynaptic terminals. An open question, but one which is crucial to the plasticity of the nervous system, is to understand how receptors for neurotransmitters (AMPA, NMDA, GABA, glycine, etc.) reach proper synaptic sites, and how their number is regulated at synapses during neuronal activity [100, 101]. Single QD measurements have been extremely fruitful to record the membrane trafficking of these receptors. Experiments at the single-molecule level have revealed a diffusion dynamics that was not anticipated from conventional imaging methods [3, 4]. In extrasynaptic domains, receptors diffuse rapidly and freely (Figure 4.4c–f), but at synaptic sites – where they are stabilized through interactions with scaffolding proteins – their diffusion is significantly slowed down and they remain temporarily confined. More importantly, receptors can exchange dynamically between the extrasynaptic and synaptic domains. After a couple of seconds or minutes, those receptors stabilized at synapses can escape and rapidly explore the extrasynaptic membrane until, by random diffusion, they reach another synaptic site. When using photostable QDs, the kinetic parameters of this process – that is, the residence time in each domain – can be quantified (Figure 4.4d–g) and related to important cellular properties such as the cytoskeleton organization or the neuronal activity [100]. Altogether, these experiments demonstrate that both diffusion and interaction mechanisms can be addressed in the membrane, one molecule (QD) at a time [83]. The ability to study biochemical reactions *in situ* is particularly interesting, since *in vitro* studies cannot necessarily take into account parameters such as the 2-D nature of the membrane, the inhomogeneities of protein concentrations, or interfering interactions with other molecules such as lipids.

One frequently asked question is whether the labeling of a membrane protein with a QD significantly affects its lateral mobility. Although, the additional friction due to the extracellular QD is generally negligible compared to the friction of the viscous membrane [102], side effects can alter the diffusion dynamics. In particular, the use of divalent antibodies can induce artificial crosslinking, making the preparation of monofunctional probes all the more important [70]. Also, the QD size can create steric hindrance when exploring confined environment such as a synapse [103]. However, there is no general rule of the QD effect, and this should be tested case by case.

4.7.3
Visualizing Internalization Pathways

Understanding the mechanisms by which membrane proteins are internalized and transported upon activation is crucial to the analysis of signaling pathways. Several experiments have shown that the brightness and photostability of QDs are very useful for the visualization of internalization pathways. Once labeled with a QD, ligands can be detected as they bind to their target receptors. The membrane diffusion and subsequent endocytosis of the ligand/receptor complex can then be recorded [96]. The resultant endosomes can be tracked as they are transported along, possibly using cytoskeleton-dependent mechanisms. One prominent example of this approach concerns the response to EGF, a ligand which initiates signaling cascades involved in many cellular functions [96]. By labeling biotinylated EGF, it was possible to decipher the early stages of the signaling process: binding to receptor tyrosine-kinases followed by cellular uptake and active transport.

Following these pioneering experiments, many other measurements have taken advantage of the possibility to monitor QDs (functionalized or not) as they traffic from the membrane to the cytoplasm through internalization processes [104–109]. Of note, whilst some of these experiments do not require single-molecule detection, the number of nanoparticles in an endosome is often limited to a couple of units, which means that measurements are carried out in a regime of few QDs.

4.7.4 Tracking Intracellular Motor Molecules

Moving beyond the membrane and directly targeting intracellular proteins is an important goal, as many biochemical reactions take place directly in the cytoplasm or the nucleus. However, the single QD imaging of intracellular proteins remains a challenge, and experiments in this direction are still in their infancy. Indeed, reaching single-molecule sensitivity in the detection of biomolecules within live cells raises several difficulties. First, the QDs must enter the cell cytoplasm (without being trapped in endosomal compartments) and reach their molecular target. Second, the fluorescence of individual QDs must be detected in a noisy environment due to the autofluorescence of intracellular compartments and organelles. Finally, the motion in the cytosol is likely to be three-dimensional, compared to 2-D diffusion in the membrane.

Several approaches have been tested to insert QDs in the cytoplasm of living cells [17, 110]. In all these techniques, a major challenge is to effectively release QDs into the cytosol and to avoid sequestering QDs in organelles. The most efficient method is probably mechanical microinjection [111]; for example, QDs were successfully microinjected into *Xenopus* embryos [65] and used for cell-lineage tracing. However, in this experiment, the number of injected QDs was large (far above the single-molecule regime) and did not permit a careful analysis of the behavior of individual nanoparticles. Despite many advantages, microinjection is a delicate and time-consuming method which does not permit the labeling of a high number of cells, and for these reasons other techniques have been explored to insert objects into the cytosol. Methods based on the transient membrane permeabilization with electroporation [112] or on transfection agents [113] have been tested, but often seemed to produce QD aggregates within the cells. A more sophisticated approach is based on a chemical modification of the particle surface. By making use of concepts developed for drug delivery [114], the surface of QDs can be modified in such a way so that, after incorporation by the cell via endocytotic pathways, the vesicular structures in which the QDs are trapped are dissolved so that the QDs are released into the cytosol. In an even more sophisticated approach, the QDs are coated with peptides isolated from viruses (e.g., the TAT peptide), which possibly allow crossing of the cell membrane into the cytosol [115, 116]. In all of the above-described techniques, it seems that the efficiency of the internalization depends on the QD charge and surface coating. A general rule here is that a good knowledge and control of the colloidal parameters of the nanoparticles is essential for understanding their intracellular behavior.

A single QD assay has been recently implemented to probe, for the first time, the transport properties of individual kinesin and myosin V motors in live cells [6, 117] (Figure 4.5a,b). The properties of these molecular motors have been abundantly investigated with single-molecule *in vitro* experiments [118]. By means of micromanipulation techniques and fluorescence assays, important parameters such as the velocity, the processivity and the stall force have been determined. However, the intrinsic limitation of these experiments is that the motors are

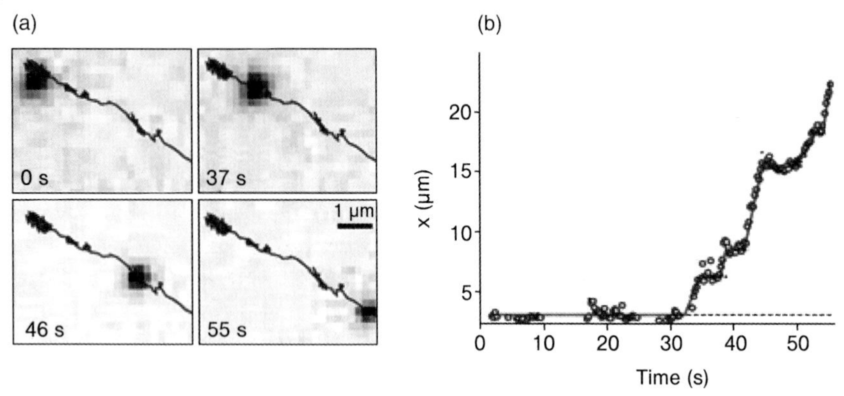

Figure 4.5 (a) Motion of an individual kinesin motor labeled with a QD in the cytoplasm of a living HeLa cell. Adapted from Ref. [6]. (b) Position along the x-axis versus time, indicating a succession of phases of directed and diffusive motion.

observed out of their natural cellular habitat. In fact, the cell is a complex environment where molecular parameters (pH, ionic strength, ATP concentration, etc.) are actively regulated, and crowding is extreme. All of these aspects can potentially affect the motor motility, and are difficult to faithfully reproduce *in vitro*. To visualize single motors in live cells, purified kinesins or myosins V were labeled *in vitro*, their activity tested in standard motility assays, and subsequently internalized into HeLa cells using a method based on the osmotic release of pynocytic vesicles. An analysis of the movement of individual motors gave access to their velocity and processivity in cells. Interestingly, the results (velocity ~0.5–1 $\mu m\, s^{-1}$ and processivity ~ 1–2 s) were relatively unchanged compared to their *in vitro* values. In the case of myosin V, it was also possible to observe individual steps of the motors [117].

4.7.5
From Single Molecules to Populations

A specificity of QDs for single-molecule experiments in live cells is that many nanoparticles can be easily detected and tracked in parallel. For instance, when using a UV lamp as the excitation source (rather than a laser) and an intensified CCD, field of views of ~$100 \times 100\,\mu m^2$ containing tens to hundreds of discernable QDs can be recorded at frame rate of ~10–30 images per second. Once analyzed with an appropriate algorithm [80, 81], these sequences of images yield a large number of individual trajectories. From a practical point of view, this means that experiments have a higher throughput of single-molecule events, and that statistically meaningful results can be more rapidly obtained. The fact that many QDs can be simultaneously tracked has also conceptual implications on the way in which the spatiotemporal dynamics of molecular populations can be investigated.

Indeed, the subpopulation of QD-labeled molecules, which appear as individual fluorescence spots in the images (Figure 4.4a), corresponds to sampling (presumably random) of the whole population. By analyzing the dynamics of these QD spots, the properties of the population can be inferred with a sensitivity that, often, cannot be achieved using ensemble labeling. For instance, the first moments of the spatial distribution, as well as spatiotemporal correlations between individual spots, can be computed with high spatial and temporal resolution. Using QDs as tracers within populations of molecules of interest will in fact reinforce the conceptual link between the analysis of molecular systems in cells and the methods used in soft-condensed matter to investigate the dynamics of complex systems.

This aspect of single QD experiments was recently used to probe the dynamics of chemoreceptors in nerve growth cones [90]. When submitted to an external gradient of chemoattractants, the QD-labeled chemoreceptors were redistributed asymmetrically towards the gradient source (this is indicative of an amplification step in gradient sensing) [119]. The analysis of individual receptor trajectories showed that the symmetry-breaking resulted from a microtubule-dependent directed transport. Furthermore, by investigating the mean position of the QD spots, the formation of polarity at the cell surface could be quantitatively described and modeled as a positive-feedback loop involving receptor asymmetric activation and microtubule dynamics [90].

In general, understanding how cell organization and morphogenesis dynamically emerge, and how they are related to molecular events, should greatly benefit from ultrasensitive QD imaging in live cells. It is anticipated that the approach described above will be fruitful for deciphering the complex processes involved in the self-organization of biological systems [120, 121]. It should be especially useful in order to develop multiscale descriptions of biological systems for which molecular-level information (determined by examining single trajectories) are integrated into a view at a systems-level.

4.8 Conclusions and Perspectives

Quantum dots represent a new class of inorganic fluorophores with unique optical properties. When detected at the single-molecule level, they offer exciting possibilities to visualize the real-time motion of individual biomolecules in their cellular environment. Several reported examples have illustrated the interest of single QD experiments in cell biology to decipher the molecular processes that contribute to cell organization and dynamics. However, making QD imaging a standard tool in cell biology still requires a significant effort, in order to improve the colloidal and biochemical properties of the probes.

From the synthesis and bioconjugation points of view, two improvements can be expected for the future. Clearly, experiments would benefit from QDs with reduced blinking [34], and the use of elongated rather than spherical QDs might represent a step in this direction [122]. The conjugation of QDs with biological ligands, in particular with proteins, can also be improved. While currently most experiments are performed with an uncontrolled number of ligand molecules per QD, it is today possible to synthesize QDs with an exactly controlled number of binding sites per QD [51, 70]. The way in which ligands are attached to the QD can also be improved. At present, for most of the time, the geometry of the linkage between the ligand and the QD is completely undefined. However, if the proteins were to be linked via their amino-groups to the carboxy-groups on the QD surface with crosslinker molecules, the binding geometry would be undefined, as each protein can have several amino-groups. Enzymes or antibodies might thus be linked in a way that they are inactivated. However, by fusing specific linking domains to proteins, they could be linked in more defined ways to the particle surface [57].

An improved QD technology should permit the investigation of new classes of scientific question. So far, single-particle tracking has been dedicated almost exclusively to tracing the positions of labeled molecules, although in many cases the orientation of molecules with respect to their binding partners has also played a crucial role. Rotational motion can be observed by measuring polarized fluorescence [123] or by defocused microscopy [124]. Although, rod-shaped QD rods have a polarized emission, and thus should be suited to this type of experiment [125], today most experiments using QDs are based on spherical particles. However, due to their changed blinking behavior and polarized emission, a much wider use of asymmetric QD structures is predicted in the future.

References

1 Schutz, G.J., Kada, G., Pastushenko, V.P., and Schindler, H. (2000) *EMBO J.*, **19**, 892.
2 Harms, G.S., Cognet, L., Lommerse, P.H.M., Blab, G.A., Kahr, H., Gamsjäger, R., Spaink, H.P., Soldatov, N.M., Romanin, C., and Schmidt, T. (2001) *Biophys. J.*, **81**, 2639.
3 Dahan, M., Levi, S., Luccardini, C., Rostaing, P., Riveau, B., and Triller, A. (2003) *Science*, **302**, 442.
4 Tardin, C., Cognet, L., Bats, C., Lounis, B., and Choquet, D. (2003) *EMBO J.*, **22**, 4656.
5 Kural, C., Kim, H., Syed, S., Goshima, G., Gelfand, V.I., and Selvin, P.R. (2005) *Science*, **308**, 1469.
6 Courty, S., Luccardini, C., Bellaiche, Y., Cappello, G., and Dahan, M. (2006) *Nano Lett.*, **6**, 1491.
7 Murase, K., Fujiwara, T., Umemura, Y., Suzuki, K., Iino, R., Yamashita, H., Saito, M., Murakoshi, H., Ritchie, K., and Kusumi, A. (2004) *Biophys. J.*, **86**, 4075.
8 Yu, J., Xiao, J., Ren, X.J., Lao, K.Q., and Xie, X.S. (2006) *Science*, **311**, 1600.
9 Shav-Tal, Y., Darzacq, X., Shenoy, S.M., Fusco, D., Janicki, S.M., Spector, D.L., and Singer, R.H. (2004) *Science*, **304**, 1797.
10 Douglass, A.D., and Vale, R.D. (2005) *Cell*, **121**, 937.
11 Lommerse, P.H.M., Snaar-Jagaiska, B.E., Spaink, H.P., and Schmidt, T. (2005) *J. Cell Sci.*, **118**, 1799.
12 Giepmans, B.N.G., Adams, S.R., Ellisman, M.H., and Tsien, R.Y. (2006) *Science*, **312**, 217.
13 Chen, I., and Ting, A.Y. (2005) *Curr. Opin. Biotechnol.*, **16**, 35.
14 Zhang, J., Campbell, R.E., Ting, A.Y., and Tsien, R.Y. (2002) *Nat. Rev. Mol. Cell Biol.*, **3** (**12**), 906–918.

15 Alivisatos, P. (2004) *Nat. Biotechnol.*, **22**, 47.

16 Parak, W.J., Pellegrino, T., and Plank, C. (2005) *Nanotechnology*, **16**, R5.

17 Michalet, X., Pinaud, F.F., Bentolila, L.A., Tsay, J.M., Doose, S., Li, J.J., Sundaresan, G., Wu, A.M., Gambhir, S.S., and Weiss, S. (2005) *Science*, **307**, 538.

18 Pellegrino, T., Manna, L., Kudera, S., Liedl, T., Koktysh, D., Rogach, A.L., Keller, S., Rädler, J., Natile, G., and Parak, W.J. (2004) *Nano Lett.*, **4**, 703.

19 Pellegrino, T., Kudera, S., Liedl, T., Javier, A.M., Manna, L., and Parak, W.J. (2005) *Small*, **1**, 48.

20 Gerion, D., Pinaud, F., Williams, S.C., Parak, W.J., Zanchet, D., Weiss, S., and Alivisatos, A.P. (2001) *J. Phys. Chem. B*, **105**, 8861.

21 Bruchez, M.J., Moronne, M., Gin, P., Weiss, S., and Alivisatos, A.P. (1998) *Science*, **281**, 2013.

22 Chan, W.C.W. and Nie, S. (1998) *Science*, **281**, 2016.

23 Parak, W.J., Manna, L., and Nann, T. (2008) *Nanotechnology: Volume 1: Principles and Fundamentals* (ed. G. Schmid), Wiley-VCH Verlag GmbH, Weinheim, p. 73.

24 Dabbousi, B.O., Rodriguez-Viejo, J., Mikulec, F.V., Heine, J.R., Mattoussi, H., Ober, R., Jensen, K.F., and Bawendi, M.G. (1997) *J. Phys. Chem. B*, **101**, 9463.

25 Wu, M.X., Liu, H., Liu, J., Haley, K.N., Treadway, J.A., Larson, J.P., Ge, N., Peale, F., and Bruchez, M.P. (2003) *Nat. Biotechnol.*, **21**, 41.

26 Medintz, I.L., Uyeda, H.T., Goldman, E.R., and Mattoussi, H. (2005) *Nat. Mater.*, **4**, 635.

27 Fernández-Argüelles, M.T., Yakovlev, A., Sperling, R.A., Luccardini, C., Gaillard, S., Medel, A.S., Mallet, J.-M., Brochon, J.-C., Feltz, A., Oheim, M., and Parak, W.J. (2007) *Nano Lett.*, **7**, 2613.

28 Dahan, M., Laurence, T., Pinaud, F., Chemla, D.S., Alivisatos, A.P., Sauer, M., and Weiss, S. (2001) *Opt. Lett.*, **26**, 825.

29 Giepmans, B.N., Deerinck, T.J., Smarr, B.L., Jones, Y.Z., and Ellisman, M.H. (2005) *Nat. Methods*, **2**, 743.

30 Kim, S., Lim, Y.T., Soltesz, E.G., De Grand, A.M., Lee, J., Nakayama, A., Parker, J.A., Mihaljevic, T., Laurence, R.G., Dor, D.M., Cohn, L.H., Bawendi, M.G., and Frangioni, J.V. (2004) *Nat. Biotechnol.*, **22**, 93.

31 Nirmal, M., Dabbousi, B.O., Bawendi, M.G., Macklin, J.J., Trautman, J.K., Harris, T.D., and Brus, L.E. (1996) *Nature*, **383**, 802.

32 Shimizu, K.T., Neuhauser, R.G., Leatherdale, C.A., Empedocles, S.A., Woo, W.K., and Bawendi, M.G. (2001) *Phys. Rev. B*, **63**, 205316.

33 Brokmann, X., Hermier, J.P., Messin, G., Desbiolles, P., Bouchaud, J.P., and Dahan, M. (2003) *Phys. Rev. Lett.*, **90**, 120601.

34 Mahler, B., Spinicelli, P., Buil, S., Quelin, X., Hermier, J.-P., and Dubertret, B. (2008) *Nat. Mater.*, **7**, 659.

35 Chen, Y., Vela, J., Htoon, H., Casson, J.L., Werder, D.J., Bussian, D.A., Klimov, V.I., and Hollingsworth, J.A. (2008) *J. Am. Chem. Soc.*, **130**, 5026.

36 Lin, C.-A.J., Liedl, T., Sperling, R.A., Fernández-Argüelles, M.T., Costa-Fernández, J.M., Pereiro, R., Sanz-Medel, A., Chang, W.H., and Parak, W.J. (2007) *J. Mater. Chem.*, **17**, 1343.

37 Kudera, S., Carbone, L., Manna, L., and Parak, W.J. (2008) *Semiconductor Nanocrystal Quantum Dots* (ed. A.L. Rogach), Springer, Vienna, p. 1.

38 Peng, X., Wickham, J., and Alivisatos, A.P. (1998) *J. Am. Chem. Soc.*, **120**, 5343.

39 Peng, X., Manna, L., Yang, W., Wickham, J., Scher, E., Kadavanich, A., and Alivisatos, A.P. (2000) *Nature*, **404**, 59.

40 Mokari, T., Rothenberg, E., Popov, I., Costi, R., and Banin, U. (2004) *Science*, **304**, 1787.

41 Manna, L., Scher, E.C., and Alivisatos, A.P. (2002) *J. Cluster Sci.*, **13**, 521.

42 Cozzoli, P.D., Pellegrino, T., and Manna, L. (2006) *Chem. Soc. Rev.*, **35**, 1195.

43 Kumar, S. and Nann, T. (2006) *Small*, **2**, 316.

44 Kudera, S., Carbone, L., Zanella, M., Cingolani, R., Parak, W.J., and Manna, L. (2006) *Phys. Stat. Sol. (c)*, **203**, 1329.

45 Kudera, S., Carbone, L., Carlino, E., Cingolani, R., Cozzoli, P.D., and Manna, L. (2007) *Physica E*, **37**, 128.

46 Derfus, A.M., Chan, W.C.W., and Bhatia, S.N. (2004) *Nano Lett.*, **4**, 11.

47 Kirchner, C., Liedl, T., Kudera, S., Pellegrino, T., Javier, A.M., Gaub, H.E., Stölzle, S., Fertig, N., and Parak, W.J. (2005) *Nano Lett.*, **5**, 331.

48 Sperling, R.A., Liedl, T., Duhr, S., Kudera, S., Zanella, M., Lin, C.-A.J., Chang, W., Braun, D., and Parak, W.J. (2007) *J. Phys. Chem. C*, **111**, 11552.

49 Kanaras, A.G., Kamounah, F.S., Schaumburg, K., Kiely, C.J., and Brust, M. (2002) *Chem. Commun.*, **2002**, 2294.

50 Ballou, B., Lagerholm, B.C., Ernst, L.A., Bruchez, M.P., and Waggoner, A.S. (2004) *Bioconjug. Chem.*, **15**, 79.

51 Sperling, R.A., Pellegrino, T., Li, J.K., Chang, W.H., and Parak, W.J. (2006) *Adv. Funct. Mater.*, **16**, 943.

52 Rojo, J., Diaz, V., Fuente, J.M.D.L., Segura, I., Barrientos, A.G., Riese, H.H., Bernad, A., and Penades, S. (2004) *ChemBioChem*, **5**, 291.

53 von Holt, B., Kudera, S., Weiss, A., Schrader, T.E., Manna, L., Parak, W.J., and Braun, M. (2008) *J. Mater. Chem.*, **18**, 2728.

54 Aldana, J., Wang, Y.A., and Peng, X. (2001) *J. Am. Chem. Soc.*, **123**, 8844.

55 Kim, S., Kim, S., Tracy, J., Jasanoff, A., and Bawendi, M. (2005) *J. Am. Chem. Soc.*, **127**, 4556.

56 Li, Z., Jin, R., Mirkin, C.A., and Letsinger, R.L. (2002) *Nucleic Acids Res.*, **30**, 1558.

57 Mattoussi, H., Mauro, J.M., Goldman, E.R., Anderson, G.P., Sundar, V.C., Mikulec, F.V., and Bawendi, M.G. (2000) *J. Am. Chem. Soc.*, **122**, 12142.

58 Parak, W.J., Gerion, D., Zanchet, D., Woerz, A.S., Pellegrino, T., Micheel, C., Williams, S.C., Seitz, M., Bruehl, R.E., Bryant, Z., Bustamante, C., Bertozzi, C.R., and Alivisatos, A.P. (2002) *Chem. Mater.*, **14**, 2113.

59 Alejandro-Arellano, M., Ung, T., Blanco, A., Mulvaney, P., and Liz-Marzan, L.M. (2000) *Pure Appl. Chem.*, **72**, 257.

60 Pinaud, F., King, D., Moore, H.-P., and Weiss, S. (2004) *J. Am. Chem. Soc.*, **126**, 6115.

61 Nikolic, M.S., Krack, M., Aleksandrovic, V., Kornowski, A., Forster, S., and Weller, H. (2006) *Angew. Chem., Int. Ed.*, **45**, 6577.

62 Lin, C.-A.J., Sperling, R.A., Li, J.K., Yang, T.-Y., Li, P.-Y., Zanella, M., Chang, W.H., and Parak, W.J. (2008) *Small*, **4**, 334.

63 Yu, W.W., Chang, E., Falkner, J.C., Zhang, J., Al-Somali, A.M., Sayes, C.M., Johns, J., Drezek, R., and Colvin, V.L. (2007) *J. Am. Chem. Soc.*, **129**, 2871.

64 Luccardini, C., Tribet, C., Vial, F., Marchi-Artzner, V., and Dahan, M. (2006) *Langmuir*, **22**, 2304.

65 Dubertret, B., Skourides, P., Norris, D.J., Noireaux, V., Brivanlou, A.H., and Libchaber, A. (2002) *Science*, **298**, 1759.

66 Carion, O., Mahler, B., Pons, T., and Dubertret, B. (2007) *Nat. Protoc.*, **2**, 2383.

67 Boulmedais, F., Bauchat, P., Brienne, M.J., Arnal, I., Artzner, F., Gacoin, T., Dahan, M., and Marchi-Artzner, V. (2006) *Langmuir*, **22**, 9797.

68 Mattoussi, H., Mauro, J.M., Goldman, E.R., Green, T.M., Anderson, G.P., Sundar, V.C., and Bawendi, M.G. (2001) *Phys. Stat. Sol. B*, **224**, 277.

69 Hermanson, G.T. (1996) *Bioconjugate Techniques*, Academic Press, San Diego.

70 Howarth, M., Liu, W.H., Puthenveetil, S., Zheng, Y., Marshall, L.F., Schmidt, M.M., Wittrup, K.D., Bawendi, M.G., and Ting, A.Y. (2008) *Nat. Methods*, **5**, 397.

71 Pons, T., Uyeda, H.T., Medintz, I.L., and Mattoussi, H. (2006) *J. Phys. Chem. B*, **110**, 20308.

72 Parak, W.J., Boudreau, R., Gros, M.L., Gerion, D., Zanchet, D., Micheel, C.M., Williams, S.C., Alivisatos, A.P., and Larabell, C.A. (2002) *Adv. Mater.*, **14**, 882.

73 Pradhan, N. and Peng, X.G. (2007) *J. Am. Chem. Soc.*, **129**, 3339.

74 Pradhan, N., Battaglia, D.M., Liu, Y.C., and Peng, X.G. (2007) *Nano Lett.*, **7**, 312.

75 Weiss, S. (1999) *Science*, **283**, 1676.

76 Moerner, W.E. and Orrit, M. (1999) *Science*, **283**, 1670.

77 Kulzer, F., Kummer, S., Matzke, R., Bräuchle, C., and Basche, T. (1997) *Nature*, **387**, 688.

78 Bobroff, N. (1986) *Rev. Sci. Instrum.*, **57**, 1152.

79 Thompson, R.E., Larson, D.R., and Webb, W.W. (2002) *Biophys. J.*, **82**, 2775.

80 Bonneau, S., Dahan, M., and Cohen, L.D. (2005) *IEEE Trans. Image Process.*, **14**, 1384.

81 Sergé, A., Bertaux, N., Rigneault, H., and Marguet, D. (2008) *Nat. Methods*, **5**, 687.

82 Jaqaman, K., Loerke, D., Mettlen, M., Kuwata, H., Grinstein, S., Schmid, S.L., and Danuser, G. (2008) *Nat. Methods*, **5**, 695.

83 Ehrensperger, M.V., Hanus, C., Vannier, C., Triller, A., and Dahan, M. (2007) *Biophys. J.*, **92**, 3706.

84 Qian, H. (2000) *Biophys. J.*, **79**, 137.

85 Saxton, M.J. and Jacobson, K. (1997) *Annu. Rev. Biophys. Biomol. Struct.*, **26**, 373.

86 Bouzigues, C. and Dahan, M. (2007) *Biophys. J.*, **92**, 654.

87 Meilhac, N., Le Guyader, L., Salome, L., and Destainville, N. (2006) *Phys. Rev. E*, **73**, 011915.

88 Simson, R., Sheets, E.D., and Jacobson, K. (1995) *Biophys. J.*, **69**, 989.

89 Huet, S., Karatekin, E., Tran, V.S., Fanget, I., Cribier, S., and Henry, J.P. (2006) *Biophys. J.*, **91**, 3542.

90 Bouzigues, C., Morel, M., Triller, A., and Dahan, M. (2007) *Proc. Natl Acad. Sci. USA*, **104**, 11251.

91 Groc, L., Heine, M., Cognet, L., Brickley, K., Stephenson, F.A., Lounis, B., and Choquet, D. (2004) *Nat. Neurosci.*, **7**, 695.

92 Groc, L., Heine, M., Cousins, S.L., Stephenson, F.A., Lounis, B., Cognet, L., and Choquet, D. (2006) *Proc. Natl Acad. Sci. USA*, **103**, 18769.

93 Nechyporuk-Zloy, V., Stock, C., Schillers, H., Oberleithner, H., and Schwab, A. (2006) *Am. J. Physiol. Cell Physiol.*, **291**, C266.

94 Haggie, P.M., Kim, J.K., Lukacs, G.L., and Verkman, A.S. (2006) *Mol. Biol. Cell*, **17**, 4937.

95 Crane, J.M., Hoek, A.N.V., Skach, W.R., and Verkman, A.S. (2008) *Mol. Biol. Cell*, **19**, 3369.

96 Lidke, D.S., Nagy, P., Heintzmann, R., Arndt-Jovin, D.J., Post, J.N., Grecco, H.E., Jares-Erijman, E.A., and Jovin, T.M. (2004) *Nat. Biotechnol.*, **22**, 198.

97 Pinaud, F., Michalet, X., Iyer, G., Margeat, E., Moore, H.P., and Weiss, S. (2009) *Traffic*, **10**, 691–712.

98 Andrews, N.L., Lidke, K.A., Pfeiffer, J.R., Burns, A.R., Wilson, B.S., Oliver, J.M., and Lidke, D.S. (2008) *Nat. Cell Biol.*, **10**, 955.

99 Singer, S.J. and Nicolson, G.L. (1972) *Science*, **175**, 720.

100 Triller, A. and Choquet, D. (2008) *Neuron*, **59**, 359.

101 Choquet, D. and Triller, A. (2003) *Nat. Rev. Neurosci.*, **4**, 251.

102 Saffman, P.G. and Delbrück, M. (1975) *Proc. Natl Acad. Sci. USA*, **72**, 3111.

103 Groc, L., Lafourcade, M., Heine, M., Renner, M., Racine, V., Sibarita, J.B., Lounis, B., Choquet, D., and Cognet, L. (2007) *J. Neurosci.*, **27**, 12433.

104 Rajan, S.S., Liu, H.Y., and Vu, T.Q. (2008) *ACS Nano*, **2**, 1153.

105 Cui, B.X., Wu, C.B., Chen, L., Ramirez, A., Bearer, E.L., Li, W.P., Mobley, W.C., and Chu, S. (2007) *Proc. Natl Acad. Sci. USA*, **104**, 13666.

106 Nabiev, I., Mitchell, S., Davies, A., Williams, Y., Kelleher, D., Moore, R., Gun'ko, Y.K., Byrne, S., Rakovich, Y.P., Donegan, J.F., Sukhanova, A., Conroy, J., Cottell, D., Gaponik, N., Rogach, A., and Volkov, Y. (2007) *Nano Lett.*, **7**, 3452.

107 Cambi, A., Lidke, D.S., Arndt-Jovin, D.J., Figdor, C.G., and Jovin, T.M. (2007) *Nano Lett.*, **7**, 970.

108 Ruan, G., Agrawal, A., Marcus, A.I., and Nie, S. (2007) *J. Am. Chem. Soc.*, **129**, 14759.

109 Tekle, C., van Deurs, B., Sandvig, K., and Iversen, T.G. (2008) *Nano Lett.*, **8**, 1858.

110 Luccardini, C., Yakovlev, A., Gaillard, S., Alberola, M.V.T., Hoff, A.P., Mallet, J.-M., Parak, W.J., Feltz, A., and Oheim, M. (2007) *J. Biomed. Biotechnol.*, **2007**, Article ID, 68963.

111 Tada, H., Higuchi, H., Wanatabe, T.M., and Ohuchi, N. (2007) *Cancer Res.*, **67**, 1138.

112 Derfus, A., Chan, W., and Bhatia, S. (2004) *Adv. Mater.*, **16**, 961.

113 Chen, F. and Gerion, D. (2004) *Nano Lett.*, **4**, 1827.

114 Plank, C., Oberhauser, B., Mechtler, K., Koch, C., and Wagner, E. (1994) *J. Biol. Chem.*, **269**, 12918.

115 de la Fuente, J.M. and Berry, C.C. (2005) *Bioconjug. Chem.*, **16**, 1176.

116 Lewin, M., Carlesso, N., Tung, C.H., Tang, X.W., Corry, D., Scadden, D.T., and Weissleder, R. (2000) *Nat. Biotechnol.*, **18**, 410.

117 Pierobon, P., Achouri, S., Courty, S., Dunn, A.R., Spudich, J.A., Dahan, M., and Cappello, G. (2009) *Biophys. J.*, **96**, 4268–4275.

118 Howard, J. (2001) *Mechanics of Motor Proteins and the Cytoskeleton*, Sinauer.

119 Mortimer, D., Fothergill, T., Pujic, Z., Richards, L.J., and Goodhill, G.J. (2008) *Trends Neurosci.*, **31**, 90.

120 Karsenti, E. (2008) *Nat. Rev. Mol. Cell. Biol.*, **9**, 255.

121 Misteli, T. (2001) *J. Cell Biol.*, **155**, 181.

122 Wang, S., Querner, C., Emmons, T., Drndic, M., and Crouch, C.H. (2006) *J. Phys. Chem. B*, **110**, 23221.

123 Hugel, T., Michaelis, J., Hetherington, C.L., Jardine, P.J., Grimes, S., Walter, J.M., Faik, W., Anderson, D.L., and Bustamante, C. (2007) *PLoS Biol.*, **5**, 558.

124 Brokmann, X., Ehrensperger, M.V., Hermier, J.P., Triller, A., and Dahan, M. (2005) *Chem. Phys. Lett.*, **406**, 210.

125 Hu, J., Yang, L.-W., Wang, L.-S., Li, W., and Alivisatos, A.P. (2002) *J. Phys. Chem. B*, **106**, 2447.

Part II
Energy Transfer on the Nanoscale

Single Particle Tracking and Single Molecule Energy Transfer
Edited by Christoph Bräuchle, Don C. Lamb, and Jens Michaelis

ISBN: 978-3-527-32296-1

5
Single-Pair FRET: An Overview with Recent Applications and Future Perspectives

Don C. Lamb

5.1
Introduction

In recent years, Förster resonance energy transfer (FRET), which is also known as "fluorescence resonance energy transfer," has become a popular method for biological, chemical, biochemical, and biophysical investigations. As FRET requires no direct contact with the sample, it can be easily applied to live-cell measurements; moreover, as an optical method it can be detection with high sensitivity, even down to the single donor–acceptor pair level.

The first experimental detection of energy transfer was performed by Franck and Cario in 1922 [1–3]. (For an excellent review on the history of FRET, see the review by Clegg [4].) The phenomenon of FRET was first correctly described by Förster during the 1940s [5–8]. However, it was only after an experimental verification of the distance-dependence of FRET (which was reported in a seminal study by Stryer and Haugland [9]) that the technique began to be used as a "molecular ruler." The main advantage here is that, as FRET is sensitive to the distance between the donor and acceptor molecule, it can be used to investigate distances on the 2–10 nm scale.

Ha and coworkers were the first to perform FRET using a single donor and a single acceptor [10]. This so-called "single-molecule FRET" (or, more generally, single-pair FRET; spFRET) combines the ability of FRET to measure distances on the nanometer scale with the advantages of single-molecule detection. As single-molecule measurements are not inherently ensemble averaged, information is made available on the heterogeneity of the system, in addition to average values. Thus, subpopulations, conformational changes, and brief, transient fluctuations can all be detected with spFRET [11].

In this chapter, following a brief description of the fundamentals of FRET, the two current approaches to spFRET measurements are outlined: (i) *solution measurements*, where snapshots of the individual molecules are taken as the molecule diffuses through the focus of the confocal microscope; and (ii) measurements on *immobilized molecules*, where the spFRET trajectories can be

Single Particle Tracking and Single Molecule Energy Transfer
Edited by Christoph Bräuchle, Don C. Lamb, and Jens Michaelis

ISBN: 978-3-527-32296-1

followed over time. For solution measurements, results relating to protein folding and the role of Chaperon proteins are detailed, while for the immobilized molecules the results describe the mechanism of helicases and the TATA box binding protein. Finally, a brief discussion regarding of the future directions of single-pair FRET is provided. Whilst it impossible to acknowledge the extent of the spFRET studies currently being conducted, several excellent reviews have been prepared, to which the reader is referred [11–19].

5.2
Principles of FRET

Förster resonance energy transfer occurs via dipole–dipole interactions of the excited-state of one fluorophore (the FRET donor, D) with the ground state of a nearby chromophore (the FRET acceptor, A). The rate of energy transfer depends on the spectral overlap of the of the fluorescence emission of the donor fluorophore and the absorption spectrum of the acceptor molecule (spectral overlap integral, J; see Figure 5.1a), the relative orientation of the two dipoles in three-dimensional (3-D) space (the orientational factor, κ^2; Figure 5.1b) and the distance between the donor and acceptor, R (Figure 5.1c):

$$k_T = \frac{9000(\ln 10)\kappa^2 \varphi_D J}{128\pi^5 N_A n^4 \tau_D R^6} = \left(\frac{1}{\tau_D}\right)\left(\frac{R_0}{R}\right)^6 \tag{5.1}$$

where φ_D is the quantum yield of the donor fluorophore in the absence of the acceptor, n is the index of refraction of the sample (typically approximated as 1.4 for protein solutions), τ_D is the fluorescence lifetime of the donor in the absence of the acceptor, and R_0 is the Förster radius. R_0 is given by:

$$R_0 = \left(\frac{9000(\ln 10)\kappa^2 \varphi_D J}{128\pi^5 N_A n^4}\right)^{1/6} \tag{5.2}$$

The FRET efficiency is defined as the fraction of excitations that undergo energy transfer from the donor molecule to the acceptor molecule. Alternately, the FRET efficiency can be defined as the probability of the energy of an excited donor molecule being transferred to the acceptor molecule. The FRET efficiency is given by:

$$E = \frac{k_T}{k_T + k_D} \tag{5.3}$$

where $k_D = 1/\tau_D$ is the rate of fluorescence decay of the donor in the absence of an acceptor, and k_T is given by Equation 5.1. Combining Equations 5.1 and 5.3 yields a direct relationship between the FRET efficiency and the donor–acceptor separation:

$$E = \frac{R_0^6}{(R_0^6 + R^6)} \tag{5.4}$$

(a) Spectral overlap

$$J = \frac{\int d\lambda\, F_D(\lambda)\varepsilon_A(\lambda)\lambda^4}{\int d\lambda\, F_D(\lambda)}$$

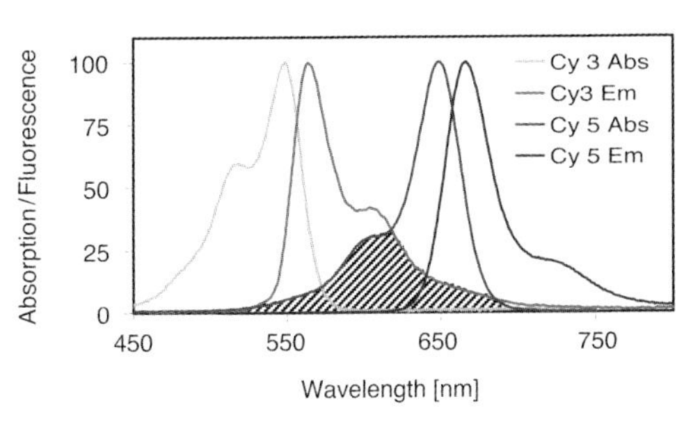

(c) D-A separation

Dansyl-(pro)$_n$-linker-napthyl

$$E = \frac{R_0^6}{\left(R_0^6 + R^6\right)} \qquad R_0 = 34\ \text{Å}$$

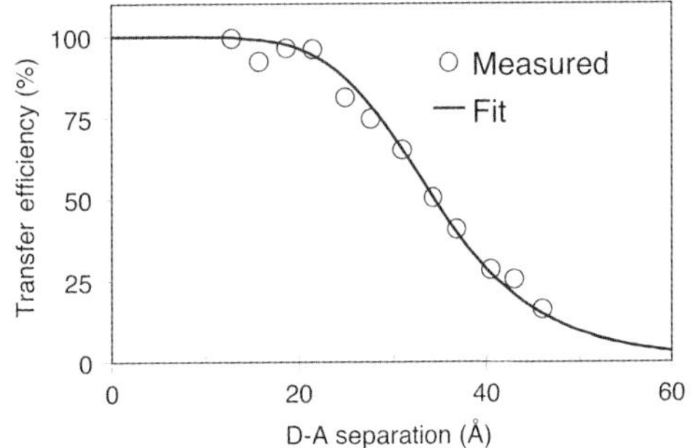

(b) Orientation factor

$$\kappa^2 = \left(\cos\theta_T - 3\cos\theta_D \cos\theta_A\right)^2$$

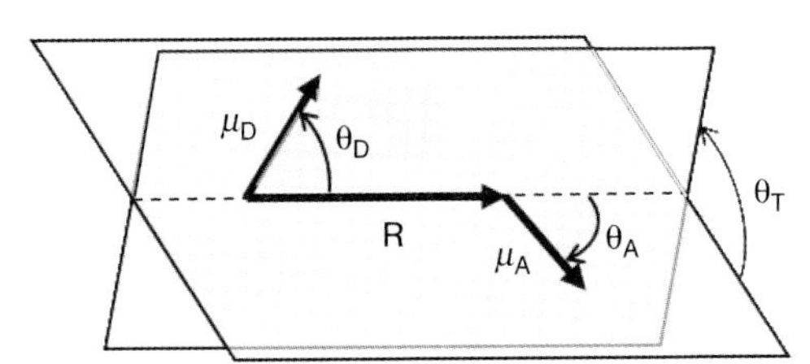

Figure 5.1 Factors that influence the rate of Förster resonance energy transfer. (a) The absorption and emission spectra of Cy3 and Cy5. The spectral overlap integral, J, gives a measure of how "resonant" the donor and acceptor fluorophores are. By selecting the FRET pair used for spFRET measurements, the Förster radius can be varied from ~45 Å to ~80 Å. (b) The relative orientation of the dipole moments of the two fluorophores in three dimensions influences the rate of energy transfer. The orientation factor, κ^2, can range from 0 to 4, but it is often sufficient to average over all possible orientations and use the value of 2/3. (c) The distance between the donor and acceptor fluorophores strongly influences the rate of energy transfer. This R^6 dependence of the energy transfer efficiency with distance was experimentally verified by Stryer and Haugland, using a poly-proline linker and dansyl and naphthyl as donor and acceptor molecules. The data shown here have been taken from this seminal work [9]. Because of the high sensitivity of FRET on the distance scale of a few Angstroms, FRET is often referred to as a "molecular ruler."

The amount of energy transfer can be calculated from donor quenching in the presence of acceptor, by enhanced acceptor fluorescence in the presence of a donor, or from the ratio of intensity from the donor and acceptor channels. The latter approach is what is typically used in spFRET experiments. Here, the FRET efficiency can be calculated using:

$$E = \frac{F_A}{\gamma F_D + F_A} \tag{5.5}$$

where F_A and F_D are the fluorescence intensity of the acceptor and donor channel, respectively, and

$$\gamma = \frac{\varphi_A \eta_A}{\varphi_D \eta_D} \tag{5.6}$$

is the detection correction factor which compensates for differences in sensitivity of the system to donor and acceptor fluorescence, φ_i is the quantum yield, and η_i is the detection efficiency of the apparatus to the donor and acceptor fluorophores. Equation 5.5 is correct in the ideal case without spectral crosstalk or direct excitation. To correct for these effects, Equation 5.5 can be modified as

$$E = \frac{F_{FRET}}{\gamma F_D + F_{FRET}} \tag{5.7}$$

where

$$F_{FRET} = F_A - \beta F_D - \alpha F_A^{Dir} \tag{5.8}$$

β is the crosstalk correction factor giving the fraction of donor signal detected in the acceptor channel with respect to the donor signal in the donor channel, and α is the correction factor for direct excitation of the acceptor and the ratio of acceptor intensity with donor excitation (in the absence of a donor) to the acceptor intensity when excited at the wavelength used to measure F_A^{Dir}.[1] The γ factor can be measured on single FRET pairs when there is a dynamic change in the FRET signal, or when the acceptor bleaches before the donor molecule. The value of γ varies from molecule to molecule, and the distribution is typically broad. Hence, the best results for measuring histograms of FRET efficiency are obtained when the γ factor is determined individually for each molecule or complex.

The orientation factor depends on the relative orientation of the donor and acceptor dipoles in 3-D space (Figure 5.1b). In spFRET experiments, there is often concern when the anisotropy of the individual fluorophores is too high (typically above 0.2), that an averaged κ^2 value of 2/3 cannot be assumed. In a recent study, Ha and Lilley measured the FRET efficiency between Cy3 and Cy5 attached to DNA. From NMR studies, it was shown that the fluorophores stacked on the end of the DNA and thus had a well-defined orientation. The end result of this study was that whilst orientational effects are observable, they play a smaller role than might be expected [20].

5.3 spFRET in Solution

In solution-based spFRET measurements, the fluorescence emitted from molecules diffusing freely through the small observation volume is detected in bursts.

1) An equivalent approach is discussed in Chapter 8, where Equation 5.5 is used and the definition of γ is changed.

The concentration of fluorescently marked molecules is sufficiently low (typically <50 pM) that the probability of having a single molecule in the probe volume is much less than one. The photons that arrive within a time bin or a burst are summed together and analyzed; hence, the method is often referred to as a "burst analysis." The first experiments that efficiently detected photon burst from individual fluorophores where performed at Los Alamos National Laboratory around 1990; the first studies used pulsed laser excitation and time-gated detection [21], but this was later followed by single-molecule detection using a continuous-wave laser [22]. Rigler and coworkers subsequently adapted the method for confocal microscopy [23]. In order to extract more information from single-molecule bursts, the group of Seidel has developed an apparatus and methods of analysis to determine the fluorescence lifetime of molecules [24] and to detect polarization during a burst [25]. The first spFRET investigations to be performed in solution using burst analysis were performed by Weiss and coworkers [26], who demonstrated the strength of spFRET methods to detect subpopulations in solution. Since these pioneering studies, solution spFRET has been used by many groups to investigate the conformation and dynamics of biomolecules.

5.3.1 Experimental Considerations

When performing spFRET measurements in solution, a compromise must be determined between the number of molecules detected and the percentage of coincident events. For burst analysis, dilute concentrations of the fluorescently labeled sample are used (typically in the range of 20–100 pM). Burst analysis operates on the assumption that only one molecule is in the volume at a time. For typical experiments, the probability of having at least one molecule in the probe volume is ≤1%. Then, assuming a Poisson distribution with an average of 0.01 molecules in the probe volume, the probability of having at least two molecules in the volume simultaneously is 5×10^{-5}. Considering that an event is only analyzed when at least one molecule is in the volume, the number of events with more than one molecule will be ~0.5%. For measurements with thousands of events, there will always be a few bursts with multiple molecules, which are normally easy to recognize as outliers on a spFRET histogram. To avoid coincidence, the use of more dilute samples is better, but the measurement time for obtaining sufficient data for statistical analysis is increased. However, increasing the measurement time is not always an option for biological samples that are unstable at room temperature. Hence, the optimal parameters will vary, depending on the sample being measured.

The average number of molecules in the probe volume depends on the size of the probe volume. Although a smaller focus would allow higher concentrations to be used in the measurements, the number of photons in a burst will also depend on the volume size. Thus, the larger the volume, the more time a molecule will spend in that volume, increasing the number of photons measured in a burst. Therefore, volumes much larger than the diffraction-limited spot size of a high numerical aperture (NA) objective are typically used for burst analysis measurements. Measurements using volumes as large as several femtoliters can be performed, provided that the buffers are free from fluorescent impurities and that the appropriate filters are used to suppress Raleigh and Raman scattered light.

Another important consideration regarding burst analysis is that of *laser power*. The quality of the measurement depends on the number of photons detected in the individual bursts, and increasing the laser power typically increases the number of detected photons. However, at higher laser powers the probability of a molecule photobleaching while traversing the probe volume is significant. A burst analysis with complexes where the acceptor has photobleached during the burst leads to distortion of the data. One possibility of using high laser powers and minimizing artifacts from photobleaching is to check the individual bursts for photobleaching and removing any bursts that appear to have undergone photobleaching. Such procedures have been suggested by the groups of Seidel [27], Weiss [28] and Schuler [29].

5.3.2
Protein-Folding Kinetics

One of the most prominent applications of spFRET measurement in solution was reported by Eaton and coworkers, who investigated the free-energy surface of protein-folding for the cold shock protein in *Themotoa maritime* (Csp*Tm*) [30]. Cysteines were engineered into the Csp*Tm* at the N terminus (after Met 1) and at the C terminus, and labeled with Alexa488 and Alexa594. Burst analysis experiments were then carried out on Csp*Tm* at different denaturant (GdmCl) concentrations (Figure 5.2). For control measurements, two polyproline peptides of different lengths were synthesized and labeled with the same fluorophores. Under native conditions, two populations were observable in the spFRET histograms; there was a high FRET population which corresponded to the folding protein, and a low FRET population which corresponded to the donor-only species. As the concentration of denaturant was increased, an additional population was observable with a FRET efficiency of ~0.5. This intermediate FRET state corresponded to the denatured protein. The mean FRET values of the Csp*Tm* subpopulations and the control proline 20-mer, as well as the widths of the unfolded state, are shown in Figure 5.2b. No significant changes were observed in the FRET efficiency of the polyproline chains, whereas a strong shift in FRET efficiency was observed for the unfolded state. Hence, there was a compaction or collapse of the polypeptide chain in the unfolding state, and an increase in the population of the folded state when passing from high to low GdmCl concentrations.

For single-molecule measurements, the width of the distribution also provides useful information. The authors noted that, although both the unfolded protein and the control polyproline sample had widths significantly above the shot-noise limit, there was no discernable difference in the width of the two distributions [30]. It was concluded, therefore, that the polypeptide chain of the unfolded protein was dynamically averaged during transit of the probe volume, and a maximum

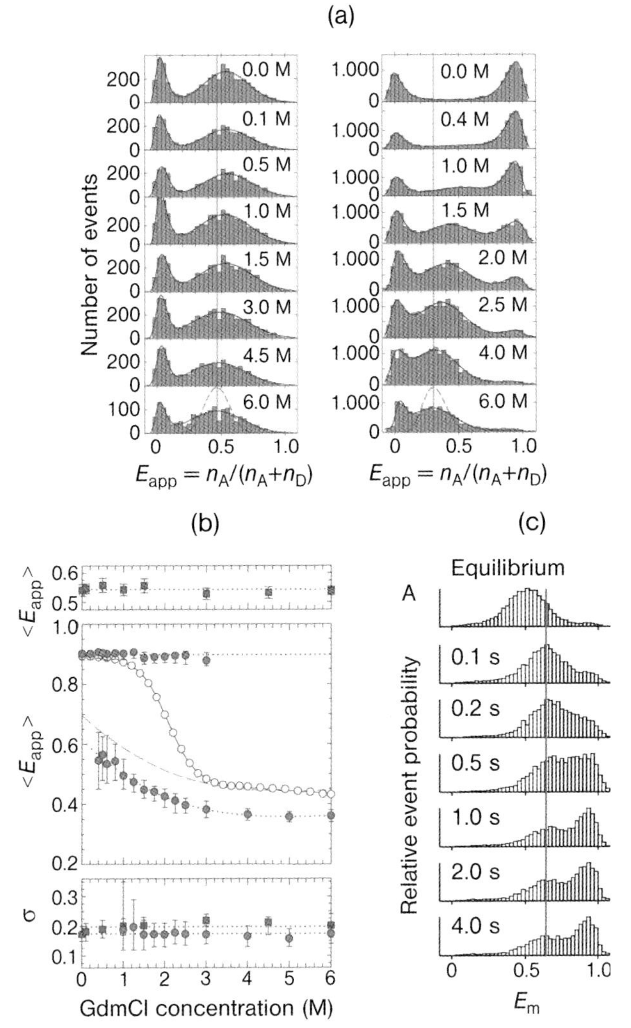

Figure 5.2 Folding of CspTm using spFRET. (a) SpFRET histograms of labeled poly-proline 20-mers (left panel) and Csp*Tm* (right panel) as a function of GdmCl concentration. A zero FRET efficiency peak is observable due to the presence of a donor-only species. The left panel shows essential no change in the peak FRET efficiency with GdmCl concentration. The right panel shows two states: a folded state at high FRET efficiency, and an unfolded state with intermediate FRET efficiency. The presence of the unfolded state is clearly observable at ~1 M GdmCl concentration and increase in amplitude with increasing denaturant concentration. The dashed line shows the expected shot-noise limit distribution for the measured bursts. Reproduced with permission from Ref. [30]. (b) The mean FRET efficiency for the poly-proline 20-mer (upper panel, squares) and the folded (upper circles) and unfolded population (lower circles) of Csp*Tm* (middle panel) and width of the poly-proline 20-mer (squares) and unfolded state (circles) as a function of GdmCl concentration. Reproduced with permission from Ref. [30]. (c) spFRET histograms of Csp*Tm* at equilibrium in 4.0 M GdmCl (top panel) and at various time points after dilution of the complex from the denaturing buffer. The line indicates the peak FRET efficiency of the unfolded state under refolding conditions. Reproduced with permission from Ref. [31].

configuration time of 0.1 ms was estimated. A minimum reconfiguration time of 40 ns was estimated by treating the polypeptide chain as a Gaussian chain. By using Kramer's theory and the extrema determined for the reconfiguration times, it was possible to estimate the free-energy barrier height for folding ($11k_BT > \Delta > 2k_BT$).

In subsequent experiments, Lipman *et al.* measured the kinetics of protein folding using spFRET with the aid of a microfluidic mixer [31]. These authors designed and constructed a microfabricated laminar-flow mixer to allow the fast mixing (<50 ms) of denatured samples with a refolding buffer. By performing burst analysis experiments at different distances from the mixing region of the microfluidic mixer, the spFRET histogram at different time intervals was measured (Figure 5.2c) and, from the ratio of folded to unfolded proteins at the different time points, the kinetics of refolding was determined. At the earliest time point, a clear shift in the FRET efficiency of the unfolded state was observed, from 0.51 in equilibrium at 4 M GdmCl to 0.64 after mixing. This increase in FRET corresponded to the rapid (often submillisecond) collapse of the unfolded state [32]. Here, the collapse resulted in an approximately 20% decrease in the radius of gyration. As these experiments were performed using single molecules, changes in average FRET efficiency due to a collapse of the unfolded state could be distinguished from refolding. Moreover, the incorporation of microfluidic mixers with spFRET measurements is opening new possibilities in the investigation of protein kinetics [33].

5.3.3 Chaperon-Assisted Protein Folding

In collaboration with the group of F. Ulrich Hartl, the present author and colleagues applied spFRET to investigate the role of chaperon proteins in protein folding [34]. Until recently, chaperons were believe to take a passive role in protein folding, simply providing an environment for the nascent polypeptide chain to fold where it is protected from the remainder of the cell. In particular, investigations were conducted into the GroEL/GroES chaperon system from *Escherichia coli* [35]. GroEL (see Figure 5.3a) [36] is a chaperonin containing two internal cavities in which nascent polypeptide chains can fold. As a substrate, a double-mutant of the maltose binding protein (MBP) was chosen (Figure 5.3b) that folds spontaneously with a rate of $t_{1/2} \sim 1200$ s, but its refolding is accelerated 13-fold in the presence of GroEL, GroES, and ATP. Two cysteine residues were engineered into MBP at multiple positions (residues 52 and 298 for the histograms shown in Figure 5.3) and labeled with donor (Atto532) and acceptor fluorophores (Atto647N). The refolded and denatured histogram of FRET efficiencies are shown in Figure 5.3c. When denatured MBP was renatured in the presence of 3 μM GroEL, a bimodal distribution of FRET efficiencies was observed in spFRET measurements (Figure 5.3d, left panel). At this concentration of GroEL, all substrates were bound to a GroEL molecule, and refolding could not occur without the addition of the co-chaperon GroES and ATP. As these data were collected using pulsed-interleaved excitation (PIE) [37, 38]/alternating laser excitation (ALEX) (see Refs [39–41] and Chapter 6), only molecules with an active donor–acceptor pair were incorporated in the analysis. Hence, both FRET subpopulations corresponded to different conformations of the substrate, namely: (i) a broad subpopulation with an intermediate FRET efficiency of 0.66; and (ii) a low-FRET subpopulation ($E = 0.06$) where the donor and acceptor were far from each other.

In the denatured state, the average FRET efficiency for this construct of MBP was 0.10, as compared to 0.06 for the low-FRET subpopulation of MBP bound to GroEL. This difference may have been due to an additional stretching of MBP upon binding to GroEL, or perhaps attributed to the polymer dynamics of the denatured protein. In order to distinguish between these two possibilities, MBP was allowed to fold spontaneously for 200 s (Figure 5.3e, left panel). Although the spFRET histogram appears very similar to that of the folded protein (Figure 5.3c, left panel) – which indicates that the donor–acceptor separation is similar for the two measurements – only approximately 10% of the MBPs had reached the native state at this time, as determined by its ability to bind maltose [34]. After 200 s of spontaneous refolding, GroEL was added to the solution; the resultant spFRET histogram is shown in Figure 5.3e (right panel). A bimodal distribution was again observed, indicating that a fraction of substrates had been stretched upon binding to GroEL.

Upon the addition of ATP to the GroEL complex, the low-FRET population disappeared and a broad intermediate FRET peak, $E \sim 0.55$, was observed (Figure 5.3d, middle panel). By using GroEL mutants that were incapable of either ATP binding (D87K) or ATP hydrolysis (D398A), it could be shown that ATP binding was responsible for inducing this change. Upon the addition of EDTA, the bound nucleotide was removed from GroEL and the low-FRET subpopulation was observable, while the intermediate-FRET population reverted to higher FRET efficiencies (Figure 5.3d, right panel). ATP binding to GroEL is known to induce a conformational change in GroEL and to alter the affinity of GroEL for the substrate. This ATP-induced reversible conformational change of GroEL led to a partial release of the substrate and a transient stretching of the remaining bound fraction of MBP. These results were consistent with molecular dynamic simulations performed by Stan *et al.*, which showed substrate unfolding during binding to GroEL and during the ATP-induced conformational change in GroEL [42]. By combining the single-molecule results with those of kinetic ensemble FRET experiments, the chaperons were shown to be much more actively involved in the folding of MBP.

The group of Schuler has also used spFRET to investigate the interactions of another GroEL substrate, rhodanese, with GroEL [29]. For this substrate, no bimodal population was observed, but rather a single, broad spFRET distribution. When Schuler's group measured the time-resolved anisotropy to include orientation effects in the description of the spFRET distribution, their results suggested that rhodanese existed in a well-defined ensemble of partially folded conformations. Taken together, the results of these two investigations showed that the interaction of GroEL with a substrate may depend heavily on the substrate itself.

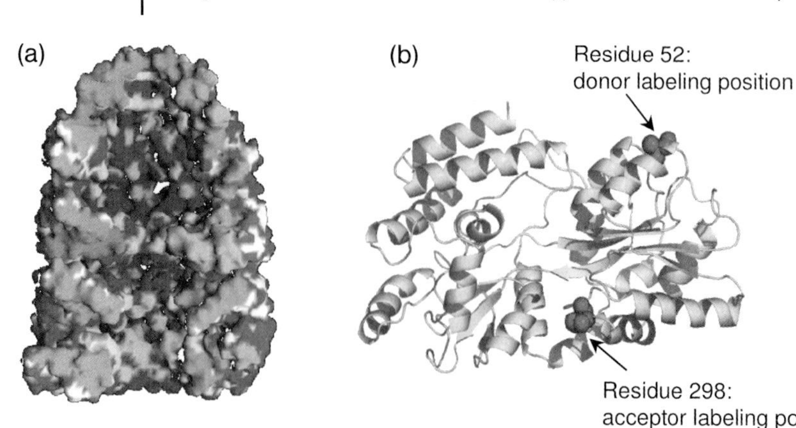

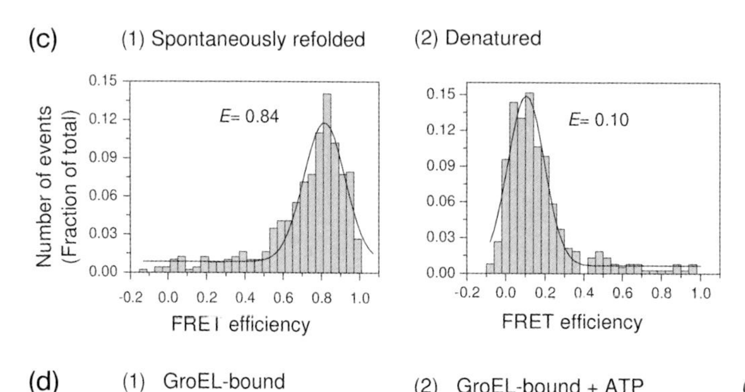

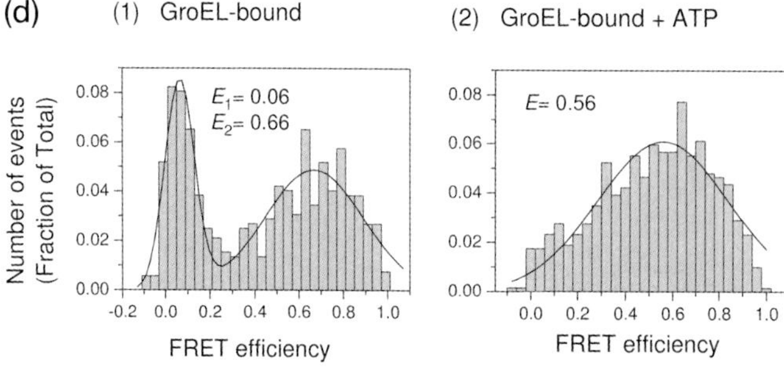

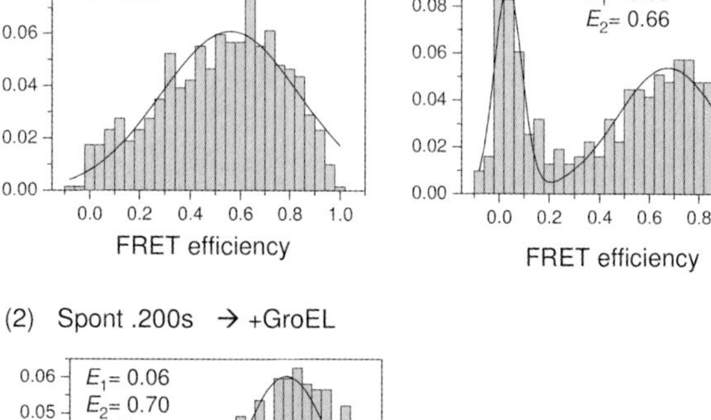

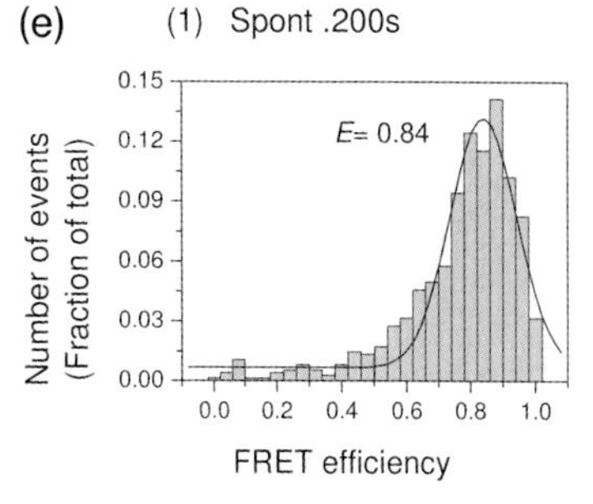

Figure 5.3 (caption see p. 109)

As unfolded proteins are inhomogeneous, methods that are able to detect and investigate heterogeneities will be powerful tools for investigating of the role of chaperon proteins [43] or the function of intrinsically disordered proteins.

5.3.4
Photon Distribution Analysis (PDA)

One (albeit currently underutilized) advantage of single-molecule studies is the availability of information relating to the heterogeneity of a sample, which can be obtained from the width of the distribution. In order to correctly interpret the significance of the width of the distribution, it is essential to recognize which sources can lead to a broadening of the spFRET signal. The width of a spFRET measurement can be affected by shot noise, background, blinking, acceptor photobleaching (and various other photophysical phenomena), misalignment of the donor and acceptor detection channels, incorrect γ-factors, distribution of orientations, conformational dynamics, and actual distributions in the donor–acceptor separation. In the past, a number of groups have investigated the statistics of spFRET distributions [28, 44–46] and, assuming that the experimentally controllable sources of error have been avoided (e.g., alignment, photobleaching, etc.), it is possible to extract information from the width of the distribution. The broadening due to shot-noise can be calculated as the photon-counting statistics are collected during an experiment. Any additional width over the shot-noise-limited distribution is attributable to a distribution in the FRET signal.

Figure 5.3 spFRET analysis of maltose binding protein (MBP) during chaperonin-assisted protein folding. (a) Structure of the GroEL/GroES complex cut away in the middle to show the cavity for protein folding. Hydrophobic resides are displayed in light grey and polar and charged side groups are shown in dark grey. Figure modified with permission from Ref. [36]. (b) The structure of MBP, the substrate used for these investigations. The positions of residues 52 and 298 which were mutated to cysteines and labeled with the donor and acceptor fluorophores respectively are shown as space-filling structures. (c) The spFRET histogram of the 52–298 MBP mutant that has folded spontaneously with a peak FRET efficiency of $E = 0.84$ (left panel) and in the unfolded state (3.0 M GdmCl) with a peak FRET efficiency of $E = 0.10$ (right panel). (d) The spFRET histogram of MBP bound to GroEL in the absence of ATP (left panel), in the presence of 2 mM ATP (middle panel), and after the addition of 10 mM EDTA 10 min following the addition of ATP (right panel). A bimodal distribution ($E_1 = 0.06$, $E_2 = 0.66$) is observed in the absence of ATP. The low-FRET efficiency population disappears upon the binding of ATP and is reversible, reappearing upon the addition of EDTA. (e) The spFRET histogram of MBP at approximately 200 s after being diluted in renaturing buffer (left panel). Approximately 10% of the sample has folded to the native state during this time. A single, high-FRET peak ($E = 0.84$) is observed. MBP bound to GroEL after 200 s of spontaneous folding in renaturing buffer (right panel). The bimodal distribution with the low-FRET efficiency peak ($E = 0.06$) is again observed.

The group of Claus Seidel has taken an elegant approach to extracting information from spFRET histograms [44]. In their approach, they have chosen to analyze the ratio of donor signal (F_D) to acceptor signal (F_A), although the analysis is also possible using FRET efficiency or proximity ratio histograms. The probability of detecting a signal ratio of F_D/F_A is given by:

$$P\left(\frac{F_D}{F_A}\right)_i = \sum_{\substack{\text{all } F_D, F_A, B_D, B_A \\ \text{which yield } (F_D/F_A)_i}} P(F)P(F_{FRET}|F)P(B_D)P(B_A) \tag{5.9}$$

where F is the total number of detected signal photons ($F = F_D + F_{FRET}$), $P(F_{FRET}|F)$ is the conditional probability of detecting F_{FRET} photons in the FRET channel (as defined in Equation 5.8) given a total of F detected signal photons, and $P(B_i)$ is the probability of detecting B_i background photons for the respective channel. The probability of detecting background photons follows a Poisson distribution:

$$P(B_i) = \frac{\langle B_i \rangle^{B_i} e^{-\langle B_i \rangle}}{B_i!} \tag{5.10}$$

where $\langle B_i \rangle$ is the average number of background photons in the ith channel. Assuming a fixed probability, ε, of a fluorescent photon being detected in the acceptor channel, the conditional probability $P(F_{FRET}|F)$ follows a binomial distribution:

$$P(F_{FRET}|F) = \frac{F!}{F_{FRET}!(F - F_{FRET})!} \varepsilon^{F_{FRET}} (1-\varepsilon)^{(F-F_{FRET})} \tag{5.11}$$

To complete Equation 5.9, an expression is needed for $P(F)$. One alternative is to approximate $P(F)$ with $P(N)$, where N is the total number of detected photons in a bin (this is reasonable when the background is low). $P(N)$ is given directly from the experimentally determined normalized distribution of photon counts, such that the only remaining unknown is ε. In the simplest case, a homogeneous population with a single FRET efficiency E can be assumed, and ε can be related to the FRET efficiency:

$$\varepsilon = \frac{F_{FRET}}{F_D + F_{FRET}} = 1 - \frac{1}{1 + \gamma\left(\frac{E}{1-E}\right) + \alpha} \tag{5.12}$$

With an estimate for the FRET efficiency, the signal-ratio histogram (Equation 5.9) can be calculated and compared to the real data (Figure 5.4). For a distribution of FRET efficiencies, the corresponding distribution for ε can be calculated and weighted appropriately. As the number of detected photons is quantized, the bins in the signal-ratio histogram may have higher or lower probabilities, which are reflected in the analysis method and allow the spikes and valleys of the histogram to be reproduced. Other approaches have been developed for approximating $P(F)$ more accurately, and the method has been expanded to include anisotropy data [47] and multiple molecular states [48].

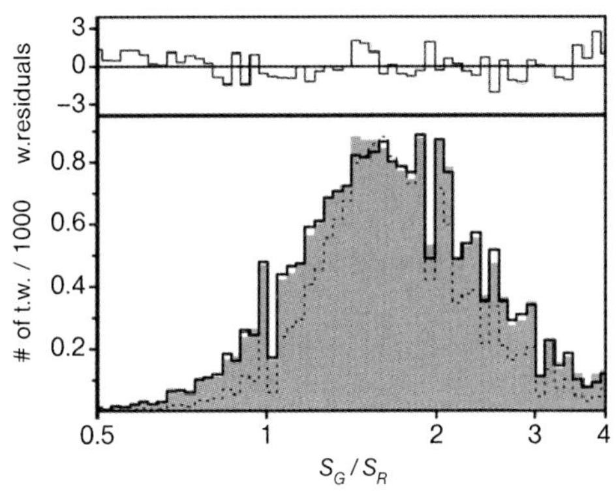

Figure 5.4 The photon distribution analysis (PDA). A histogram of signal ratio for spFRET measurements with a FRET-labeled DNA where the donor and acceptor are separated by 13 base pairs (dark gray). The dashed line shows the predicted signal-ratio histogram for a fixed FRET efficiency ($F_D/F_A = 1.61$); the solid black line shows the results assuming a Gaussian distance distribution peaked at 53.9 Å ($F_D/F_A = 1.58$) and a width of 2.3 Å. The reduced residuals for the fit to a Gaussian distribution of distances are shown in the upper portion of the graph. The spikes and valleys of the histogram are well reproduced by PDA. Reproduced with permission from Ref. [44].

5.4 spFRET on Immobilized Molecules

One disadvantage of burst analysis is the limited timescale over which the FRET efficiency can be measured on an individual complex, due to the short transit time of the molecule through the probe volume (typically 1–10 ms). In a series of challenging experiments conducted by Börsch and coworkers, the rotation of the F0-F1 ATPase in liposomes was investigated, with occasional bursts in excess of 100 ms being reported [49]. However, this proved to be an exception and, as the liposomes diffused so slowly, it was difficult to collect sufficient data for a statistical analysis. When dynamics occurred on the transit timescale of the molecule (or faster), the dynamics could be investigated using fluorescence correlation spectroscopy (FCS) [50, 51]. When using spFRET to investigate the conformational transitions of single molecules or complexes on the 10 ms timescale or longer, it is necessary first to immobilize the complexes. The spFRET signal is then measured until the fluorophores photobleach, using either total internal reflection fluorescence (TIRF) microscopy or nonfocal microscopy. The advantage of TIRF is that it has a lower background signal, since the excitation depth of the evanescent wave is only 100–200 nm. In addition, a sensitive camera can be used [e.g., an electron-multiplying charge-coupled device (EMCCD)], and data from hundreds of single molecules can be collected simultaneously. In confocal microscopy, although only one molecule can be measured at a time, the data can be collected with a much higher time-resolution.

5.4.1
Immobilization Methods

One of the main challenges in performing spFRET experiments on immobilized molecules is to ensure that the immobilization method does not affect the functionality of the biomolecules. Several excellent reviews have been produced of the various immobilization methods available [52–54]. The initial experiments were performed on nucleic acids that had been labeled with a single donor–acceptor pair and immobilized using biotinylated bovine serum albumin (BSA) and a biotin–streptavidin–biotin linkage to a biotin-labeled DNA or RNA strand (Figure 5.5a) [10, 55]. It is known that BSA adsorbs to hydrophobic glass surfaces and anchors the nucleic acids to the coverslip. However, although BSA works well with nucleic acids, its surface coverage is often incomplete and many proteins interact nonspecifically with the surface, even in the presence of BSA. In order to investigate the interaction of proteins with DNA, the group of Greene used the adsorption of neutravidin to a glass coverslip as an anchor for biotinylated DNA, and then passivated the surface using a supported lipid bilayer of zwitterionic lipids [56]. For spFRET studies with proteins, polyethylene glycol (PEG) is often used to cover the surface (Figure 5.5b) [57] and, depending on the affinity of the protein to the surface, higher fractions of PEG may be necessary to pacify the surface [58]. A small fraction of the PEG (typically ~1%) that has been modified with a reactive group such as *N*-hydroxysuccinimide ester (NHS), or has been labeled with biotin, is then used to attach the biomolecules to the surface. Branched polymers of different compositions have also been used successfully to immobilize biomolecules on surfaces [59–61]. Another immobilization method uses a His-tag and chelated Ni^{2+} or Cu^{2+} groups that are attached to the surface (Figure 5.5c) [62, 63]. This may be advantageous, as many proteins have already been cloned with His-tags in regions that do not affect the functionality of the protein, for purification purposes. Another approach to immobilization is to trap the single molecules in a nanocontainer, which is then immobilized. In this case, the biomolecules do not need to be physically attached to the surface, which can often cause deleterious effects on the functionality of the protein. The group of Ha has developed a method for trapping molecules in vesicles and immobilizing the vesicles to the surface (Figure 5.5d) [64]. Objects can also be immobilized away from surfaces by using optical or magnetic tweezers. For example, aqueous nanodroplets have been formed in an immiscible fluorocarbon liquid and the nanodroplets trapped with optical tweezers, whereby spFRET signal measurements could be performed over several seconds until the fluorophores were photobleached [65]. Another method which has proved capable of keeping molecules within the focus of a confocal microscope for extended periods of time, was developed by Cohen while working in the group of W.E. Moerner [66–68]. By using microfabrication, the electrodes could be placed in a sample holder near the focus of the confocal laser system. The Brownian motion of the molecule could then be detected and countered by applying electrokinetics forces through the electrodes in a feedback loop. This device is referred to as an anti-Brownian electrokinetic (ABLE) trap.

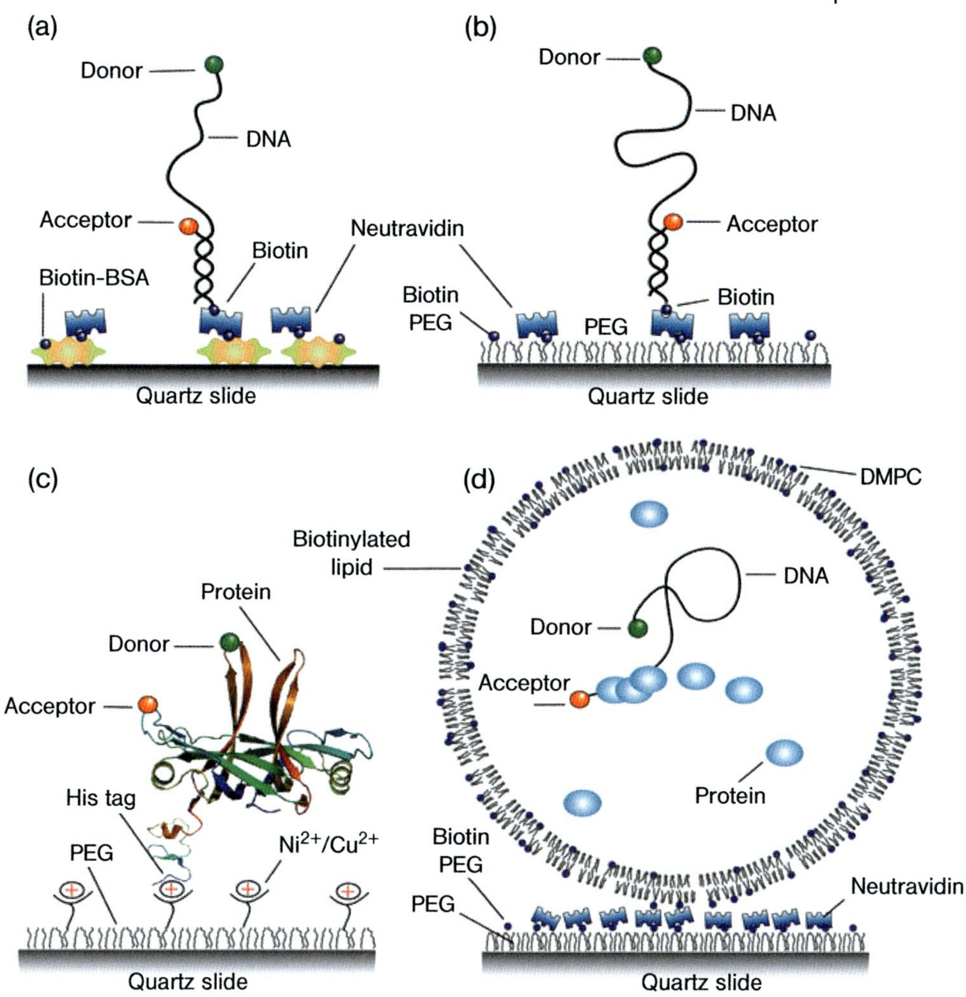

Figure 5.5 Immobilization methods. Different approaches to the immobilization of molecules for spFRET experiments are shown. (a) Biotinylated bovine serum albumin (BSA) molecules attach non-specifically to the coverslip and are used to anchor biotin-containing complexes via a multivalent-avidin molecule such as neutravidin. This approach works well for experiments with nucleic acids. (b) Complexes are immobilized using the same biotin–avidin–biotin linkage as in panel (a), but in this case, the surface is covered with polyethylene glycol (PEG) where a small fraction of the PEG contains a biotin. (c) Proteins are tethered to the surface using a His-tag and Ni^{2+} or Cu^{2+} chelated groups attached to a PEG surface. (d) Proteins can be encapsulated in ~100 nm-sized vesicles, and the vesicles immobilized to a biotinylated PEG surface via avidin and biotinylated lipids. Reproduced with permission from Ref. [53].

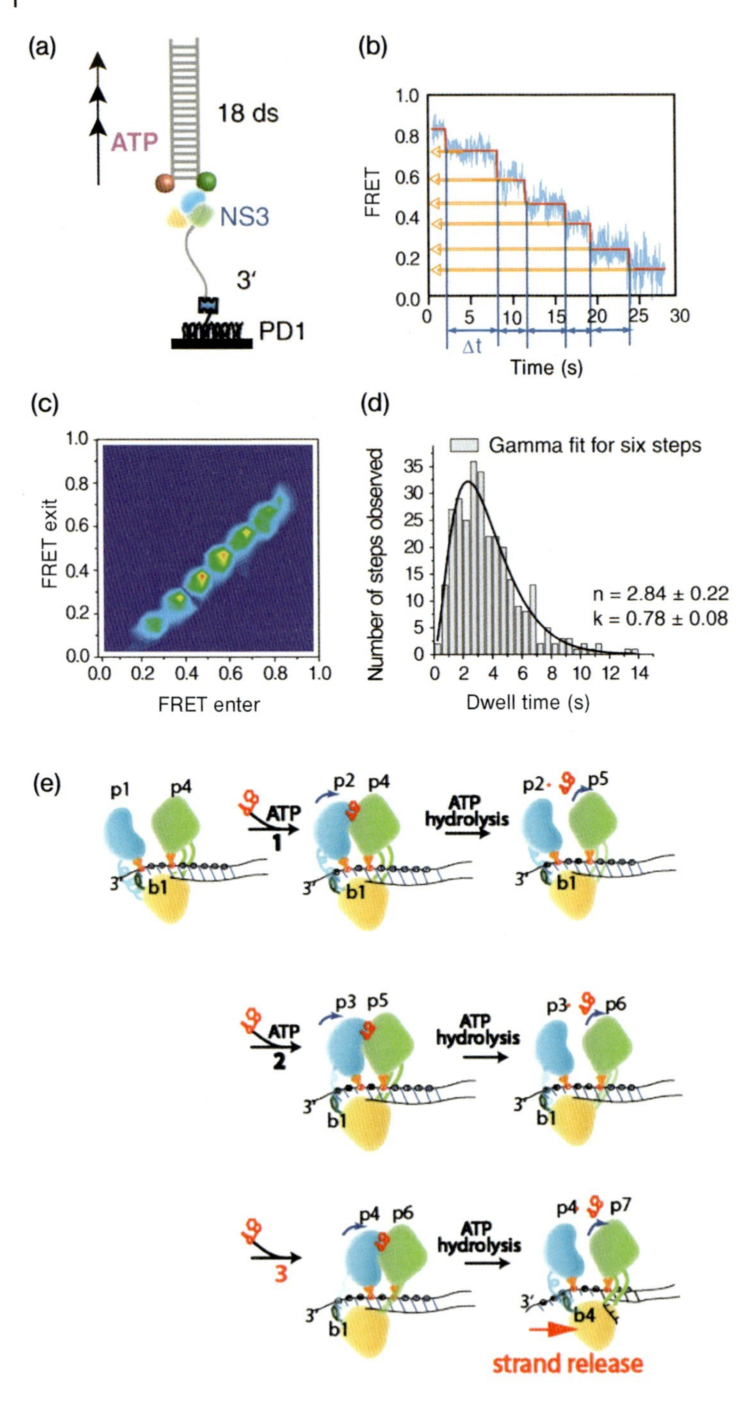

Figure 5.6 (caption see p. 115)

5.4.2
Mechanism of DNA Unwinding by the NS3 Helicase

As a demonstration of the power of spFRET measurements for elucidating mechanistic information relating to protein function, a recent report from the group of Taekjip Ha on the spring-loaded mechanism of DNA unwinding by the hepatitis C virus NS3 helicase, is highlighted below [69]. The NS3 helicase forms a complex with the viral polymerase and other cofactors, and is essential for virus replication. In addition to knowledge of the crystal structure, a significant quantity of biochemical data is also available for NS3 [70], which makes it an excellent model system for investigating the mechanical function of helicase enzymes. The main function of NS3 appears to be the unwinding of the highly structured hepatitis C virus genome, although it is capable also of unwinding both DNA and RNA substrates. Ensemble biochemical studies have shown that NS3 is able to unwind short RNA strands in steps of ~18 bp [71]. Single-molecule optical-tweezers experiments conducted in the laboratory of Bustamante have shown that NS3 takes rapid steps of 3 to 4 bp [72].

As a substrate, Myong *et al.* used a partial-duplex DNA (18 bp) with a 3′ single-stranded DNA (ss-DNA) tail of 20 bases that was attached to the surface via biotin at the 3′-end (Figure 5.6a). The short DNA strand was labeled with the acceptor using a 5′ modified C6 linker, whereas the donor (Cy3) was attached internally to a substituted aminodeoxythymidine base so that the backbone structure of the DNA would not be disturbed. The fluorophores were placed across from each other such that the initial FRET efficiency was ~0.9. Subsequently, NS3 was incubated with the DNA at 25 nM concentration for 15 min before being attached to the surface of the flow chamber. Upon the addition of 4 mM ATP, the helicase began to unwind the DNA, and several drops in the FRET efficiency were observed (Figure 5.6b). The decrease in FRET efficiency corresponded to an increase in time-averaged donor–acceptor separation. Six plateaus were observable with this DNA substrate. The experiment was repeated with an identical DNA construct,

Figure 5.6 The mechanism of DNA unwinding by the NS3 helicase. (a) Schematic representation of the experimental system including DNA, fluorescent labels, and starting position for the NS3 helicase. (b) An individual FRET trace showing the different steps observed as the NS3 helicase unwinds the DNA. The orange lines represent the results of the step-finding algorithm used in this study. (c) A 2-D transition plot showing the initial FRET values (ordinate) and final FRET values (abscissa) for the individual transitions from 75 molecules. Transitions are only observed from higher to low FRET efficiencies. (d) A histogram of delay times between steps in the spFRET data. The line represents a fit of the histogram to a Gamma function assuming *n* hidden, irreversible steps with identical rates, *k*. Values for *n* and *k* are giving in the lower right-hand corner. (e) A schematic model of the proposed mechanism of unwinding for NS3. Domains 1, 2, and 3 are labeled in blue, green, and yellow respectively. As domains 1 and 2 translocate along the DNA, strain is built up in domain 3. Eventually, the "spring-loaded" domain moves forward in a burst, unwinding the nucleotide strands by three base pairs. Figure adapted with permission from Ref. [69].

where the fluorophores were labeled in the middle of the DNA duplex. In these experiments, only three plateaus were observed, which was consistent with the 3-bp step-size for unwinding. A simple algorithm developed by Kerssemakers *et al.* [73] was used to quantify the individual steps in the FRET trajectory. Once the dynamics of the spFRET traces had been modeled, detailed information was obtained with regards to the kinetics of the protein. For example, a 2-D transition plot was calculated (Figure 5.6c), where the ordinate indicated the initial FRET value before a transition, the abscissa provided the FRET value after transition, and the color indicated how often this transition had been observed. There were seven, easily observed FRET states with transitions passing from higher to lower FRET values. Transitions were only observed between adjacent states, which was consistent with the processive helicase activity of NS3. In addition, a histogram of dwell times could be determined from the kinetics (Figure 5.6d); in this case, the dwell-time distribution showed at least two phases – a rising phase followed by a decay – which indicated that multiples processes were involved in the kinetics. Assuming n hidden, irreversible steps with identical kinetics, the histogram of dwell times was fitted with a Gamma function:

$$g(t) = t^{n-1}e^{-kt} \tag{5.13}$$

where k is the rate of the individual steps. For the two DNA complexes measured, the dwell-time histograms could be described with an n-value of approximately 3 and a rate in the range of 0.78–0.89 s^{-1}. This rate compared well with the rate of unwinding determined from ensemble measurements on unlabeled DNA (0.66 s^{-1} bp^{-1}) [74]. Hence, the data were consistent with each 3 bp step being composed of three hidden steps, most likely coupled to ATP hydrolysis events. Following the hydrolysis of three ATPs, a conformational change occurred which resulted in the separation of 3 bp from the DNA. Based on the structure, the authors proposed a model where domains 1 and 2 of NS3 moved along the DNA, one nucleotide at a time. The third domain remained anchored to the DNA. When three such hidden steps had been taken, the "spring-loaded" lagging domain moved forward in a burst, unzipping 3 bp of the DNA (Figure 5.6e).

5.4.3
Dynamics of the TBP–NC2–DNA Complex

Previously, spFRET has been used to investigate the dynamics of the TATA box binding protein (TBP)–Negative Cofactor 2 (NC2) complex on DNA, in collaboration with the group of Meisterernst [58]. As a general initiation factor, TBP is involved in the early steps of gene transcription. Upon binding to the core-promoter TATA boxes on DNA, TBP first induces a sharp bend in the DNA (Figure 5.7a), and then recruits other transcription factors before initiating assembly of the preinitiation complex. The preinitiation complex remains bound to the core-promoter until RNA polymerase II clears the promoter site, and is then disassembled. This early step in the transcription process serves as a popular target

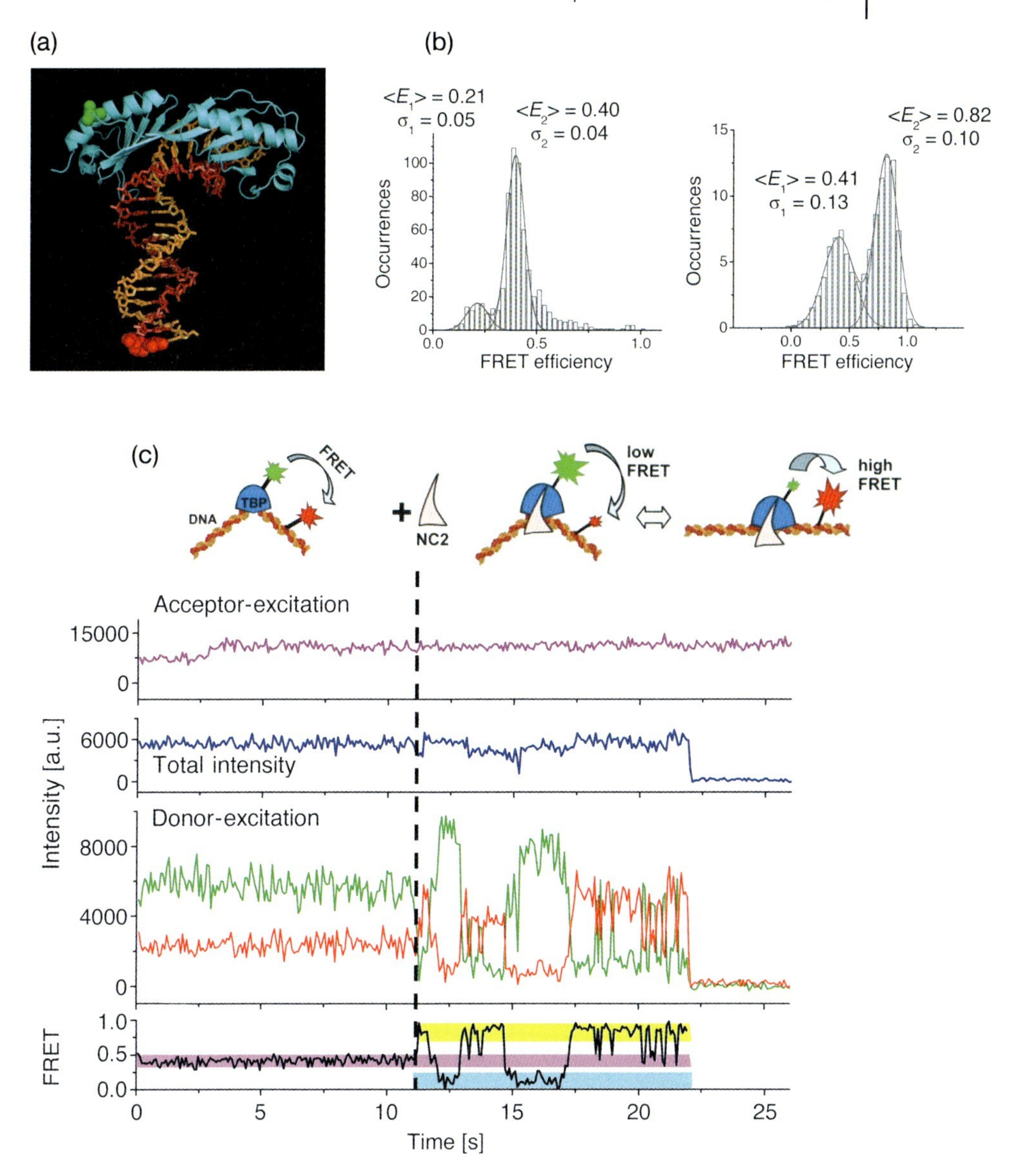

Figure 5.7 The dynamics of the TBP–NC2 complex on DNA. (a) The crystal structure of the TBP–DNA complex, with the attachment positions of the donor and acceptor molecules shown in green and red, respectively. (b) The molecule-wise (631) spFRET histogram of the TBP–TFIIA–DNA complex before the addition of NC2 (left panel). A dominant FRET population at $E = 0.4$ is observed. The frame-wise spFRET histogram for 110 molecules after addition of NC2 is shown (right panel). Two dominant populations are observable in the histogram. The peak FRET efficiencies and widths of the Gaussian fits are given in the plots. (c) A schematic and a spFRET trace showing the transition from a steady to a dynamic FRET signal upon addition of NC2. The FRET efficiency (black), donor (green), and acceptor (red) intensities, the total intensity $F = \gamma F_D + F_A$ (blue) and the acceptor intensity with direct excitation (F_A^{Dir}) (magenta), are shown as a function of time. The FRET efficiency of different conformations are highlighted in the spFRET trace in yellow, magenta, and cyan.

for gene regulation; therefore, both positive and negative cofactors may interact with TBP to either up regulate or down regulate transcription.

NC2 binds to the TBP–DNA complex from the bent underside of the DNA, and forms a ring-like structure with TBP [75]. The crystal structure of the TBP–DNA–NC2 complex suggests that the presence of NC2 may sterically hinder the binding of other transcription factors, and thereby repress gene transcription. However, footprinting measurements performed by Meisterernst and colleagues showed that the footprint of TBP vanished upon the addition of NC2, despite its still being bound to the DNA [58]. These findings suggest that dynamics might also play an important role in gene regulation.

In order to test the hypothesis that TBP becomes mobile along the DNA in the presence of NC2, a spFRET experiment was designed for which a single-cysteine mutant (S61C) of TBP from yeast [76, 77] (where the cysteine is located just outside the highly conserved region of the protein) was kindly provided by Tony Weil. The TBP was labeled with a donor fluorophore (Atto532), which did not influence its functionality [58]. DNA containing the Adeno major-late promoter site (AdML) and an acceptor fluorophore (Atto748N) was obtained commercially. To ensure the correct orientation of TBP on the TATA box, the general transcription factor TFIIA was used in preparing the TBP–DNA complexes [78, 79]. TBP (5–10 nM) was incubated with the DNA (10 nM) and TFIIA (20 nM) for 15 min at 28 °C before being diluted and added to a flow chamber. The dynamics of the initial binding of TBP to DNA has been investigated elsewhere [79–84]. When the TBP had become bound stably to the DNA, more than 93% of the complexes demonstrated steady spFRET signals until either the donor or acceptor fluorophore was photobleached. A histogram of spFRET efficiencies is shown in Figure 5.7b (left panel); here, two populations are observed, a dominant population with a peak FRET efficiency of 0.4, and a minor population with a FRET efficiency of 0.21. The low-FRET population was most likely due to a different binding position of the TBP on the DNA, whereas the major FRET population had a peak FRET efficiency that corresponded to the expected donor–acceptor separation observed in the TBP–DNA crystal structure.

After establishing that TBP on the AdML promoter exhibited a steady FRET signal of ~0.4 for the construct in the absence of NC2, NC2 was flowed through the sample chamber during the measurement (Figure 5.7c). Upon the addition of NC2, the behavior of TBP on the DNA changed dramatically, with the FRET signal switching from steady to dynamic and fluctuating between distinct states on the 100 ms to second timescale. A histogram of FRET values per frame collected on over 100 molecules showed two major subpopulations with values of 0.41 and 0.82, respectively (Figure 5.7b, right panel). From the time course of the FRET signal shown in Figure 5.7c, a third subpopulation with a FRET efficiency of ~0.20 was observed. A detailed analysis of the dynamics using a Hidden Markov Modeling (HMM) analysis on data collected at a higher time resolution (5 ms per frame) revealed a fourth state with a FRET efficiency of ~0.6 (these studies were conducted in collaboration with N. Zarrabi and M. Börsch, unpublished results).

To interpret the fluctuations in FRET signal as real changes in the separation between donor and acceptor, it was necessary to perform control measurements so as to exclude other possible causes for the observed dynamics [58]. For example, blinking of the acceptor molecule could lead to fluctuations in donor and acceptor intensities, without any change in donor–acceptor separation. Experiments performed using millisecond ALEX (see also Chapter 6 and Ref. [39]) such as the data displayed in Figure 5.7c verified that the photophysics of the acceptor did not play any role in the observed dynamics. The intensity of the directly excited acceptor molecule (as displayed in the uppermost trace of Figure 5.7c) did not show any correlation with the dynamics observed in the FRET signal.

The 0.41 FRET efficiency subpopulation shown in Figure 5.7b (right panel) corresponded to a conformation similar to the TBP–DNA in the absence of NC2 (Figure 5.7a). The second dominant FRET population, at 0.82, was attributed to a DNA-extended conformation. As the DNA is highly distorted in the TBP–DNA complex, it is energetically favorable for the DNA to extend when the interaction between TBP and DNA is weakened by the binding of NC2. However, the 0.20 FRET efficiency peak which was observed transiently in the FRET traces of individual molecules (see Figure 5.7c) was attributed not only to conformation changes in the TBP–DNA–NC2 complex, but involved movement of the TBP–NC2 complex along the DNA.

Measurements on DNA with various promoter sites indicated that the observed dynamics was a general phenomenon observed with different TATA-containing promoters, although the details of the dynamics differed with the promoter sequence ([58] and D.C. Lamb, unpublished results). The biological consequences of the NC2-induced dynamics of TBP on the DNA is currently under investigation. An alteration of the TBP–DNA configuration upon binding of NC2 inhibits the association of TFIIB and results in a repression of transcription. Relocation of the TBP on the DNA may also enhance repression. Alternatively, it is conceivable that a structural rearrangement of the TBP–DNA complex and a shuttling of the transcription factors over short distances may optimize the positioning of the complex for initiation, and lead to an enhanced transcription under *in vivo* conditions.

5.4.4 Hidden Markov Modeling

As demonstrated in the previous two subsections, kinetic information can be collected from spFRET measurements. However, to extract the kinetic information, it is necessary to model the individual traces and to assign various levels or states as a function of time to the spFRET data. In simple systems, such an assignment can be performed either visually or by applying simple algorithms, such as a threshold analysis where a user-defined cut-off is used to distinguish the state of the system [85]. During the past few years, many maximum-likelihood approaches have been developed to investigate single-molecule dynamics [86–90]. Although, HMM was initially developed for speech recognition, it has found application in several other fields, including the analysis of single-ion channel measurements

[91, 92] and spFRET data [93]. Indeed, several reviews have been produced describing HMM and its application to single-molecule data [94, 95].

A *Markov process* is one where transitions occur between discrete states, and where the probabilities of the various transitions are constants. Hence, the system is termed *stationary*, which means that the rates are not evolving in time. A Markov process is considered to be "hidden" when the noise of the measurement is significant enough that the transitions between different states are no longer discernable by eye. For the evaluation of spFRET data, a Markov model is generated with a given number of states and a probability transition matrix, giving the probability of a transition between each of the states. The model is optimized by determining the parameters that yield the maximum likelihood for the given trace or traces. When the number of states is known for a given system, the appropriate number of states can be entered into the model. However, when the number of states is unknown, either hidden Markov models can be performed for various numbers of states and the results compared, or a large number of states can be given (e.g. 10) and the results analyzed to determine how many states are significantly populated. Multiple traces can be fitted by using a single, global model, or individual traces with a cluster analysis being used to identify the different states. An excellent description of the details of one HMM approach, and its application to single-molecule data, was provided by McKinney *et al.* [93]. The optimized HMM provides the most probable assignment of the states for the individual spFRET trajectories.

A spFRET trace showing the dynamics of TBP–DNA in the presence of NC2 is shown in Figure 5.8a, along with the optimized global fit to approximately 500 traces. When the states have been assigned for the individual traces, a wealth of information can be extracted from the spFRET data. For example, Figure 5.8b shows the number of observed transitions from an initial FRET state S_i to a final FRET state S_f. Transitions are observable between the 0.4, 0.6, and 0.8 FRET subpopulations, and between the 0.20 and 0.4 FRET populations, but very few transitions are seen between the 0.20 FRET subpopulation and the higher FRET efficiency states (0.6 or 0.8). The lifetime that the complex spends in the different states can also be extracted from the data, and further evaluated.

Although HMM represents an excellent tool for analyzing single-molecule data, care must be taken that the data fulfill the assumptions of a Markov process. When the transition rates or available FRET states are time-dependent, as in the case of measurements with enzymes that exhibit a memory effect (see also Chapter 11) [96, 97], the results of an HMM analysis must be interpreted with care.

5.5 Future Prospects

The examples described in the previous sections have highlighted the way in which spFRET measurements are currently being applied to biological systems. Two main strategies exist for measuring individual complexes, namely *burst analysis*, where photons are collected as a molecule freely diffuses through the confocal

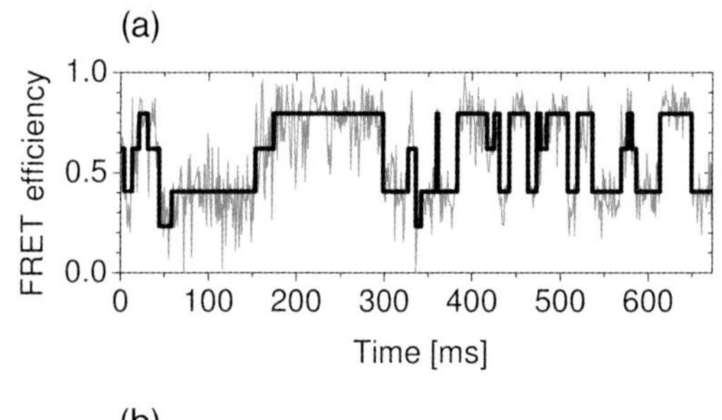

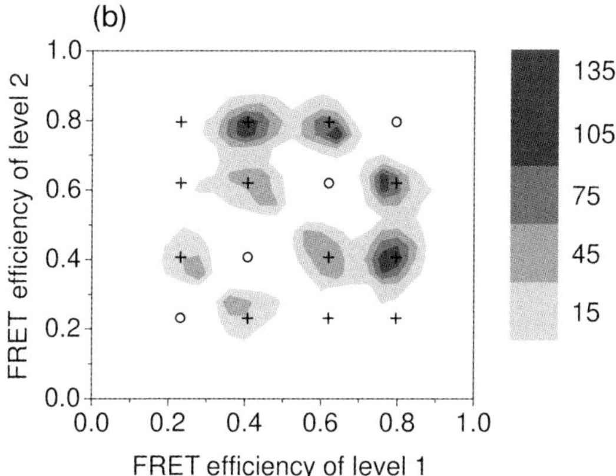

Figure 5.8 Hidden Markov analysis. The dynamics of TBP–NC2 on DNA were measured with 5 ms time resolution and approximately 500 traces were fitted globally to a HMM with four states. (a) A single FRET trace along with the results from the global HMM analysis. (b) A transition plot showing the probability for the various transitions between the four different states. The "+"s mark possible transitions between the four FRET states determined from the global HMM fit to the data. The "o"s mark the diagonal elements of the transition matrix. Transitions are observed between the 0.4 FRET state and all other states, whereas transitions between the 0.2 state and the 0.6 and 0.8 FRET states are only rarely observed, if at all.

volume of the microscope; and spFRET with *immobilized molecules*. Whilst many other excellent spFRET experiments have been conducted by a variety of groups, a limitation of space in this chapter has prevented the inclusion of many studies worthy of mention. Nonetheless, the quality of spFRET experiments continues to improve, with methods being developed to extract additional information from single-molecule data. In the remainder of the chapter, attention is focused on two areas where a significant potential exists to advance the future use of spFRET measurements.

5.5.1
Three-Color FRET

The main advantage of FRET is that it provides information over distances and changes in distances on the Angstrom scale. With the addition of a third fluorophore, it is possible to observe *three* distances simultaneously (Figure 5.9a).

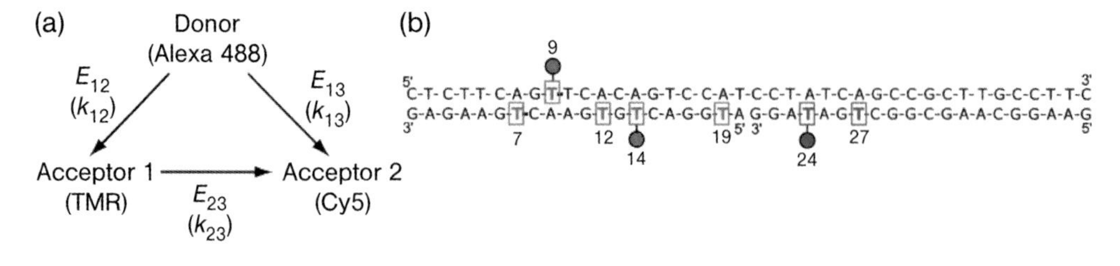

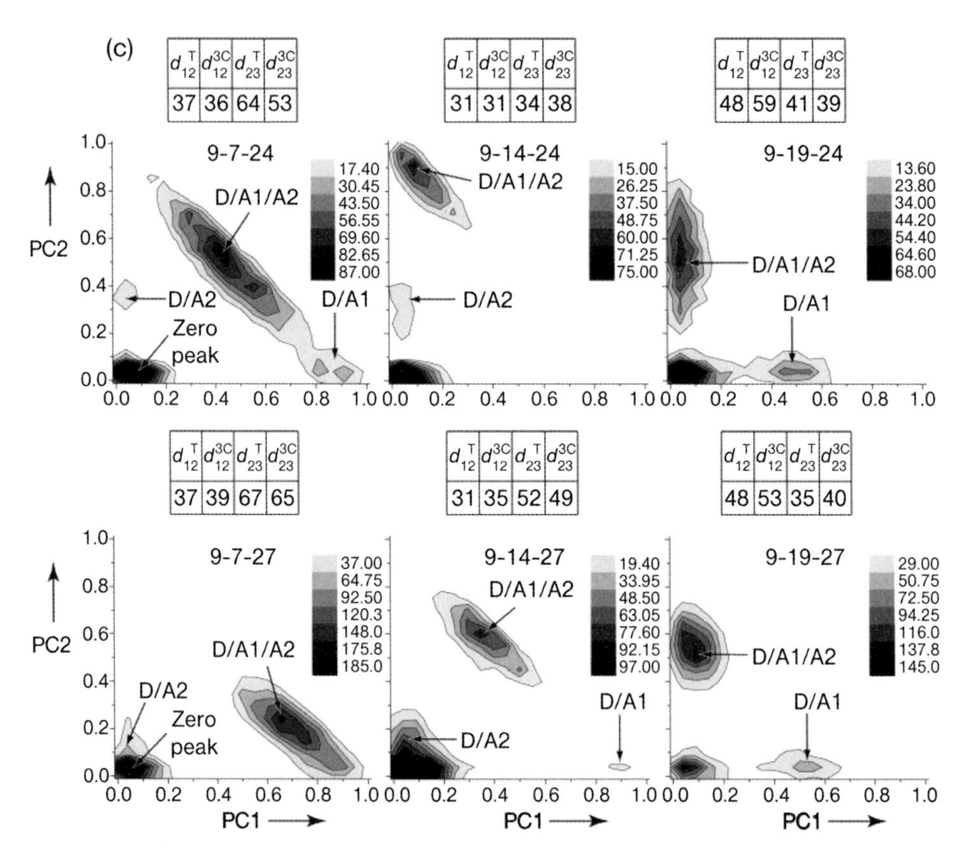

Figure 5.9 Three-color FRET. (a) When using three fluorophores, the energy transfer efficiency between the three fluorophores, and hence three different distances, can be calculated; (b) The DNA construct used by Clamme and Deniz to demonstrate the capabilities of 3cFRET in solution. The labeling position of the donor (9), various positions for acceptor 1/donor 2 (7, 12, 14, and 19) and the labeling positions of acceptor 2 (24 and 27) are shown; (c) Two-dimensional proximity coordinate plots for the various constructs are shown. The labeling positions measured in the different graphs are given above the plot (in the order D_1-A_1/D_2-A_2). The different subpopulations are labeled in the plot. Above the graph, a table gives the theoretical and calculated distances between donor 1 and acceptor 1 and acceptor 1/donor 2 and acceptor 2, as determined from the measurements. Figure adapted with permission from Ref. [104].

Three-color FRET (3cFRET) was first applied in ensemble measurements as a means of extending the length scale over which FRET is observable [98, 99], and shortly afterwards was used as a structural tool for measuring multiple distances [100, 101]. Three-color FRET measurements performed on single complexes were first performed on immobilized complexes [102, 103], and later in solution [104]. For example, in studies conducted by Clamme and Deniz, DNA was labeled specifically at three locations (Figure 5.9b). A ratiometric analysis was performed [26] using two proximity coordinates (PCs):

$$PC1 = \frac{F_{A_1/D_2}}{F_{D_1} + F_{A_1/D_2} + F_{A_2}}$$
$$PC2 = \frac{F_{A_2}}{F_{D_1} + F_{A_1/D_2} + F_{A_2}} \tag{5.14}$$

where F represents the fluorescence intensity of the appropriate channel: the first donor (D_1), the first acceptor, which can also act as a second donor (A_1/D_2), and the second acceptor (A_2). From the 2-D histograms obtained (Figure 5.9c), the various subpopulations are observable and the distances between the different fluorophores can be calculated. A comparison of the theoretical distances and the distances determined from the experiment are given above the appropriate histograms in Figure 5.9c. The agreement is typically within 5 Å for these measurements.

Having three distances is particularly beneficial for measuring polymer conformations, such as in protein-folding dynamics [105]. Here, in addition to knowing the distance between the end fluorophores, information is also available on the conformation of the polymer in between the end-labeling positions. Three-color FRET can also be used for measuring binding-induced conformational changes, where one component is used to determine the presence of the binding partner, while the other two fluorophores can monitor the conformation of one of the components. For mobile systems, such as motor proteins moving along a track, the directionality of motion can be investigated using 3cFRET. For this, the two acceptors can be attached to the ends of the track (e.g., DNA) and the complex labeled with the donor fluorophore. Depending on the direction in which the complex moves, a distinct FRET signal will be observable. Three-color FRET experiments can be improved with the aid of three-color ALEX or PIE ([106]; see also Chapter 6). One of the main challenges in 3cFRET is the simultaneous determination of all three distances; however, by alternating the laser excitation it is possible to determine the distance between one FRET pair, independent of the other distances.

5.5.2
FRET of Tracked Particles

A second possibility that will undoubtedly be of immense value in the future is the ability to measure FRET while tracking single particles. In this case, it should be possible to visualize interactions or conformational changes that occur in a single molecule as it goes about its function. One variation of this would be to trap particles using, for example, the ABLE trap (as mentioned above) [67]. When using electrokinetic forces, the molecule is maintained at the center of the optical device, while the FRET signal can be measured for as long as the molecule remains in the focus. Another possibility would be to use optical or magnetic tweezers to immobilize and apply forces to the complex, while investigating its conformation or interactions via the FRET efficiency [107, 108].

It should be noted that, in all of the above examples, the complexes are either trapped or immobilized. Yet, when investigating the function of biomolecules within their native environment – the cell – it is necessary that they are first allowed to move freely, and then are tracked. Recently, Hellriegel and Gratton demonstrated an ability to measure spectroscopic parameters of individual complexes during tracking [109], which means that the FRET efficiency of a complex can be measured while the complex is tracked. If this proves to be feasible, it will open up the possibility of following interactions or conformational fluctuations of a complex along its trajectory via FRET, in a living cell.

Abbreviations

ALEX	Alternating laser excitation
FRET	Förster resonance energy transfer
HMM	Hidden Markov model
PDA	Photon distribution analysis
PIE	Pulsed interleaved excitation
TIRF	Total internal reflection fluorescence

References

1 Cario, G. (1922) Über Entstehung wahrer Lichtabsorption und scheinbare Koppelung von Quantensprüngen. *Z. Physik.*, **10**, 185–199.

2 Cario, G. and Franck, J. (1922) Über Zerlegugen von Wasserstoffmolekülen durch angeregte Quecksilberatome. *Z. Physik.*, **11**, 161–166.

3 Franck, J. (1922) Einige aus der Theorie von Klein und Rosseland zu ziehende Folgerungen über Fluoreszenz, photochemische Prozesse und die Elektronenemission glühender Körper. *Z. Physik.*, **9**, 259–266.

4 Clegg, R.M. (2006) The history of FRET, in *Reviews in Fluorescence 2006* (eds C.D. Geddes and J.R. Lakowicz), Springer, USA, pp. 1–45.

5 Förster, T. (1946) Energiewanderung und Fluoreszenz. *Naturwissenschaften*, **6**, 166–175.

6 Förster, T. (1947) Fluoreszenzversuche an Farbstoffmischungen. *Angew. Chem. A*, **59**, 181–187.

7 Förster, T. (1948) Zwischenmolekulare Energiewanderung und Fluoreszenz. *Ann. Phys. (Leipzig)*, **2**, 55–75.

8 Förster, T. (1951) Fluoreszenz organischer Verbindungen. Vandenhoeck & Ruprecht, Göttingen, p. 315.

9 Stryer, L. and Haugland, R.P. (1967) Energy transfer: a spectroscopic ruler. *Proc. Natl Acad. Sci. USA*, **58** (2), 719–726.

10 Ha, T., *et al.* (1996) Probing the interaction between two single molecules:

fluorescence resonance energy transfer between a single donor and a single acceptor. *Proc. Natl Acad. Sci. USA*, **93** (13), 6264–6268.

11 Weiss, S. (1999) Fluorescence spectroscopy of single biomolecules. *Science*, **283** (5408), 1676–1683.

12 Borgia, A., Williams, P.M., and Clarke, J. (2008) Single-molecule studies of protein folding. *Annu. Rev. Biochem.*, **77** (1), 101–125.

13 Joo, C., *et al.* (2008) Advances in single-molecule fluorescence methods for molecular biology. *Annu. Rev. Biochem.*, **77** (1), 51–76.

14 Deniz, A.A., Mukhopadhyay, S., and Lemke, E.A. (2008) Single-molecule biophysics: at the interface of biology, physics and chemistry. *J. R. Soc. Interface*, **5** (18), 15–45.

15 Nienhaus, G.U. (2006) Exploring protein structure and dynamics under denaturing conditions by single-molecule FRET analysis. *Macromol. Biosci.*, **6** (11), 907–922.

16 Michalet, X., Weiss, S., and Jager, M. (2006) Single-molecule fluorescence studies of protein folding and conformational dynamics. *Chem. Rev.*, **106** (5), 1785–1813.

17 Schuler, B. (2005) Single-molecule fluorescence spectroscopy of protein folding. *ChemPhysChem*, **6** (7), 1206–1220.

18 Haustein, E. and Schwille, P. (2004) Single-molecule spectroscopic methods. *Curr. Opin. Struct. Biol.*, **14** (5), 531–540.

19 Weiss, S. (2000) Measuring conformational dynamics of biomolecules by single molecule fluorescence spectroscopy. *Nat. Struct. Mol. Biol.*, **7** (9), 724–729.

20 Iqbal, A., *et al.* (2008) Orientation dependence in fluorescent energy transfer between Cy3 and Cy5 terminally attached to double-stranded nucleic acids. *Proc. Natl Acad. Sci. USA*, **105** (32), 11176–11181.

21 Shera, E.B., *et al.* (1990) Detection of single fluorescent molecules. *Chem. Phys. Lett.*, **174** (6), 553–557.

22 Soper, S.A., *et al.* (1991) Single-molecule detection of rhodamine 6G in ethanolic solutions using continuous wave laser excitation. *Anal. Sci.*, **63** (5), 432–437.

23 Edman, L., Mets, U., and Rigler, R. (1996) Conformational transitions monitored for single molecules in solution. *Proc. Natl Acad. Sci. USA*, **93** (13), 6710–6715.

24 Zander, C., *et al.* (1996) Detection and characterization of single molecule in aqueous solution. *Appl. Phys. B*, **63**, 517–523.

25 Kühnemuth, R. and Seidel, C.A.M. (2001) Principles of single molecule multiparameter fluorescence spectroscopy. *Single Mol.*, **2** (4), 251–254.

26 Deniz, A.A., *et al.* (1999) Single-pair fluorescence resonance energy transfer on freely diffusing molecules: observation of Forster distance dependence and subpopulations. *Proc. Natl Acad. Sci. USA*, **96** (7), 3670–3675.

27 Eggeling, C., *et al.* (2006) Analysis of photobleaching in single-molecule multicolor excitation and Förster resonance energy transfer measurements. *J. Phys. Chem. A*, **110** (9), 2979–2995.

28 Nir, E., *et al.* (2006) Shot-noise limited single-molecule FRET histograms: comparison between theory and experiments. *J. Phys. Chem. B*, **110** (44), 22103–22124.

29 Hillger, F., *et al.* (2008) Probing protein-chaperone interactions with single-molecule fluorescence spectroscopy. *Angew. Chem., Int. Ed.*, **47** (33), 6184–6188.

30 Schuler, B., Lipman, E.A., and Eaton, W.A. (2002) Probing the free-energy surface for protein folding with single-molecule fluorescence spectroscopy. *Nature*, **419** (6908), 743–747.

31 Lipman, E.A., *et al.* (2003) Single-molecule measurement of protein folding kinetics. *Science*, **301** (5637), 1233–1235.

32 Chan, C.K., *et al.* (1997) Submillisecond protein folding kinetics studied by ultrarapid mixing. *Proc. Natl Acad. Sci. USA*, **94** (5), 1779–1784.

33 Hamadani, K.M. and Weiss, S. (2008) Nonequilibrium single molecule protein folding in a coaxial mixer. *Biophys. J.*, **95** (1), 352–365.

34 Sharma, S., *et al.* (2008) Monitoring protein conformation along the pathway of chaperonin-assisted folding. *Cell*, **133** (1), 142–153.

35 Hartl, F.U. and Hayer-Hartl, M. (2002) Molecular chaperones in the cytosol: from nascent chain to folded protein. *Science*, **295** (5561), 1852–1858.

36 Xu, Z.H., Horwich, A.L., and Sigler, P.B. (1997) The crystal structure of the asymmetric GroEL-GroES-(ADP)7 chaperonin complex. *Nature*, **388** (6644), 741–749.

37 Müller, B.K., *et al.* (2005) Pulsed interleaved excitation. *Biophys. J.*, **89** (5), 3508–3522.

38 Laurence, T.A., *et al.* (2005) Probing structural heterogeneities and fluctuations of nucleic acids and denatured proteins. *Proc. Natl Acad. Sci. USA*, **102** (48), 17348–17353.

39 Kapanidis, A.N., *et al.* (2005) Alternating-laser excitation of single molecules. *Acc. Chem. Res.*, **38** (7), 523–533.

40 Kapanidis, A.N., *et al.* (2004) Fluorescence-aided molecule sorting: analysis of structure and interactions by alternating-laser excitation of single molecules. *Proc. Natl Acad. Sci. USA*, **101** (24), 8936–8941.

41 Lee, N.K., *et al.* (2005) Accurate FRET measurements within single diffusing biomolecules using alternating-laser excitation. *Biophys. J.*, **88** (4), 2939–2953.

42 Stan, G., *et al.* (2007) Coupling between allosteric transitions in GroEL and assisted folding of a substrate protein. *Proc. Natl Acad. Sci. USA*, **104** (21), 8803–8808.

43 Mickler, M., *et al.* (2009) The large conformational changes of Hsp90 are only weakly coupled to ATP hydrolysis. *Nat. Struct. Mol. Biol.*, **16** (3), 281–286.

44 Antonik, M., *et al.* (2006) Separating structural heterogeneities from stochastic variations in fluorescence resonance energy transfer distributions via photon distribution analysis. *J. Phys. Chem. B*, **110** (13), 6970–6978.

45 Schuler, B., *et al.* (2005) Polyproline and the "spectroscopic ruler" revisited with single-molecule fluorescence. *Proc. Natl Acad. Sci. USA*, **102** (8), 2754–2759.

46 Gopich, I. and Szabo, A. (2005) Theory of photon statistics in single-molecule Forster resonance energy transfer. *J. Chem. Phys.*, **122** (1), 14707.

47 Kalinin, S., *et al.* (2007) Probability distribution analysis of single-molecule fluorescence anisotropy and resonance energy transfer. *J. Phys. Chem. B*, **111** (34), 10253–10262.

48 Kalinin, S., *et al.* (2008) Characterizing multiple molecular states in single-molecule multiparameter fluorescence detection by probability distribution analysis. *J. Phys. Chem. B*, **112** (28), 8361–8374.

49 Zimmermann, B., *et al.* (2006) Subunit movements in membrane-integrated EF0F1 during ATP synthesis detected by single-molecule spectroscopy. *Biochim. Biophys. Acta – Bioenerg.*, **1757** (5–6), 311–319.

50 Mukhopadhyay, S., *et al.* (2007) A natively unfolded yeast prion monomer adopts an ensemble of collapsed and rapidly fluctuating structures. *Proc. Natl Acad. Sci. USA*, **104** (8), 2649–2654.

51 Slaughter, B.D., *et al.* (2004) Single-molecule resonance energy transfer and fluorescence correlation spectroscopy of calmodulin in solution. *J. Phys. Chem. B*, **108** (29), 10388–10397.

52 Ha, T. (2001) Single-molecule fluorescence resonance energy transfer. *Methods*, **25** (1), 78–86.

53 Roy, R., Hohng, S., and Ha, T. (2008) A practical guide to single-molecule FRET. *Nat. Methods*, **5** (6), 507–516.

54 Selvin, P.R. and Ha, T. (eds) (2008) *Single-Molecule Techniques: A Laboratory Manual*, Cold Spring Harbor Laboratory Press, Cold Spring Harbor, p. 507.

55 Kim, H.D., *et al.* (2002) Mg^{2+}-dependent conformational change of RNA studied by fluorescence correlation and FRET on immobilized single molecules. *Proc. Natl Acad. Sci. USA*, **99** (7), 4284–4289.

56 Graneli, A., *et al.* (2006) Organized arrays of individual DNA molecules tethered to supported lipid bilayers. *Langmuir*, **22** (1), 292–299.

57 Ha, T., *et al.* (2002) Initiation and re-initiation of DNA unwinding by the *Escherichia coli* rep helicase. *Nature*, **419** (6907), 638–641.

58 Schluesche, P., *et al.* (2007) A dynamic model for class II gene promoter targeting through mobile TBP-NC2 complexes. *Nat. Struct. Mol. Biol.*, **14** (12), 1196–1201.

59 Amirgoulova, E.V., *et al.* (2004) Biofunctionalized polymer surfaces exhibiting minimal interaction towards immobilized proteins. *ChemPhysChem*, **5** (4), 552–555.

60 Heyes, C.D., *et al.* (2007) Synthesis, patterning and applications of star-shaped poly(ethylene glycol) biofunctionalized surfaces. *Mol. Biosyst.*, **3** (6), 419–430.

61 Heyes, C.D., *et al.* (2004) Biocompatible surfaces for specific tethering of individual protein molecules. *J. Phys. Chem. B*, **108** (35), 13387–13394.

62 Cha, T., Guo, A., and Zhu, X.Y. (2005) Enzymatic activity on a chip: the critical role of protein orientation. *Proteomics*, **5** (2), 416–419.

63 Manon, J.W., Ludden, A.M., Schulze, K., Subramaniam, V., Tampé, R., and Huskens, J. (2008) Anchoring of histidine-tagged proteins to molecular printboards: self-assembly, thermodynamic modeling, and patterning. *Chemistry*, **14** (7), 2044–2051.

64 Okumus, B., *et al.* (2004) Vesicle encapsulation studies reveal that single molecule ribozyme heterogeneities are intrinsic. *Biophys. J.*, **87** (4), 2798–2806.

65 Reiner, J.E., *et al.* (2006) Optically trapped aqueous droplets for single molecule studies. *Appl. Phys. Lett.*, **89** (1), 013904–013903.

66 Cohen, A.E. and Moerner, W.E. (2005) Method for trapping and manipulating nanoscale objects in solution. *Appl. Phys. Lett.*, **86** (9), 093109–093103.

67 Cohen, A.E. and Moerner, W.E. (2006) Suppressing Brownian motion of individual biomolecules in solution. *Proc. Natl Acad. Sci. USA*, **103** (12), 4362–4365.

68 Cohen, A.E. and Moerner, W.E. (2008) Controlling Brownian motion of single protein molecules and single fluorophores in aqueous buffer. *Opt. Express*, **16** (10), 6941–6956.

69 Myong, S., *et al.* (2007) Spring-loaded mechanism of DNA unwinding by hepatitis C virus NS3 helicase. *Science*, **317** (5837), 513–516.

70 Cho, H.-S., *et al.* (1998) Crystal structure of RNA helicase from genotype 1b hepatitis C virus. A feasible mechanism of unwinding duplex RNA. *J. Biol. Chem.*, **273** (24), 15045–15052.

71 Serebrov, V. and Pyle, A.M. (2004) Periodic cycles of RNA unwinding and pausing by hepatitis C virus NS3 helicase. *Nature*, **430** (6998), 476–480.

72 Dumont, S., *et al.* (2006) RNA translocation and unwinding mechanism of HCV NS3 helicase and its coordination by ATP. *Nature*, **439** (7072), 105–108.

73 Kerssemakers, J.W.J., *et al.* (2006) Assembly dynamics of microtubules at molecular resolution. *Nature*, **442** (7103), 709–712.

74 Pang, P.S., *et al.* (2002) The hepatitis C viral NS3 protein is a processive DNA helicase with cofactor enhanced RNA unwinding. *EMBO J.*, **21** (5), 1168–1176.

75 Kamada, K., *et al.* (2001) Crystal structure of negative cofactor 2 recognizing the TBP-DNA transcription complex. *Cell*, **106** (1), 71–81.

76 Banik, U., *et al.* (2001) Fluorescence-based analyses of the effects of full-length recombinant TAF130p on the interaction of TATA box-binding protein with TATA box DNA. *J. Biol. Chem.*, **276** (52), 49100–49109.

77 Gumbs, O.H., Campbell, A.M., and Weil, P.A. (2003) High-affinity DNA binding by a Mot1p-TBP complex: implications for TAF-independent transcription. *EMBO J.*, **22** (12), 3131–3141.

78 Kays, A.R. and Schepartz, A. (2000) Virtually unidirectional binding of TBP to the AdMLP TATA box within the quaternary complex with TFIIA and TFIIB. *Chem. Biol.*, **7** (8), 601–610.

79 Schluesche, P., Heiss, G., Meisterernst, M. and Lamb, D.C. (2008) Dynamics of TBP binding to the TATA box, in *Single Molecule Spectroscopy and Imaging* (eds J. Enderlein, Z.K. Gryczynski and R. Erdmann), SPIE, San Jose, USA, pp. 6862E-1–6862E-8.

80 Tolić-Nørrelykke, S.F., *et al.* (2006) Stepwise bending of DNA by a single TATA-box binding protein. *Biophys. J.*, **90** (10), 3694–3703.

81 Parkhurst, K.M., Brenowitz, M., and Parkhurst, L.J. (1996) Simultaneous binding and bending of promoter DNA by the TATA binding protein: real time kinetic measurements. *Biochemistry*, **35** (23), 7459–7465.

82 Parkhurst, K.M., *et al.* (1999) Intermediate species possessing bent DNA are present along the pathway to formation of a final TBP-TATA complex. *J. Mol. Biol.*, **289** (5), 1327–1341.

83 Powell, R.M., *et al.* (2001) Marked stepwise differences within a common kinetic mechanism characterize TATA-binding protein interactions with two consensus promoters. *J. Biol. Chem.*, **276** (32), 29782–29791.

84 Powell, R.M., Parkhurst, K.M., and Parkhurst, L.J. (2002) Comparison of TATA-binding protein recognition of a variant and consensus DNA promoters. *J. Biol. Chem.*, **277** (10), 7776–7784.

85 McKinney, S.A., *et al.* (2003) Structural dynamics of individual Holliday junctions. *Nat. Struct. Biol.*, **10** (2), 93–97.

86 Brand, L., *et al.* (1997) Single-molecule identification of coumarin-120 by time-resolved fluorescence detection: comparison of one- and two-photon excitation in solution. *J. Phys. Chem. A*, **101** (24), 4313–4321.

87 Edel, J.B., Eid, J.S., and Meller, A. (2007) Accurate single molecule FRET efficiency determination for surface immobilized DNA using maximum likelihood calculated lifetimes. *J. Phys. Chem. B*, **111** (11), 2986–2990.

88 Enderlein, J. (1995) Maximum-likelihood criterion and single-molecule. *Appl. Opt.*, **34** (3), 514–526.

89 Watkins, L.P. and Yang, H. (2004) Information bounds and optimal analysis of dynamic single molecule measurements. *Biophys. J.*, **86** (6), 4015–4029.

90 Yang, H. and Xie, X.S. (2002) Probing single-molecule dynamics photon by photon. *J. Chem. Phys.*, **117** (24), 10965–10979.

91 Qin, F., Auerbach, A., Sachs, F., and Sachs, H. (2000) Markov modeling for single channel kinetics with filtering and correlated noise. *Biophys. J.*, **79** (4), 1928–1944.

92 Qin, F., Auerbach, A., and Sachs, F. (2000) A direct optimization approach to hidden Markov modeling for single channel kinetics. *Biophys. J.*, **79** (4), 1915–1927.

93 McKinney, S.A., Joo, C., and Ha, T. (2006) Analysis of single-molecule FRET trajectories using hidden Markov modeling. *Biophys. J.*, **91** (5), 1941–1951.

94 Eddy, S.R. (2004) What is a hidden Markov model? *Nat. Biotechnol.*, **22** (10), 1315–1316.

95 Talaga, D.S. (2007) Markov processes in single molecule fluorescence. *Curr. Opin. Colloid Interface Sci.*, **12** (6), 285–296.

96 Lu, H.P., Xun, L., and Xie, X.S. (1998) Single-molecule enzymatic dynamics. *Science*, **282** (5395), 1877–1882.

97 Edman, L. and Rigler, R. (2000) Memory landscapes of single-enzyme molecules. *Proc. Natl Acad. Sci. USA*, **97** (15), 8266–8271.

98 Haustein, E., Jahnz, M., and Schwille, P. (2003) Triple FRET: a tool for studying long-range molecular interactions. *ChemPhysChem*, **4** (7), 745–748.

99 Tong, A.K., *et al.* (2001) Triple fluorescence energy transfer in covalently trichromophore-labeled DNA. *J. Am. Chem. Soc.*, **123** (51), 12923–12924.

100 Liu, J. and Lu, Y. (2002) FRET study of a trifluorophore-labeled DNAzyme. *J. Am. Chem. Soc.*, **124** (51), 15208–15216.

101 Watrob, H.M., Pan, C.P., and Barkley, M.D. (2003) Two-step FRET as a structural tool. *J. Am. Chem. Soc.*, **125** (24), 7336–7343.

102 Hohng, S., Joo, C., and Ha, T. (2004) Single-molecule three-color FRET. *Biophys. J.*, **87** (2), 1328–1337.

103 Ross, J., *et al.* (2007) Multicolor single-molecule spectroscopy with alternating laser excitation for the investigation of interactions and dynamics. *J. Phys. Chem. B*, **111** (2), 321–326.

104 Clamme, J.P. and Deniz, A.A. (2005) Three-color single-molecule fluorescence resonance energy transfer. *ChemPhysChem*, **6** (1), 74–77.

105 Ting, C.L. and Makarov, D.E. (2008) Two-dimensional fluorescence resonance energy transfer as a probe for protein folding: a theoretical study. *J. Chem. Phys.*, **128** (11), 115102.

106 Lee, N.K., *et al.* (2007) Three-color alternating-laser excitation of single molecules: monitoring multiple interactions and distances. *Biophys. J.*, **92** (1), 303–312.

107 Hohng, S., *et al.* (2007) Fluorescence-force spectroscopy maps two-dimensional reaction landscape of the Holliday junction. *Science*, **318** (5848), 279–283.

108 Tarsa, P.B., *et al.* (2007) Detecting force-induced molecular transitions with fluorescence resonant energy transfer. *Angew. Chem., Int. Ed.*, **46** (12), 1999–2001.

109 Hellriegel, C. and Gratton, E. (2009) Real-time multi-parameter spectroscopy and localization in three-dimensional single-particle tracking. *J. R. Soc. Interface*, **6**, S3–S14.

6
Alternating-Laser Excitation and Pulsed-Interleaved Excitation of Single Molecules

Seamus J. Holden and Achillefs N. Kapanidis

6.1
Introduction

Since the pioneering studies that led to its first observation in 1996 [1], single-molecule fluorescence resonance energy transfer (smFRET) has significantly advanced our understanding of biomolecular mechanisms, and has been employed in an ever-expanding array of single-molecule methods that survey single biological machines at work. FRET has been discussed extensively in several book chapters and reviews [2–5]; here, only a brief introduction will be presented.

Fluorescence resonance energy transfer (FRET) is the process of nonradiative energy transfer from an excited "donor" fluorophore to an "acceptor" chromophore in close proximity [6]. Changes in donor–acceptor separation within the 2–10 nm range alter the efficiency of FRET and the fluorescence properties of the donor and acceptor. By taking advantage of the FRET sensitivity to nanoscale distances, biomolecules can be labeled with a donor–acceptor pair, and FRET then used as a "nanoscale ruler" to monitor intramolecular distances and the conformational changes that occur in timescales from nanoseconds to minutes. Single-molecule FRET, wherein FRET is measured between donor–acceptor pairs at the level of single molecules, combines the sensitivity of FRET to nanoscale distances with the ability of single-molecule techniques to resolve dynamic and static heterogeneity within a sample – properties which are inaccessible at the ensemble level [7–9]. If a single donor–acceptor pair is involved, the more specific term "single-pair FRET" is also used [1].

While the first smFRET methods employed a single excitation wavelength for direct excitation of the donor fluorophore, advanced sample-illumination schemes have added flexibility and increased the information content of smFRET measurements. Alternating laser excitation (ALEX) [10] is an illumination scheme that excites molecules by alternating between a laser that primarily excites the donor and a laser that primarily excites the acceptor, to provide direct information on the presence and state of both fluorophores. This additional information allows the virtual sorting of molecules into subpopulations. The extra information available

Single Particle Tracking and Single Molecule Energy Transfer
Edited by Christoph Bräuchle, Don C. Lamb, and Jens Michaelis

ISBN: 978-3-527-32296-1

through ALEX increases the resolution and accuracy of measurements compared with smFRET methods, which rely on single-laser excitation.

Three complementary implementations of ALEX have been developed, and these can be combined to probe a wide range of timescales that extend from nanoseconds to hours:

- Experiments on systems at chemical equilibrium (or systems undergoing kinetic changes occurring on the timescale of a few minutes [10]) can be carried out using solution-based measurements and laser modulation at the microsecond timescale using microsecond-ALEX (μsALEX), which provides snapshots of molecules as they diffuse through a femtoliter-size sample volume.
- In order to probe fast dynamics and recover fluorescence-lifetime-based observables, nanosecond-ALEX or pulsed interleaved excitation (nsALEX/PIE) [10, 11], which uses a nanosecond-timescale alternation of pulsed laser sources, can be employed.
- For nonequilibrium experiments and the simultaneous observation of dynamics of multiple molecules for milliseconds to several minutes, millisecond-ALEX (msALEX) [12], which combines total internal reflection fluorescence (TIRF) microscopy of surface-immobilized molecules with laser modulation at the millisecond timescale, may be used.

In this chapter, the principles, applications, advantages and limitations of the family of ALEX methods are described, and a detailed discussion of recent ALEX applications on biological and biomimetic systems is presented. For additional information on the subject, the reader should consult earlier reviews [13, 14].

6.2 ALEX: The Principles of Operation

Fluorescence resonance energy transfer is a photophysical interaction that occurs when the excited-state energy of a fluorophore (the FRET donor, D) is transferred nonradiatively to a nearby chromophore (the FRET acceptor, A). In a typical donor–acceptor FRET pair, the acceptor chromophore is also a fluorophore, and the two probes are selected such that the emission spectrum of the donor overlaps significantly with the absorption spectrum of the acceptor. The efficiency of energy transfer E between donor and acceptor is inversely proportional to the sixth power of the donor–acceptor separation, as given by:

$$E = \frac{1}{1+\left(\frac{R_{DA}}{R_0}\right)^6} \tag{6.1}$$

where R_{DA} is the donor–acceptor distance, and R_0 is the Förster radius, which is defined as the donor–acceptor distance at which E is 50%. The R_0 distance is related to the properties of the donor and acceptor fluorophores, and can be measured experimentally; specifically, R_0 depends on the fluorescence quantum yield Q_D of the donor, the spectral overlap between the two fluorophores, the refractive index n of the medium, and the relative dipole orientation factor κ^2.

The sensitive distance-dependence of E, and the fact that R_0 values fall typically within the 2–6 nm range [4], render FRET a powerful tool for measuring distances and distance changes on the nanometer scale – a scale which is ideal for studying average-size proteins, nucleic acids, and lipid bilayers. The factor κ^2 is particularly important when there is a need to extract distance measurements from FRET data. In the case of high rotational freedom of the donor and acceptor fluorophores, the donor–acceptor pair approaches the limit of orientational averaging, and the simplifying assumption of $\kappa^2 = 2/3$ can be safely made. In cases of restricted dipole rotation, the uncertainty for the distance measurement increases significantly. However, in most cases, FRET is used as a reporter of the presence, relative magnitude and kinetics of a *distance change* rather than as an accurate molecular ruler. In terms of labeling, smFRET measurements require the specific attachment of donor and acceptor fluorophores to the biomolecules of interest. For example, in order to measure a distance between a site on a protein and a site on a DNA fragment within a protein–DNA complex, the fluorophores must be incorporated site-specifically in the DNA and the protein. In fact, it is common for protein labeling with reactive forms of fluorophores to be preceded by a genetic modification of the protein in order to introduce a single reactive site on the protein surface, which often is a single surface-exposed cysteine residue [15].

In single-excitation smFRET experiments (Figure 6.1a–c), photons emitted in the donor- and acceptor-emission channels upon donor excitation for a smFRET pair under observation are collected; these photon counts are defined as $f_{D_{ex}}^{D_{em}}$ and $f_{D_{ex}}^{A_{em}}$, where f_X^Y is the photon count for a single molecule upon excitation at wavelength X (where D_{ex} is the wavelength of substantial excitation of the FRET donor) and detection at wavelength range Y (where D_{em}, A_{em} are wavelengths of substantial emission of FRET donor and acceptor, respectively). The extent of FRET can be evaluated using a convenient expression of FRET efficiency, E^* (which does not account for spectral crosstalks in the acceptor-emission channel):

$$E^* = \frac{f_{D_{ex}}^{A_{em}}}{f_{D_{ex}}^{A_{em}} + \gamma f_{D_{ex}}^{D_{em}}} \tag{6.2}$$

where γ is a detection-correction factor that depends on the donor and acceptor quantum yields and the detection efficiencies of the donor and acceptor emission channels; typically $0.5 < \gamma < 2$. A simpler expression that can still observe relative FRET changes assumes $\gamma = 1$; in that case, a *proximity ratio*, E_{PR}, is recovered:

$$E_{PR}^{raw} = \frac{f_{D_{ex}}^{A_{em}}}{f_{D_{ex}}^{A_{em}} + f_{D_{ex}}^{D_{em}}} \tag{6.3}$$

When accurate FRET values are required (e.g., for evaluating donor–acceptor distances within biomolecular complexes), it is necessary to measure crosstalk factors and detection-correction biases. In the case of immobilized molecules,

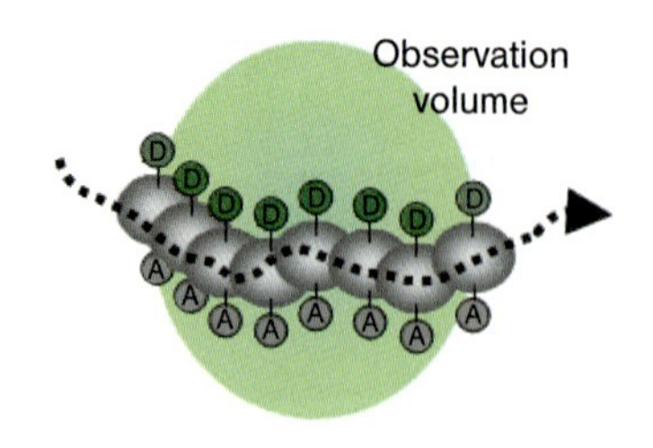

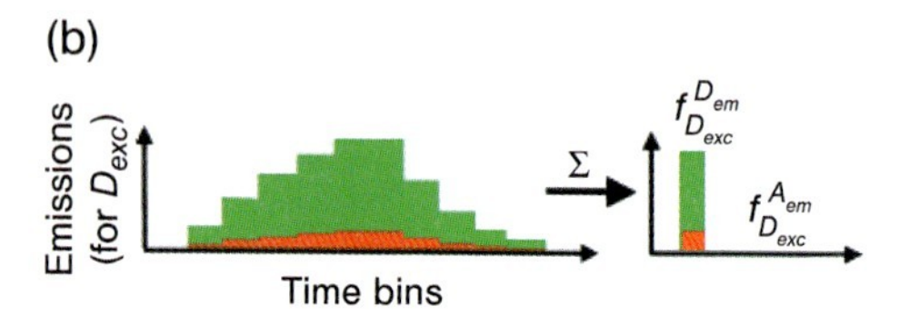

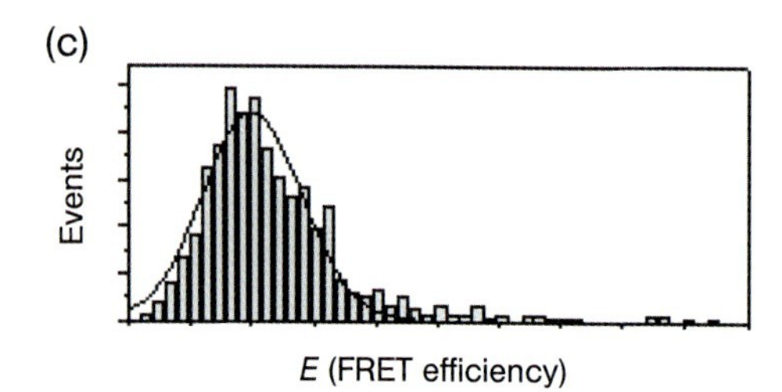

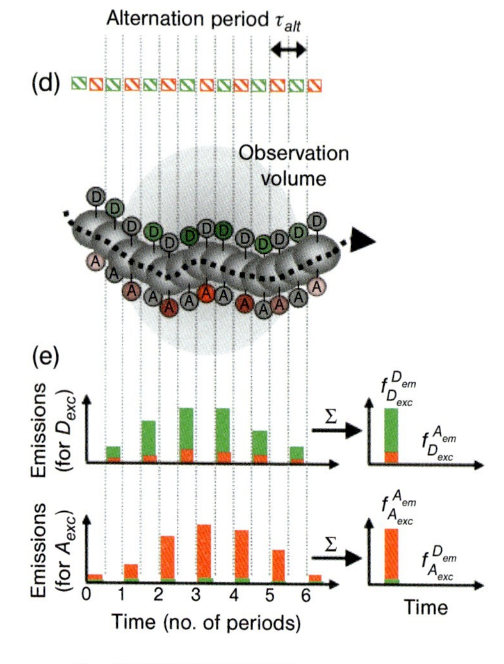

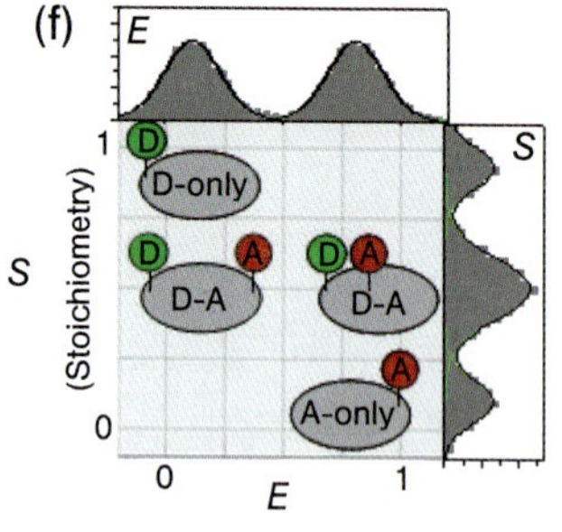

Figure 6.1 The concept of alternating laser excitation (ALEX) and comparison with single-excitation single-molecule FRET: an example for diffusing molecules. (a–c) Single-molecule FRET using single-laser excitation. A fluorescent molecule traverses a focused green-laser beam and emits photons at the donor- and acceptor-emission wavelengths (green and red bars, respectively). The sum of photon counts at these two wavelengths for each molecular transit are used to generate one-dimensional histograms of the FRET efficiency, *E*. (d–f) Single-molecule FRET using ALEX. A fluorescent molecule traverses an observation volume illuminated in an alternating fashion using focused green- and red-laser beams. By using the photons emitted in the donor- and acceptor-emission wavelengths for each laser excitation, a 2-D histogram can be generated of *E* and the relative fluorophore stoichiometry, *S*, enabling molecular sorting. Donor-only ("D-only") and acceptor-only ("A-only") molecules have high and low stoichiometry, respectively, and can be excluded from the analysis of the FRET efficiency of molecules containing both a donor and an acceptor ("D-A"), which have intermediate stoichiometry.

the methodology described in Ref. [16] is followed, extracting γ and other correction factors by direct measurement of the change in emitted donor and acceptor intensity upon acceptor photobleaching; these are determined for individual molecules where the acceptor photobleaches before the donor. This can then be applied to all molecules in the experiment. In the case of solution-based measurements, the ALEX method provides the necessary information (see Section 6.3).

Alternating laser excitation (Figure 6.1d–f) provides one extra nonzero photon count, $f_{A_{ex}}^{A_{em}}$, for acceptor emission upon *direct* acceptor excitation [10, 17]. This allows a new parameter to be calculated, the relative fluorophore stoichiometry ratio S:

$$S = \frac{f_{D_{ex}}}{f_{D_{ex}} + f_{A_{ex}}} \tag{6.4}$$

where $f_{D_{ex}} = f_{D_{ex}}^{D_{em}} + f_{D_{ex}}^{A_{em}}$ and $f_{A_{ex}} \approx f_{A_{ex}}^{A_{em}}$. The S ratio also contains information about the relative quantum yield of the donor and acceptor fluorophores, the relative ratio of donor and acceptor fluorophores on a single molecule, and the ratio of laser-excitation powers. The additional information is summarized in two-dimensional (2-D) histograms and allows virtual molecular sorting. Sorting can remove artifacts that complicate FRET (such as the presence of states with inactive donor or acceptor, and the presence of complex fluorophore stoichiometries) while introducing new observables, such as the observation of an acceptor-only population which is helpful for evaluating biomolecular interactions.

As with single-excitation smFRET [18, 19], there are two main experimental realizations of ALEX: the first approach uses confocal microscopy for point detection; and the second uses fluorescence imaging for area detection. For experimental details on the components, assembly, acquisition and analysis using ALEX, the reader should consult a recent review [20].

Confocal fluorescence microscopy (Figure 6.2a,b) allows the point detection of single fluorophores (either in solution or immobilized on a surface) by using a clever experimental format that reduces the observation volume to a diffraction-limited size [5, 21]. Excitation light is focused into a sample using a high-numerical aperture microscope objective; the fluorescence emitted from the sample is collected using the same objective and then focused through a small pinhole. The conjugation of the objective and the pinhole creates a spatial filter, which cuts the sampling volume down to a diffraction-limited size (~1 fl) [22]. This small sampling volume reduces substantially the contribution of the background and increases the signal-to-noise ratio (SNR) to the level that permits the detection of single fluorophores or fluorescently labeled molecules diffusing through the confocal volume. By extending this scheme to multiple excitation and detection wavelengths, smFRET and ALEX can be carried out.

On the other hand, TIRF microscopy uses evanescent-wave excitation within a thin layer off a surface and wide-field imaging on an ultrasensitive camera to observe surface-immobilized molecules for extended periods [23, 24]. When a light beam crosses an interface into a medium with a lower index of refraction at an

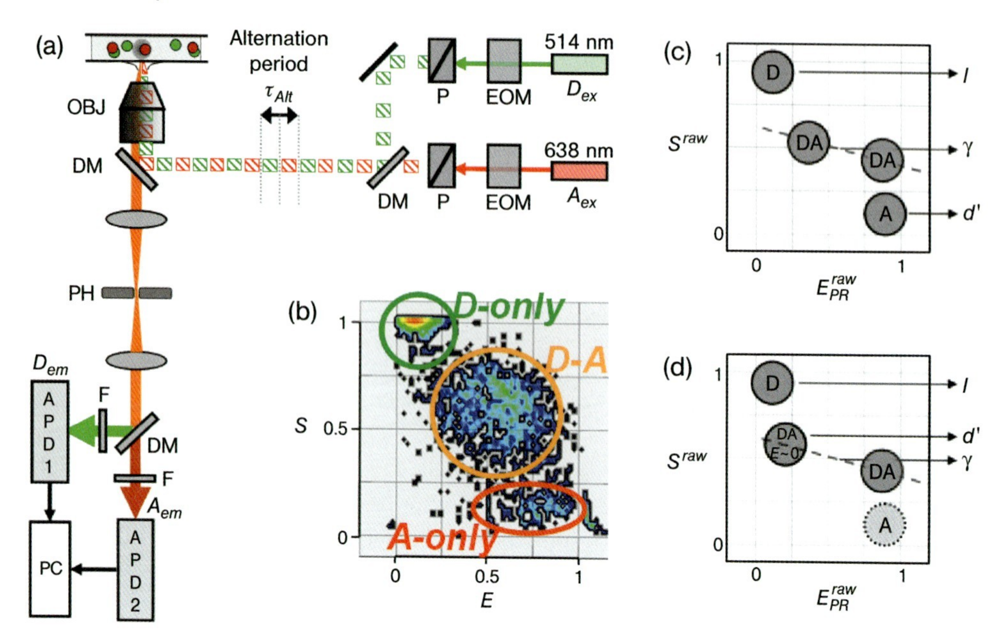

Figure 6.2 μsALEX microscopy: instrumentation and typical experimental data. (a) Set-up for μsALEX microscopy. The alternating laser source excites fluorescent diffusing molecules in a solution. The molecules then emit fluorescence (orange line) which is collected in the donor-emission and acceptor-emission channels. EOM = electro-optical modulator; P = polarizer; DM = dichroic mirror; OBJ = objective; PH = pinhole; F = optical filter; APD = avalanche photodiode. (b) An experimental *E*–*S* histogram. Fluorescence emission from diffusing molecules is classified based on excitation and emission wavelength, allowing the calculation of *E* and *S* and enabling subpopulation sorting. Molecules containing both a donor and an acceptor ("D-A") are clearly separated from donor-only ("D-only") and acceptor-only ("A-only") molecules. (c) Species required for recovering accurate FRET using a method that depends on laser-alternation characteristics. *D*-only species provides the *D*-leakage factor *l*, *A*-only species provides the *A*-direct-excitation factor *d*, and two donor–acceptor species with a large difference in *E* provide the γ-factor. (d) Species required for recovering accurate FRET using a method independent of laser-alternation characteristics. *D*-only species provides the *D*-leakage factor *l*, a donor–acceptor species with $E \sim 0$ ("simple-coincidence" control) provides the modified *A*-direct-excitation factor *d′*, and a donor–acceptor species with appreciable *E* provides the γ-factor. The use of *A*-only species is not necessary. Reproduced from Ref. [17].

angle larger than a critical angle, an evanescent wave is generated close to the interface, with the intensity of the field $I(d)$ decaying exponentially with distance: $I(d) = I_0 e^{-z/d}$, where z is the distance from the surface and d is a characteristic depth, typically on the order of 100 nm. Because the evanescent wave decays rapidly with distance, single molecules immobilized on the surface can be imaged on a sensitive camera with little background from molecules diffusing in solution at focal planes away from the surface.

6.3 μsALEX

Microsecond ALEX uses continuous-wave lasers modulated at the microsecond timescale to probe molecules in solution diffusing through a confocal volume (Figure 6.2). Modulation can be achieved using electro-optical and acousto-optical modulators; moreover, several solid-state lasers nowadays can be modulated directly using software-driven triggering pulses [20]. Detected photons are classified based on the excitation period and emission channel. For a sample with a low concentration of fluorescent species (typically 50–100 pM), there is a low probability of having more than one fluorescent molecule in the confocal volume. At this concentration regime, the diffusion of a fluorescent molecule through the confocal volume is detected as a "burst" of fluorescence photons over a low background due to scattered excitation photons and detector electronics. Diffusing molecules can thus be identified using simple search algorithms [25, 26]. The ALEX photon $f_{D_{ex}}^{D_{em}}$, $f_{D_{ex}}^{A_{em}}$, $f_{A_{ex}}^{D_{em}}$, and $f_{A_{ex}}^{A_{em}}$ are calculated for each burst and are used to calculate E^* and S values for each molecular transit; the (E^*,S) results are plotted on a 2-D histogram allowing subpopulation sorting.

Various subpopulations are clearly identifiable on the 2-D histogram (Figure 6.2b). Donor-only (*D*-only) labeled molecules appear as a population with $S \sim 1$ and $E^* \sim 0$, and acceptor-only (*A*-only) molecules as a population with $S \sim 0$ and broad E^* (since from 6.2); E^* becomes ill-defined in the absence of a donor molecule). These singly-labeled molecules can be excluded for FRET analysis. Solutions that involve fluorophores that bleach or blink during their transit through the confocal volume, or that are present at concentrations which result in subpopulation mixing, give rise to characteristic patterns in the E^*–S histogram; this diagnostic capability of the histogram is instructive for further experimental design [20, 26].

Additionally, in experiments where donor and acceptor fluorophores are placed separately on molecules that can associate to form a complex, donor–acceptor species can be examined in order to study the thermodynamics and kinetics of complex formation [10]. In this case, subpopulation sorting is central to the method. An example of this approach for the study of a protein release in gene transcription is described later in this section.

The width of each E^* subpopulation not only determines the number of resolvable subpopulations, but also can be further analyzed to "mine" for unresolved states within a single peak [26–28]. For instance, a subpopulation due to a single noninterconverting species with a constant donor–acceptor separation will still have a significant width due to the effects of shot noise (an inherent property of any stochastic process with low number of events). In the case of FRET distributions, shot noise arises from the statistical fluctuation of the number of emitted photons about an average value. Since shot noise determines the smallest width which a single static subpopulation can have (the "shot-noise-limited width"), the E^* distribution for the subpopulation of interest could be compared with a simulated shot-noise-limited distribution with the same mean E^*; the presence of a significant difference between the two distributions is an indication that the experimentally observed subpopulation contains unresolved states. These states could be due to either two or more noninterconverting (or slowly interconverting) states with similar E^*, or due to two or more rapidly interconverting states with a larger possible variation in E^* [26, 28].

Recently, Doose *et al.* demonstrated a simplified version of μsALEX known as periodic acceptor excitation (PAX) spectroscopy [29]. In PAX, the green-laser excitation is continuous, whereas the red-laser excitation is modulated on the microsecond timescale, simplifying the experimental implementation. PAX allows the measurement of all crosstalks, and (at least for a popular donor–acceptor FRET pair) has been shown to generate an E^*–S histogram identical to that generated using alternation of both lasers. The lower cost and complexity of such a set-up make this a viable option for laboratories considering building a simple ALEX set-up. On the other hand, the simultaneous direct and indirect (FRET-based) excitation of the acceptor may results in complex laser-based state transitions that may complicate FRET analysis and enhance photobleaching.

To pursue a diverting motoring analogy, μsALEX could perhaps be considered as the "Ford" of the ALEX methods: while there are more complex or expensive methods, μsALEX performs the basic tasks of molecular sorting and FRET analysis reliably and simply, and for many purposes is entirely sufficient. The solution-based nature of the method means that sample preparation is simple, fast, and free of possible surface-induced perturbations, but has the disadvantage that typically single molecules cannot be observed for more than about 1–2 ms. The continuous-wave lasers and simple detection set-up mean that the set-up is the least expensive of the three ALEX-based methods.

6.3.1 Accurate FRET Using ALEX

In order to obtain *accurate FRET* values rather than proximity ratios, the best approach is to measure the contribution of excitation and emission crosstalk terms, the detection efficiency of each detection channel, and the quantum yield of the donor and acceptor fluorophores. The accurate E value is defined as:

$$E = \frac{f^{FRET}}{f_{D_{ex}}^{D_{em}} + f^{FRET}} \tag{6.5}$$

where

$$f^{FRET} = f_{D_{em}}^{A_{em}} - Lk - Dir \tag{6.6}$$

The terms Lk (donor-leakage in the acceptor detection channel) and Dir (acceptor emission in the acceptor detection channel due to direct excitation by the donor-excitation laser) describe crosstalk contributions in the $f_{D_{em}}^{A_{em}}$ signal; such crosstalk terms must be subtracted from the $f_{D_{ex}}^{A_{em}}$ signal to recover the photon count associated only with FRET (Equation 6.6). For solution-based measurements, these terms are easily obtained using μsALEX (Figure 6.2c–d) [17]. The factor γ is directly related to the gradient on the E–S histogram of donor-acceptor labeled molecules of varying E, while Lk and Dir are also easily obtained from the D-only and A-only subpopulations. The factors l, d, and d' shown in Figure 6.2C and D are used to calculate the Lk and Dir (for details, see Ref. [17]).

Once E is obtained, Equation 6.1 can be used to obtain the interprobe distances. The major uncertainty in the conversion of E to distances comes from uncertainty in the relative orientation of the smFRET pair, described by the orientation factor κ^2 (see Section 6.2). If κ^2 changes for a given sample (e.g., due to a change in rotational freedom of one of the fluorophores), then R_0 will also change and the distance value obtained will therefore be inaccurate [30]. This can be addressed by measuring fluorescence anisotropy, a fluorescence ratio that reports on the reorientation of the fluorescence dipole during the fluorescence lifetime of a given fluorophore. For polarized excitation, fluorescence anisotropy is defined as:

$$r = \frac{I_{\parallel} - I_{\perp}}{I_{\parallel} + 2I_{\perp}} \tag{6.7}$$

where $I_{\parallel}$ is the emission intensity parallel to excitation, and $I_{\perp}$ is the emission intensity perpendicular to excitation. If a fluorophore is free to rotate between excitation and fluorescence emission, the fluorescence anisotropy will be low. If the fluorophore has restricted rotational freedom (the extreme case being a completely immobilized fluorophore), then the anisotropy will be high. This measurement may be applied at the ensemble level to check the rotational freedom of the fluorophores. For cases of low anisotropy, the rotational freedom is likely to be high and the corresponding assumption of $\kappa^2 = 2/3$ is likely to be valid and have low uncertainty. For the less-common case of high anisotropy, the rotational freedom of at least one fluorophore is likely to be restricted, and therefore the uncertainty on R_0 and any associated distances will increase; the uncertainty increases further if both fluorophores have restricted rotational freedom. Ensemble anisotropy measurements do not entirely address the possibility that rotational freedom is restricted to only some of the sample subpopulations, although in many cases this will still result in a noticeably high ensemble value which will indicate that rotation is likely to be restricted to some extent. However, the ideal solution is to measure the anisotropy on a subpopulation basis, which is possible if nsALEX/PIE is used (see Section 6.4). For surface-immobilized molecules, accurate FRET measurements may be obtained similarly using the methodology of [16] for both single-excitation smFRET and for ALEX.

6.3.2 Applications of μsALEX

The first test of the ability of ALEX spectroscopy to address biological questions arose during studies of DNA transcription by RNA polymerase (RNAP), a process that is crucial for gene regulation.

Transcription is the process of DNA-directed RNA biosynthesis. During bacterial transcription initiation, a complex of RNAP with a transcription initiation factor (a "σ factor") binds to promoter DNA (a specific DNA sequence located just upstream of gene transcription start sites) and unwinds a region of DNA around the transcription start site to form a catalytically competent "open complex" [31–33]. Subsequently, RNAP engages in cycles of synthesis of short "abortive" RNA until, at some point, it transcribes an RNA of ~9–11 nucleotides (nt) in length. At that point, RNAP successfully breaks its interactions with the promoter region and enters transcription elongation [31–34]. Several multistep transitions during this phase of transcription are still unclear, mainly due to sample heterogeneity and to the transient nature of many of the intermediates involved.

The first ALEX study addressed the fate of factor σ^{70} during transcription elongation [35]. Textbook versions of transcription routinely refer to the obligatory release of σ^{70} upon formation of the first stable elongation complex, formed upon synthesis of a 9–11 nt RNA; this conclusion was reached from studies that measured the σ^{70} content of early elongation complexes [36–40]. However, early methods for quantifying σ^{70} in elongation included separation steps that may have translated the decreased affinity of σ^{70} for elongation complexes into a release of σ^{70} from elongation complexes. This possibility became more real in 2001, when ensemble studies of elongation complexes using ensemble-FRET [41], or purification via gentle separation steps [42], were combined with earlier investigations [43–45] to identify elongation complexes that retain σ^{70}. However, the ensemble studies were complicated by the extensive heterogeneity of transcription complexes; for example, heterogeneity due to conformational transitions which change donor–acceptor distances in ways unaccounted for with reference complexes.

To address these concerns, μsALEX spectroscopy was used to measure σ^{70} retention at various points along elongation. From a single sample without additional controls, ALEX determines: (i) the fraction of active open complexes; (ii) the fraction of stalled elongation complexes able to resume transcription; and (iii) the fraction of transcription complexes that retain σ^{70}. The σ^{70} content of an elongation complex is measured by simple molecular coincidence of σ^{70} and DNA, thus bypassing the need for close donor–acceptor proximity. The ALEX-based σ release assay is based on observations of changes in intermolecular distances and binding stoichiometries upon escape to elongation. The assay uses analytical sorting to distinguish between σ^{70} retention and σ^{70} release in elongation. For a leading-edge FRET assay (which uses a donor introduced at the leading edge of RNAP and an acceptor at the downstream end of DNA; Figure 6.3), forward translocation and formation of a σ^{70}-containing elongation complex converts a donor–acceptor ($S \sim 0.5$) species with low E^* to a donor–acceptor ($S \sim 0.5$) species with high E^*; on the other hand, the formation of σ^{70}-free elongation complex with release of σ^{70} converts a donor–acceptor ($S \sim 0.5$) species with low E^* to donor-only ($S < 0.8$) and acceptor-only ($S < 0.3$) species.

As seen in Figure 6.3, in samples of open complex, two species are observed: (i) the open complex ($S \sim 0.55$, $E^* \sim 0.23$ and apparent donor–acceptor distance of 77 Å, consistent with previous studies [17, 41, 46–48]; and (ii) free promoter

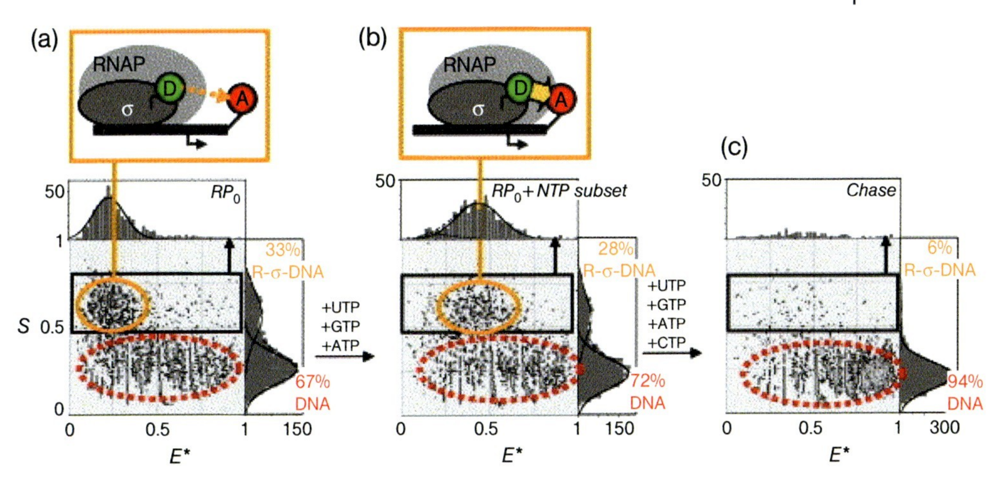

Figure 6.3 Retention of factor σ^{70} in early RNA polymerase (RNAP) elongation complexes. D-A species are marked by the orange ellipse, and A-only species by the red ellipse. Species with $S > 0.8$ (e.g., free σ^{70}) are not included, because the molecule-search criteria employed herein identify solely acceptor-containing molecules. (a) Prior to the addition of NTPs, the RNAP–DNA open complex (R-σ-DNA) is present in a low-FRET state. The red circled population indicates DNA-only molecules. (b) After the addition of a subset of NTPs that allows formation of the first stable elongation complex, the donor–acceptor population moves to a higher FRET state (species R-σ-DNA, in orange circle), which corresponds to σ^{70}-containing stalled elongation complexes. (c) On addition of CTP, the R-σ-DNA species disappears since transcription of DNA substrate is completed and RNAP has dissociated from DNA. Reproduced from Ref. [35].

DNA ($S < 0.3$; a byproduct of the disruption of nonspecific RNAP–promoter complexes). Upon adding nucleotides to form $RD_{e,11}$ and monitoring the changes in the S histogram (cf. Figure 6.3a,b), it was shown that approximately 80% of open complexes are converted to a species that exhibits the same stoichiometry ($S \sim 0.55$) but a higher FRET efficiency ($E^* \sim 0.44$ and apparent donor–acceptor distance of 58 Å); this species is a σ^{70}-containing early elongation complex. This observation led to the conclusion that nearly all open complexes can enter the elongation phase, and that about 80% retain σ^{70} in elongation. Further results confirmed that most early elongation complexes retain σ^{70}, and that a determinant for σ^{70} recognition in the initial transcribed region increases σ^{70} retention in early elongation. Moreover, by monitoring stoichiometry changes in samples containing elongation complexes on DNA fragments that lead to the synthesis of 50 nt-long RNA, it was shown that approximately 50% of the mature elongation complexes retain σ^{70}. In these complexes, the half-life of σ^{70} retention is long relative to the time-scale of elongation, which suggests that some complexes may retain σ^{70} throughout elongation.

The σ^{70}-retention study served as a benchmark for ALEX spectroscopy, and suggested the type of question addressable by this new method. To allow for a proper comparison between the ensemble-FRET study [41] and the ALEX study of σ^{70} release, identical reagents and reaction conditions were used. This provided a vital and reassuring transition from solution-based ensemble-FRET experiments to solution-based ALEX experiments, and set a new benchmark for comparison with ALEX measurements on immobilized complexes (see Section 6.5).

Following this validation of ALEX, the method was used to address the mechanism of the abortive initiation phase of transcription, during which RNAP becomes trapped in a "Catch-22" situation. Here, the strong interactions of RNAP with promoter DNA (which help RNAP associate at the correct site for starting transcription) prevent it from leaving the promoter region and entering transcription elongation. As a consequence, the polymerase goes through the process of abortive cycling, and its relation to promoter escape is a topic of much interest. However, the mechanism involved was not resolved until single-molecule methods were applied to the problem [49, 50].

Three mechanistic models [34, 40, 51–53] were proposed to explain existing ensemble work on abortive initiation: "transient excursions," "inchworming," and "DNA scrunching" (Figure 6.4a):

- The *transient excursions* model invoked transient cycles of forward and reverse translocation of RNAP, wherein RNAP translocates forward as a unit, powered by phosphodiester bond formation, and translocates backwards on the release of abortive RNA.
- *Inchworming* invokes a flexible RNAP element which connects two putative RNAP modules; the downstream module which contains the active site translocates further downstream generating abortive RNA, and then releases the RNA and returns to the initial state.
- *DNA scrunching* invokes a flexible element in DNA: during synthesis of abortive RNA, the RNAP pulls downstream DNA into itself, and extrudes the DNA on release of the abortive RNA. In this model, it is the energy stored in the "scrunched" DNA which fuels promoter escape.

The distances between different sites within RNAP–DNA complexes were monitored using smFRET (Figure 6.4b). The use of μsALEX to exclude *D*-only and *A*-only populations allowed high accuracy measurements to be performed, even at low FRET efficiencies where the *D*-only population obscures single-excitation smFRET measurements. RNAP was labeled either at its leading-edge or at its trailing edge, whereas the DNA was labeled within and around the promoter region. A map of the distance changes on the structure of the open complex clearly showed that the distance changes fell into two distinct categories. All measured distances between RNAP and upstream DNA, or between RNAP and spacer DNA (the part of the promoter region residing between the −10 and −35 promoter element sequences) remained unchanged during the abortive initiation. These results showed that there was no motion of either the leading-edge or the trailing-edge of RNAP relative to upstream DNA and relative to spacer DNA, thus ruling

(a)

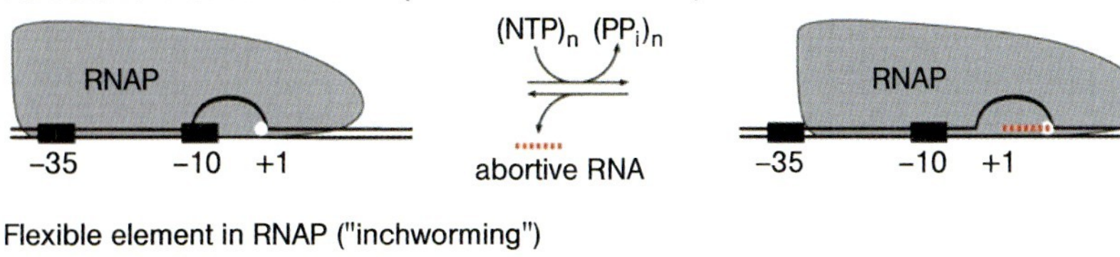

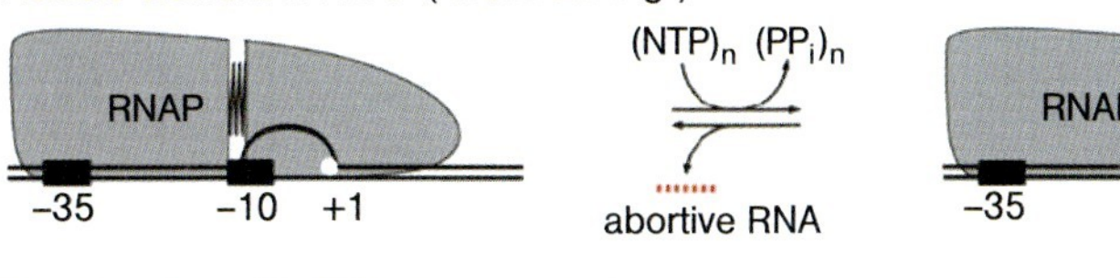

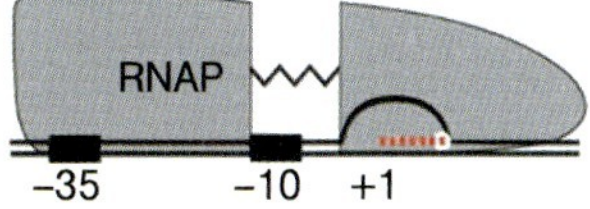

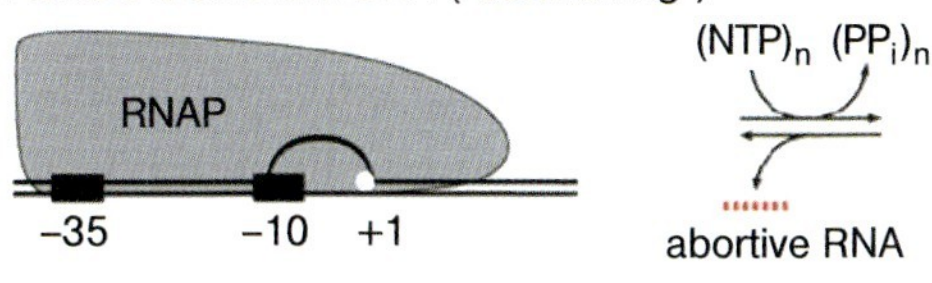

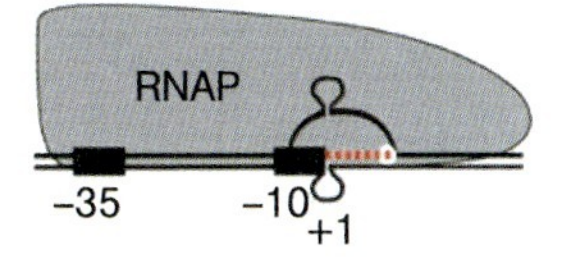

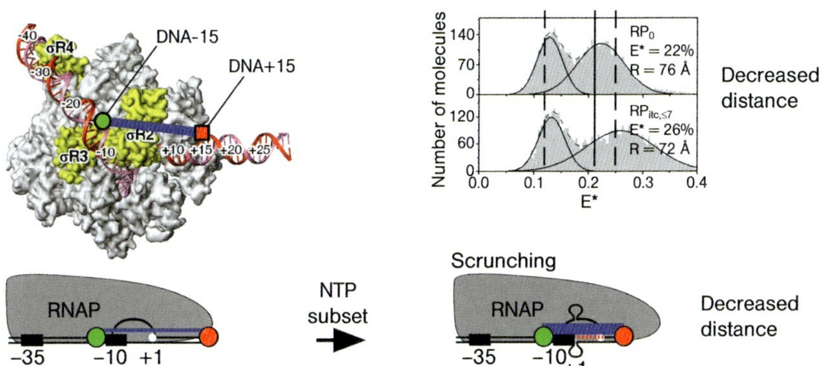

Figure 6.4 Abortive initiation mechanism of RNAP as studied using μsALEX. (a) Three models for RNAP active-center translocation during initial transcription (see Refs [34, 40, 51–53]): transient excursions, inchworming, and scrunching. White circles indicate the RNAP active center, red dashed lines the RNA, and black rectangles the promoter −10 and −35 elements. (b) Experiment showing that initial transcription involves scrunching. DNA contraction occurs between positions −15 and +15 on addition of nucleotides (Cy3B as donor at DNA position −15; Alexa647 as acceptor at DNA position +15). The two donor–acceptor species in the E^* histograms comprise free DNA (lower-E^* species) and open complex (RP_o) or initial transcribing complex ($RP_{itc,\leq 7}$). The increase in FRET efficiency of $RP_{itc,\leq 7}$ compared to RP_o indicates that DNA contracts during abortive initiation. An E–S histogram-based molecular sorting was used to exclude donor-only and acceptor-only labeled molecules from the analysis. Reproduced from Ref. [49].

out the transient-excursion and inchworming hypotheses. In contrast, all measured distances between RNAP and downstream DNA, or between the spacer DNA and downstream DNA, were decreased during abortive initiation. This strongly suggested that abortive initiation operated through a DNA-scrunching mechanism, and that DNA scrunching occurred exclusively within the DNA segment comprising positions −15 to +15.

Microsecond ALEX was also used to study DNA-based nanodevices, nanostructures with much potential for nanotechnology applications [54–56]. Goodman *et al.* used the specificity of DNA hybridization to prepare self-assembling DNA tetrahedra [57, 58] that could reversibly change shape in response to specific signals [59] (Figure 6.5). In the reconfigurable design, one of the edges of the tetrahedra contained a hairpin loop. In the absence of "fuel," the loop remained closed, producing a 10 base pair (bp)-long edge. When the "fuel" (a single-stranded DNA (ss-DNA) strand that was complementary to the hairpin loop) was added, it hybridized to both halves of the hairpin, producing a substantially extended edge (30 bp long). Importantly, the cycle was seen to be reversible, as shown using polyacrylamide gel electrophoresis (PAGE) and ensemble-FRET measurements; the addition of an "antifuel" hairpin complementary to the fuel strand was shown to bind and displace the fuel from the tetrahedra and return them to the closed state. This cycle could be repeated several times.

The FRET measurements were carried out by labeling each end of the hairpin edges with donor and acceptor fluorophores, so that the closed state would give a high FRET and the open state would give a low FRET (Figure 6.5a). However, this did not exclude the presence of either static or dynamic heterogeneity. In order to address this issue, μsALEX was used to measure FRET for the initial conversion to closed and then open tetrahedra (Figure 6.5b). Each state showed a single homogeneous population at expected FRET values, which indicated that dynamic interconversion was not present on any significant scale. This experiment illustrated well how DNA nanotechnology and smFRET methods were natural partners, as both were essentially concerned with the properties of single molecules.

Microsecond ALEX was used to study conformational changes in *lac* permease (lacY), a well-studied bacterial membrane protein that is involved in the transport of lactose [60]. By using a pair of FRET probes with short R_0 (in order to be able to probe short interprobe distances), it was shown that the binding of a specific sugar (a galactopyranoside) caused a distance decrease between the FRET probes introduced into the cytoplasmic side of the protein and a corresponding increase in the periplasmic site; this observation was consistent with an "alternating access" model for galactoside transport. The contribution of ALEX in these studies was due to its compatibility of substoichiometrically labeled proteins (e.g., one of the lacY derivatives was only 60% donor-labeled and 60% acceptor-labeled), its ability to provide accurate FRET efficiencies, and the shot-noise-limited widths. Finally, μsALEX was instrumental as an analytical tool during the development of novel site-specific protein-labeling strategies [61, 62] for preparing proteins for FRET studies of protein folding.

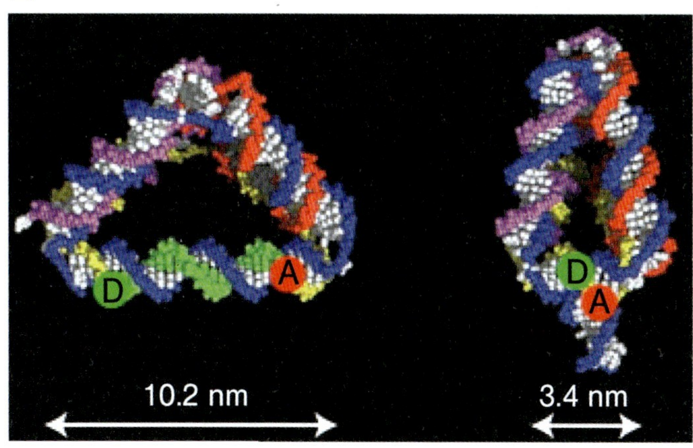

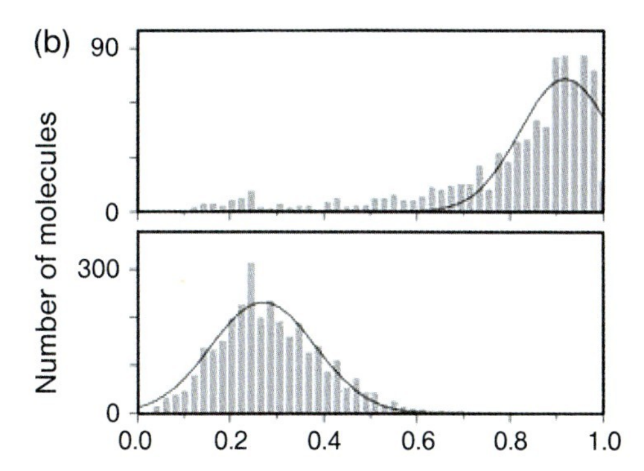

Figure 6.5 Analysis of DNA tetrahedron opening using μsALEX. (a) Open and closed tetrahedra. In the closed state, the reconfigurable edge contains a hairpin loop. In the open state, the hairpin is extended by hybridization to the fuel. Fluorophores placed at the on the reconfigurable edge (*D*, donor; *A*, acceptor) have low FRET in the open state and high FRET in the closed state. (b) μsALEX for characterization of the open and closed states. Single-molecule FRET measurements corresponding to the initial conversion of the closed tetrahedron (top panel) to the open tetrahedron (bottom panel) indicate a single, homogeneous population of fluorescent objects in each case, with no observable dynamic interconversion between the two states. Reproduced from Ref. [59].

6.4 Nanosecond-ALEX/Pulsed Interleaved Excitation (PIE)

By reducing the alternation period of ALEX to the nanosecond timescale [11, 63], the fluorescence-intensity decays can be analyzed for different smFRET subpopulations (Figure 6.6a–e). In addition to subpopulation sorting and FRET measurement, this allows measurements of time-resolved polarization anisotropy for each subpopulation, provides high temporal-resolution FRET information, and also enables a high temporal-resolution fluorescence correlation spectroscopy (FCS).

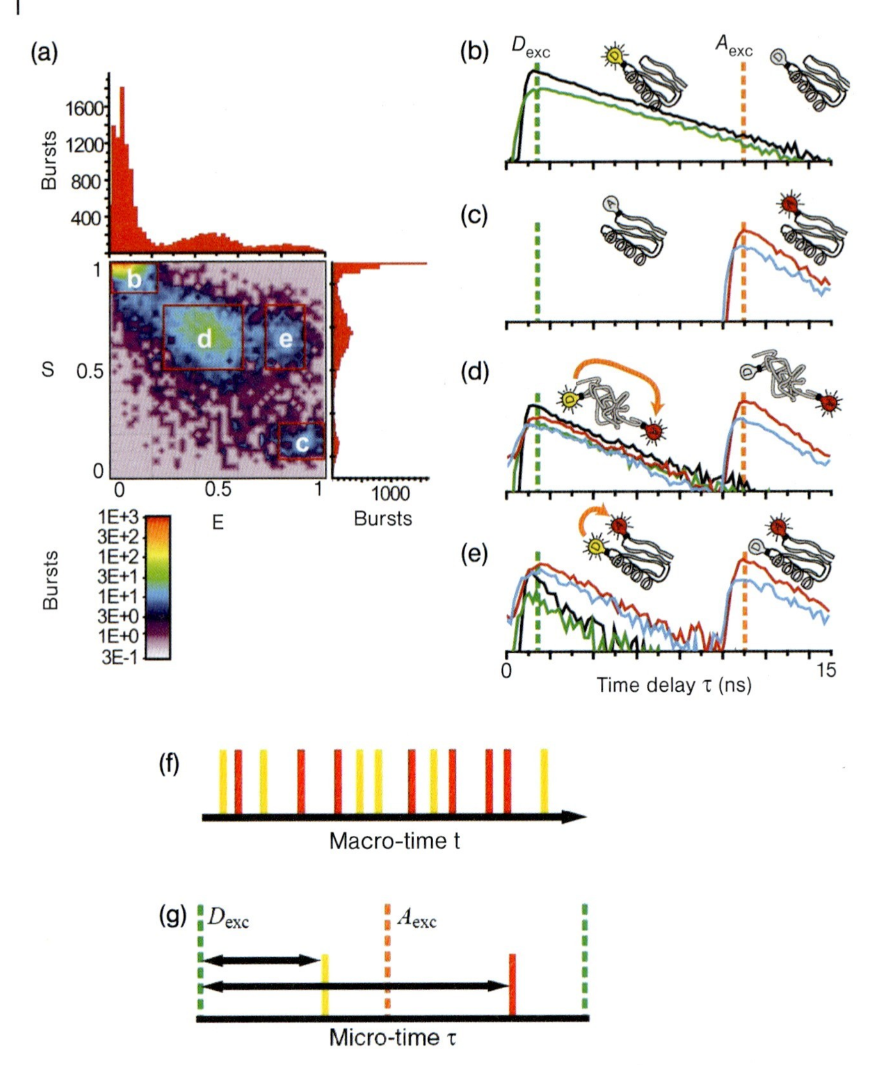

Figure 6.6 *E–S* histogram and nsALEX/PIE fluorescence-intensity decays for various subpopulations. (a) An nsALEX/PIE derived *E–S* histogram with several subpopulations with different (*E*,*S*) values. (b–e) Time-resolved fluorescence-intensity decays analyzed on a subpopulation basis. Fluorescence-intensity decays extracted for donor-only (b), acceptor only (c), donor–acceptor complex with low-FRET (d), and donor–acceptor complex with high FRET (e). The black and green decays represent donor emission, parallel and perpendicular to excitation polarization, respectively. The red and cyan decays represent acceptor emission, parallel and perpendicular to excitation polarization. (b) Donor-only molecules emit only after donor excitation (leakage of donor into acceptor channel removed for clarity). (c) Acceptor-only molecules emit only after acceptor excitation (direct excitation of acceptor by donor excitation removed for clarity). (d) Donor–acceptor low-FRET complexes emit donor and acceptor fluorescence after donor excitation pulse (ratio of intensities and lifetimes depend on FRET efficiency) and emit acceptor fluorescence after acceptor excitation.

The latter is a well-established fluorescence spectroscopy that uses the statistical analysis of fluorescence fluctuations to interrogate biomolecular dynamics [22, 64–66]. The two methods that perform alternation in the nanosecond timescale, nsALEX and PIE, were first conceived by Laurence *et al.* [63] and Muller *et al.* [11], respectively. Although the two methods are fairly similar, they were in fact developed independently and their details reported almost simultaneously–hence the surfeit of acronyms. The major difference between the original reports was one of applications: while Laurence *et al.* emphasized the applications to subpopulation fluorescence lifetime analysis, Muller *et al.* highlighted the advantages of FCS when combined with fast laser alternation. Consequently, the technique here is referred to as nsALEX/PIE–a somewhat cumbersome acronym, but one which serves the purposes of this chapter.

nsALEX/PIE is carried out using picosecond-pulsed laser sources and time-correlated single-photon counting electronics. Technically, the original PIE implementation used one pulsed laser source and one rapidly modulated continuous-wave source, which matches the outcome of two pulsed sources, albeit at a slower alternation rate [11]. The crucial difference to μsALEX is the use of pulsed rather than continuous-wave excitation (Figure 6.6f,g). During each alternation period, a short pulse excites fluorophores in the diffusing molecules, and the time difference ("microtime") between the pulse and detection of fluorescence photons is recorded with high resolution. At the same time, a coarser measurement of the photon arrival-time relative to the start of the experiment ("macrotime") is recorded and used for molecular sorting, as for μsALEX. The high-resolution arrival-time measurement can be used for fluorescence lifetime analysis.

The analysis of fluorescence lifetime decays is not possible for single diffusing molecules because only about 100 photons are detected per burst [63]. However, fluorescence lifetime analysis is possible on a subpopulation basis [67], recovering distance distributions that fluctuate on timescales down to the fluorescence lifetime [63]. For example, nsALEX/PIE can probe fluctuations occurring at the 1 μs timescale; in contrast, such a fluctuation averages out during the 1 ms-long molecular transits of μsALEX experiments, and will not add width in excess to shot-noise-limited width. Fast subpopulation analysis is useful for applications such as the study of polymer dynamics (e.g., polypeptides or oligonucleotides). The ability to measure time-resolved fluorescence anisotropy is useful for accurate distance measurements, as the rotational freedom of each fluorophore can be directly evaluated and used to estimate the orientation factor, κ^2. This is perhaps the most

(e) Donor–acceptor high-FRET complexes emit similarly to case (d), except with a higher relative intensity of acceptor compared with donor after the donor excitation pulse, and with a shorter donor lifetime, indicating a higher FRET efficiency due to shorter average distance. (f, g) Schematic diagrams of nsALEX/PIE timing, using pulsed-laser sources combined with time-correlated single photon counting (TCSPC) electronics. (f) arrival times relative to the start of experiment (macro-time) of all photons. (g) arrival times relative to the excitation time (micro-time) of each photon. Reproduced from Ref. [63].

exciting aspect of nsALEX/PIE: by measuring anisotropy on a subpopulation basis, it should be possible to check for any changes in fluorophore rotational freedom, thereby significantly decreasing the uncertainty on FRET-based distance measurements.

The rapid alternation achieved by nsALEX/PIE means that for typical FCS measurements, where the timescale of interest is order of microseconds or greater, the alternation is effectively simultaneous, whereas for μsALEX the slower laser makes FCS measurements more difficult. This advantage allows quantitative FCS measurements on complexes undergoing FRET by using direct acceptor excitation [11]. It is also likely that the application of FCS on a subpopulation basis ("purified FCS"; [68]), will benefit significantly from the ability of nsALEX/PIE to access shorter timescales.

If the spurious motoring analogy were to be continued, nsALEX/PIE would the "Ferrari" of the ALEX family: high performance, but expensive and challenging to handle. A large range of additional capabilities are available, but at much higher cost and complexity. The sophisticated instrumentation (particularly the pulsed lasers and photon-counting module) add significantly to the cost of an ALEX set-up. Moreover, the conversion of raw data into lifetime data and the implementation of lifetime fitting algorithms requires complex programming and data analysis. Therefore, nsALEX/PIE represents a versatile–but expensive and complex–tool for use in demanding experiments.

6.4.1 Applications of nsALEX/PIE

In the studies conducted by Laurence *et al.* [63], nsALEX was used to investigate the polymer flexibility of varying lengths of ss-DNA and double-stranded DNA (ds-DNA), and to study residual structures in the unfolded states of chymotrypsin inhibitor 2 (CI2) and acyl-CoA binding protein (ACBP), under conditions where the presence of folded states of the proteins would normally obscure analysis. By extracting the mean E and the variance ΔE for individual subpopulations using fluorescence lifetime measurements (as opposed to fluorescence-intensity ratios as in μsALEX), the changes in polymer flexibility could be measured under different experimental conditions, as these changes would correspond to changes in ΔE.

The studies on ss-DNA and ds-DNA could, in principle, have been performed using ensemble time resolved techniques on 100% labeled samples [63]. However, nsALEX was used to perform these studies in the presence of *D*-only and *A*-only subpopulations, which conveniently allowed an evaluation of the fluorescence lifetime and fluorescence anisotropy of the donor and acceptor labels. These studies could also be performed at concentrations where any intermolecular effects were essentially eliminated. The results demonstrated an unexpected and intriguing increased flexibility of short ds-DNA fragments compared to the theoretical predictions based on the rigid nature of DNA in length-scales of up to ~50 nm (~150 bp). Moreover, the authors were able to show, convincingly, that the electrostatic contribution to the *persistence length* (a polymer-rigidity index, defined as the mean length of a polymer at which thermal fluctuations lead to a significant deflection of the polymer long axis) of ss-DNA varied with the ionic strength to the −1/2 power ($I^{-1/2}$), again demonstrating the quantitative capabilities of the technique.

For the protein measurements, the effects of denaturant concentration on distance distributions were measured for the unfolded subpopulation of the protein, directly excluding the folded subpopulation. These measurements were compared with simulations of a Gaussian chain model, which should adequately describe an entirely denatured protein [69]. It should be noted that, although a denatured protein can be described better by the worm-like chain model [70, 71], the application of this model would not have produced sufficiently large changes to explain deviations from the model [63]. The study results showed significant deviations from the Gaussian model, thus providing strong evidence for a *transient residual structure* in the unfolded states of the proteins, with ACBP having a greater residual structure than CI2, as previously suggested. These measurements clearly indicated the potential of the nsALEX technique for studies of protein folding and dynamics.

A different aspect of protein folding, namely protein-assisted protein folding, was also probed using PIE by Scharma *et al.* [72]. It is well known that in the cytoplasm, many polypeptide chains require assistance from molecular chaperones ("folding machines") in order to reach their folded states efficiently and quickly [72, 73]. In a comprehensive ensemble and single-molecule study of protein folding by the GroEL/GroES molecular chaperon system (a cage-like multimeric protein with a hydrophobic cavity where accelerated folding occurs in an ATP-dependent manner), Sharma *et al.* produced significant new information on the mechanism and physical principles exploited to facilitate folding. In short, these studies led to the proposal that protein confinement by GroEL/GroES stretches apart strongly hydrophobic regions of a polypeptide, and thus opens up folding pathways that are inaccessible to spontaneous hydrophobic collapse.

Muller *et al.* [74] used PIE to characterize "DNA tweezers," a prototype DNA machine [75]. The principle of DNA tweezers and the labeling scheme for the PIE experiment are shown in Figure 6.7a,b. Initially, the tweezers are in the open, low-FRET state. However, upon the addition of a closing strand ("fuel") which hybridizes with the tweezers, they convert to the closed, high-FRET state. The addition of fuel may produce several DNA nanostructures other than the desired DNA tweezers; such nanostructures would have different FRET efficiencies and stoichiometries (Figure 6.7c–f) that were difficult to study using ensemble methods, as their FRET efficiencies would be expected to average out at the ensemble level. However, it was shown very elegantly in a single diffusion-based PIE measurement that, whilst open tweezers exist in a single conformation, the addition of a fuel strand generated three conformations corresponding to various states of "closed" tweezers (Figure 6.7g–i). These closed states were tentatively assigned to some of the structures in Figure 6.7c–f. Future experiments may reveal to what extent the closed-state subpopulation truly remains closed by applying the lifetime or FCS capabilities of PIE on a subpopulation basis.

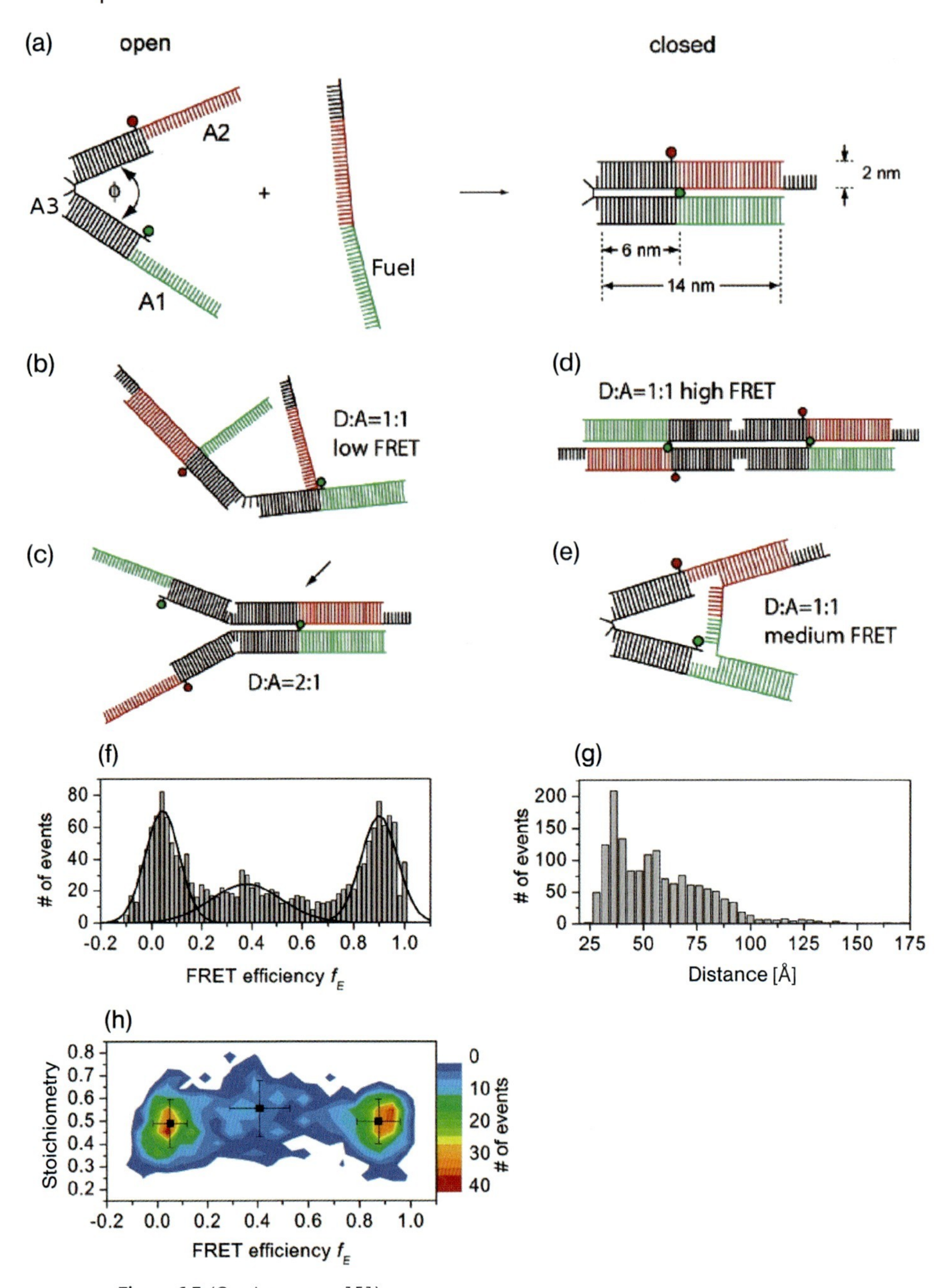

Figure 6.7 (Caption see p. 151)

6.5 msALEX

By combining TIRF microscopy with lasers alternating on the millisecond timescale, msALEX allows the extended observation of multiple single molecules simultaneously [12]. The experimental set-up (Figure 6.8) is a straightforward modification of a standard TIRF microscope to allow laser alternation and imaging of two emission channels. Once the molecules of interest have been identified in movies of a msALEX experiment, and the fluorescence intensity time traces extracted, the data analysis is similar to that for μsALEX, with the added advantage of extended FRET observations that may reveal interesting fluctuations due to conformational changes. This is particularly useful for following dynamics during a nonequilibrium reaction pathway [12], as with many celebrated single-laser smFRET measurements [76–79].

Moreover, by monitoring changes in *S* for time traces, it is possible to deconvolve FRET fluctuations due to distance variations from FRET fluctuations due to acceptor photophysics, such as acceptor blinking [12, 80], a process which is observed on popular single-molecule fluorophores such as Cy5, tetramethylrhodamine, and Alexa647. Blinking significantly increases the uncertainty on single-excitation smFRET measurements. Time traces of *S* can also report on the kinetics of assembly/disassembly reactions.

In relentlessly pursuing the automobile analogy, msALEX could be likened to a "Land Rover": slower and without the timing accuracy of the point detection methods, but with the "off-roader" advantage of being able to access temporal regimes inaccessible to the diffusion-based ALEX methods. The success of its use depends on the success of surface immobilization, which can be challenging

Figure 6.7 The use of pulsed interleaved excitation (PIE) to uncover heterogeneity in DNA tweezers. (a) Schematic depiction of the DNA tweezers. The open tweezers are formed by three strands of DNA: **A1**, **A2**, and **A3**. Upon the addition of a "fuel" strand, the two arms are pulled together. (b–e) Unwanted hybridization products resulting from the reaction of fuel strands with the tweezers. The unwanted products have a range of FRET states and stoichiometries, including contributions due to incomplete labeling (in (c), the arrow points to position of the missing label). (f, g) FRET efficiency and distance histograms for closed tweezers. Three subpopulations are observed, showing high, intermediate, and low mean FRET efficiencies. These fractions contain contributions from a variety of constructs in the "closed" tweezers sample. The high-FRET peak contains properly closed tweezers structures and perhaps dimers in which the donor and acceptor are spatially separated by ~3.5 nm (see state (d)). The low- and intermediate-FRET fractions also contain multimers, but with different ratios of tweezers and fuel strands, and/or incomplete donor and acceptor labeling. (h) An analysis of the stoichiometry of dye labeling corresponding to the FRET efficiencies in (f). The intermediate-efficiency fraction has a slightly elevated stoichiometry value, indicating contributions from dimer structures that contain more donors than acceptors. The three black squares represent the mean value and standard deviation of FRET efficiency and stoichiometry for molecules with low-, intermediate-, and high-FRET efficiency, respectively. Reproduced from Ref. [74].

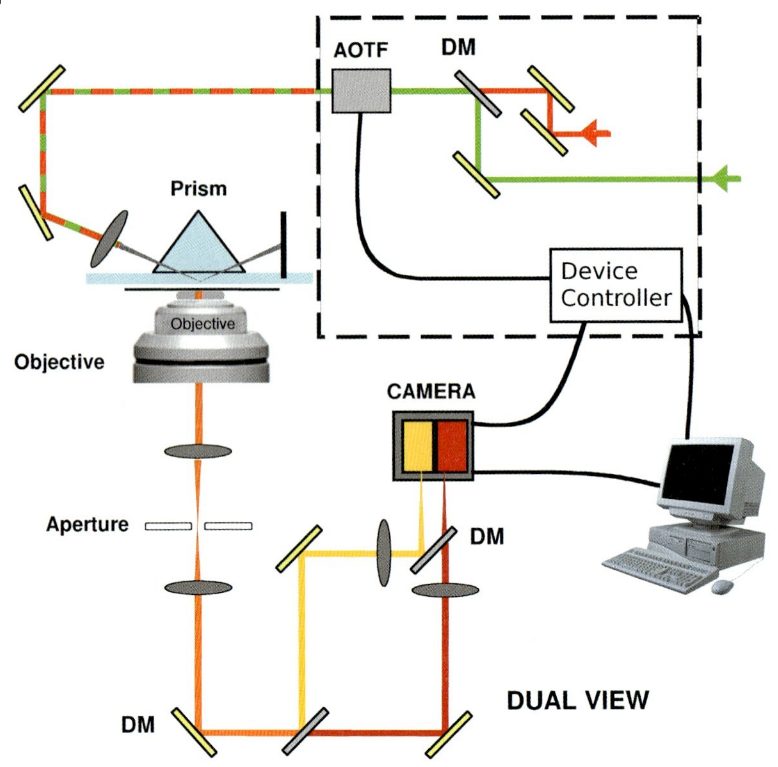

Figure 6.8 Set-up for ms-ALEX. A laser total internal reflection microscope is modified to achieve alternating-laser excitation and camera synchronization. DM = dichroic mirror, AOTF = acousto-optic tunable filter (this controls the laser alternation). Reproduced from Ref. [10].

for some biomolecules and biosystems. There are two main disadvantages with msALEX, however: (i) the surface-based nature of the technique is more challenging than diffusion-based experiments; and (ii) the current maximum refresh rate of the CCD cameras used is ~1 KHz, which sets a limit on the time resolution of experiments.

6.5.1 Applications of msALEX

The first application of msALEX was also focused on mechanistic questions in transcription initiation. ALEX studies of diffusing single molecules cannot report on the millisecond-minute kinetics of transcription because of the short dwell of molecules in the laser focus (~1 ms), and the need to accumulate significant statistical data so as to extract kinetic information. By using msALEX, Margeat *et al.* were able to detect abortive initiation and promoter escape within single immobilized transcription complexes. The transcription complexes studied were essentially identical to those used for the analysis of σ^{70} release and abortive initiation (see Section 6.3), the only difference being the addition of a biotin group at either end of a promoter DNA fragment. This format allowed the immobilization of DNA fragments or stable transcription complexes on quartz surfaces modified by hydrophilic polyethylene glycol (PEG) groups and coated with streptavidin. This immobilization strategy has been developed for the study of other DNA-processing enzymes, and used extensively [77, 78, 81].

The ALEX-based translocation assays used to study σ^{70} retention upon promoter escape (see Section 6.3) were applied to measure the fraction of immobilized open complexes able to enter elongation. Initially, open complexes for leading-edge FRET were surface-immobilized, and msALEX then used to identify donor–acceptor species and generate an E^* histogram. Upon the addition of a nucleotide subset that allowed formation of the first stable elongation complex, 50–70% of the molecules moved to elongation and retain σ^{70}. These results also settled the question of σ^{70} retention on RNAP during promoter escape versus σ^{70} dissociation before promoter escape and reassociation with RNAP after promoter escape. Since there was no free σ^{70} in the solution over the immobilized complexes, the probability of σ^{70} rebinding was negligible. Therefore, the σ^{70} molecule present in the transcription initiation was carried in elongation during promoter escape. The study results also established that, for the promoter studied, the rate-limiting steps of abortive initiation were the steps of abortive-product release and/or RNAP-active-center reverse translocation (and not the step of abortive-product synthesis and RNAP-active-center forward translocation).

The study of the transcription mechanism was also central to the second msALEX study, although in this case the biological system was the complex of a eukaryotic transcription factor (TATA-binding protein; TBP) with promoter DNA, in the presence and absence of a negative cofactor-2 (NC2) [82]. By using TIRF with a high temporal resolution (30–75 ms per frame), the authors showed, somewhat surprisingly, that the addition of NC2 to a TBP–DNA complex mobilized the complex; this conclusion was supported by the fact that the addition of NC2 caused large-scale fluctuations in the FRET signal; the application of msALEX excluded that this fluctuation was due to acceptor photophysics. These observations extended the rather "static" view of transcription initiation and provided a glimpse of exciting future developments in the study of eukaryotic gene regulation.

Continuing on the theme of DNA-binding proteins, Abbondanzieri *et al.* [83] used msALEX to study how the binding orientation of HIV reverse transcriptase to DNA and RNA substrates affected transcriptase activities. HIV reverse transcriptase (RT) is a viral polymerase that catalyzes a series of reactions to convert the single-stranded RNA genome of HIV into ds-DNA that can then be integrated into the host genome. In order to achieve this, RT has three separate functions: RNA-directed DNA synthesis; DNA-directed DNA synthesis; and DNA-directed RNA hydrolysis. Various investigations have shown how the binding orientation of RT to DNA and RNA primers was related to DNA synthesis and RNA hydrolysis activity, as the coordination of these activities and the coordination determinants were unclear.

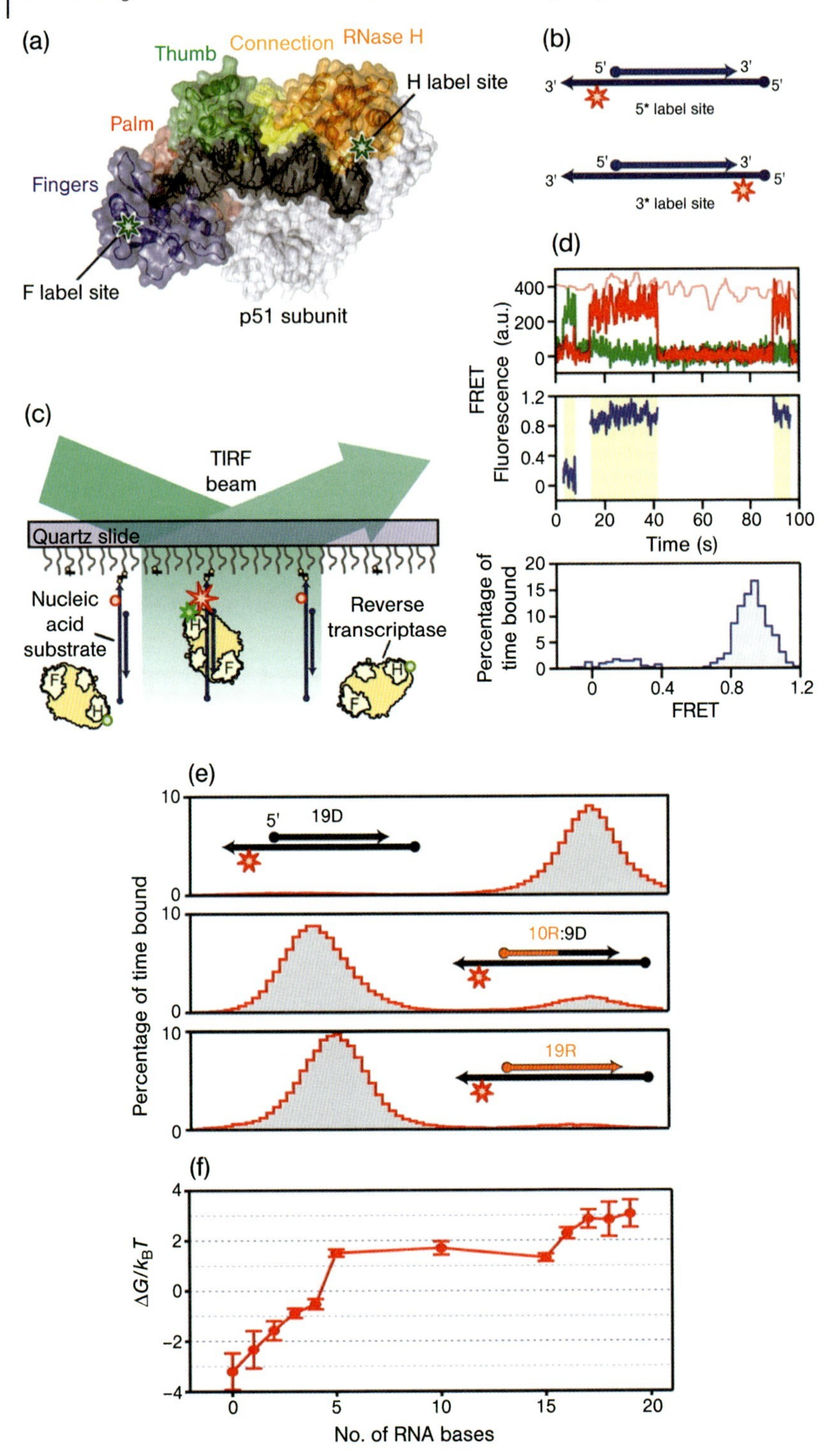

Figure 6.9 (caption see p. 155)

From the crystal structure of RT (Figure 6.9a), two possible DNA binding orientations (forwards and backwards relative to the DNA/RNA 5′ end) were identified, of which only one was observed in crystal structures of RT–nucleic acid complexes. Experiments on surface-immobilized molecules were ideal for studying the connection of binding orientation to catalytic activity. By labeling the immobilized substrate and free RT at different positions with donor and acceptor (Figure 6.9a–c), the authors were able to directly observe the binding orientation. Moreover, the ability to directly observe single molecules for extended periods (up to minutes) led to direct observation of dynamic switching between the two orientations. The use of msALEX allowed easy selection of movie frames with complexes containing both a fluorescent donor and a fluorescent acceptor, excluding sections of the FRET time traces showing significant donor- or acceptor-photophysics (Figure 6.9d). This capability allowed the authors to show, convincingly, that RT binds DNA and RNA primers in opposite directions (Figure 6.9e), and suggested that the binding orientation might determine the enzymatic activity of RT. Additionally, by using chimeric DNA/RNA substrates with different ratios of DNA/RNA, the free energy difference per extra RNA base between the two different binding orientations could be measured (Figure 6.9f). In subsequent studies, the same group showed that the binding orientation determined the

Figure 6.9 msALEX measurements investigate the effect of binding orientation on HIV-1 reverse transcriptase (RT) enzymatic activity. (a) The structure of RT bound to a DNA-primer–template complex. Labeling sites for Cy3 on the RT are shown as green stars. (b) Nucleic acid substrates. The substrates consisted of a 19–21 nt primer strand annealed to a 50 nt template strand containing a Cy5 label (red star). Cy5 was either three nucleotides from the 5′ end (circle) or four to six nucleotides from the 3′ end (arrow) of the primer. (c) Experimental format. Single-molecule detection of Cy3 (green star or sphere)-labeled RT binding to and dissociating from the surface-immobilized nucleic acid substrates labeled with Cy5 (red star or sphere). The stars and spheres indicate dyes that do and do not emit fluorescence, respectively. (d) FRET analysis for RT binding to a single primer–template complex. Top: fluorescence time traces from Cy3 (green) and Cy5 (red) under excitation at 532 nm, and that from Cy5 (pink) under excitation at 635 nm. Direct Cy5 excitation was used to confirm absence of acceptor photophysics. Middle: FRET value calculated over the duration of the binding events (yellow-shaded regions). Bottom: Computed FRET probability distribution for the binding events. Reproduced from Ref. [83]. (e) Selected FRET distributions of Cy3-labeled RT bound to Cy5-labeled substrates containing various 19 nt chimeric RNA:DNA primers hybridized to a DNA template. RT bound to entirely DNA and entirely RNA primers has opposite binding orientations, and therefore a different FRET efficiency. RT bound to chimeric RNA:DNA primers shows a mixture of the two FRET states due to switching between the binding orientations. (f) The free-energy difference ΔG between the high-FRET and low-FRET orientations plotted as a function of RNA content for xR:yD chimeras. The free-energy difference (ΔG) between the two states was most sensitive to the sugar composition of the four or five nucleotides located at each end of the 19 nt primer, suggesting that the interactions between RT and the nucleic acid at opposite ends of the primer–template binding cleft were most important in determining the binding orientation. Error bars indicate the standard error of the mean ($n = 3$). Reproduced in modified form from Ref. [83].

enzymatic activity of RT, and that cognate nucleotides and a specific RT inhibitor would have opposing effects on the switching rates. The latter observation provided a simple mechanism for the clinical effectiveness of the inhibitor in treating HIV. The fact that a single-molecule method was able to decipher the mode of action of an important pharmaceutical nicely illustrates the wide general utility of such methods.

Finally, an interesting msALEX application by Andrecka *et al.* used multiple surface-based FRET measurements to accurately triangulate the three-dimensional (3-D) position of an RNA residue within a transcription complex [84], a method similar to that of Rasnik *et al.* [85] using single-laser smFRET.

6.6 Three-Color ALEX

A recent extension of ALEX by Lee *et al.* [86] added an extra excitation wavelength (in the 470 nm region) to increase the number of excitation lasers to three and give rise to three-color ALEX (3c-ALEX). This method adds to a existing collection of three-color-based, single-molecule methods, such as three-color fluorescence cross-correlation spectroscopy (3c-FCCS; [87] and three-color smFRET in solution and on the surface [88, 89]. By adding a blue fluorophore to a biomolecule in addition to green and red, this allows up to three intermolecular distances to be monitored: blue-green FRET; blue-red FRET; and green-red FRET. In this way, 3-D histograms of both E and S may be constructed, allowing the identification of a wide range of subpopulations and the measurement of multiple distances in a single solution (Figure 6.10).

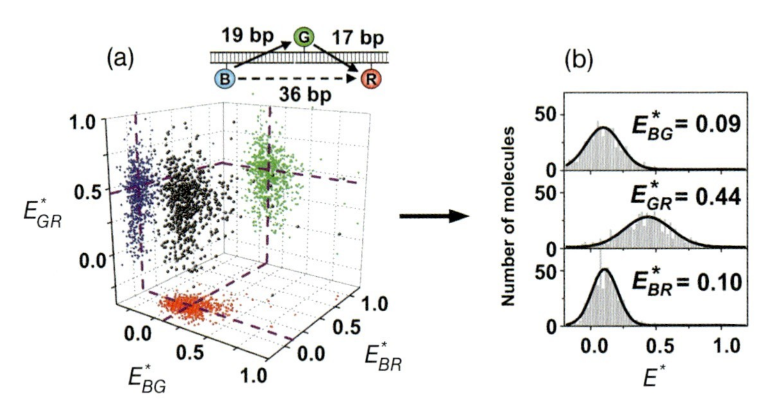

Figure 6.10 Three-color ALEX can be used to simultaneously monitor multiple distances within the same sample. (a) Three-dimensional E^* histograms for triply labeled species, *B-G-R*. The gray population on the 3-D histogram is projected onto 2-D histograms (green, blue, red) for easy visualization; (b) One-dimensional E^* histograms obtained after collapsing the 3-D histogram on each E^* axis. The reported E^* values represent the means of the fitted Gaussian distributions. Reproduced from Ref. [83].

Three-color ALEX has been used to study RNAP translocation on DNA, mainly as a demonstration of the potential of the method rather than for answering any outstanding biological questions. By labeling DNA with two fluorophores, upstream and downstream of the RNAP binding site, and by labeling RNAP with a third fluorophore, it was possible simultaneously to monitor conformational changes in DNA (e.g., to check for large-scale bending or unwinding), and the position of the RNAP relative to the two points on the DNA [86]. This allowed the unambiguous identification of the direction of RNAP translocation on DNA, and confirmed previous results regarding σ^{70} retention in elongation. Three-color ALEX was also used by Lee *et al.* [86] to study the folding characteristics of a deoxyribozyme.

A method similar to 3-color ALEX was also implemented by Ross *et al.* [90] in their investigations of three-way DNA junctions. In these studies, the alternation was implemented using an acousto-optical beam splitter and the molecules were immobilized on a glass coverslip. This format allows extended observation of multiple emission channels and complex FRET efficiencies and stoichiometries as a function of time.

Interestingly, the application of this principle to *n*-color ALEX, where *n* excitation wavelengths are used and multiple distances are simultaneously measured, offers the possibility of unambiguous determination of probe positions in 3-D space through simple triangulation [86]. However, the main drawback of such a scheme is the significantly increased experimental and analytical complexity each time that an extra excitation wavelength is added.

6.7 Conclusions and Outlook

Some five years after the first report from the laboratory of Shimon Weiss at UCLA describing its concept and first implementation [10], ALEX has become a widely applicable and useful addition to the single-molecule toolbox, accessing timescales on the range of nanoseconds to minutes, distances on the range of 2 to 10 nm, and concentrations on the range of 10 to 200 pM. A number of related approaches have been developed, and much new information has been gained through their use. The ability to exclude molecules with incorrect fluorophore stoichiometry, coupled with an ability to carry out subpopulation sorting, has significantly increased the resolution of, and confidence, in smFRET measurements. Today, it is unnecessary to stringently remove donor- and acceptor-only labeled species from samples (as is the case with some single-excitation smFRET measurements), which makes the experiments much less arduous. Finally, the ability to choose between several implementations of ALEX means there is an ALEX counterpart to every single-excitation smFRET method.

At present, the main disadvantage of ALEX appears to relate to its ease of implementation. Although a commercial instrument capable of nsALEX/PIE currently exists (the MicroTime 200; Picoquant), there remains a need to construct customized set-ups to carry out μsALEX or msALEX measurements. However, the fact that detailed construction and operation protocols for all ALEX implementations have been produced [20] should lower the barriers for building and operating an ALEX set-up. In terms of comparison with single-excitation smFRET, ALEX is more complex to implement and reduces the photon count of the measurement (since ~50% of the time is allocated to acceptor excitation). However, the backwards-compatibility of ALEX with single-excitation FRET allows the characterization of samples for subpopulations using ALEX, but then to revert to the latter method if temporal resolution is an important issue.

In future, ALEX-based data analysis will undoubtedly be developed further so as to achieve improved distance measurements within single molecules [17], and this will surely help in the studies of large, multicomponent, heterogeneous complexes of cellular machinery (especially if they are inaccessible to conventional structural biology). Benefit will also be derived from existing crystallographic and/or nuclear magnetic resonance-based structures. The use of 3c-ALEX may also lead to the real-time monitoring of coupling of conformational changes, especially when 3c-ALEX is adapted for use on immobilized molecules [12]. In the near future, excitation confinement using zero-mode waveguides [91], nanopipettes [92] and nanofluidics [93], may extend the working range of solution-based ALEX to the 1 to 1000 nM range, thus widening the applicability of the method.

Finally, ALEX is fully compatible with real-time monitoring structures and stoichiometries at the single-molecule level in living cells. Although this may seem an enormous challenge, recent successes in monitoring the diffusion properties of single *lac* repressor molecules in bacteria [94] have suggested that elegant *in vivo* experiments and ensuing discoveries may be closer to our grasp than was previously thought.

References

1 Ha, T., *et al.* (1996) Probing the interaction between two single molecules: fluorescence resonance energy transfer between a single donor and a single acceptor. *Proc. Natl Acad. Sci. USA*, **93** (13), 6264–6268.

2 Clegg, R.M. (1992) Fluorescence resonance energy transfer and nucleic acids. *Methods Enzymol.*, **211**, 353–388.

3 Selvin, P.R. (1995) Fluorescence resonance energy transfer. *Methods Enzymol.*, **246**, 300–334.

4 Selvin, P.R. (2000) The renaissance of fluorescence resonance energy transfer. *Nat. Struct. Biol.*, **7** (9), 730–734.

5 Nie, S. and Zare, R.N. (1997) Optical detection of single molecules. *Annu. Rev. Biophys. Biomol. Struct.*, **26**, 567–596.

6 Förster, T. (1948) Intramolecular energy migration and fluorescence. *Annalen der Physik*, **2**, 55–75.

7 Ha, T. (2004) Structural dynamics and processing of nucleic acids revealed by single-molecule spectroscopy. *Biochemistry*, **43** (14), 4055–4063.

8 Weiss, S. (1999) Fluorescence spectroscopy of single biomolecules. *Science*, **283** (5408), 1676–1683.

9 Weiss, S. (2000) Measuring conformational dynamics of biomolecules by single molecule fluorescence spectroscopy. *Nat. Struct. Biol.*, **7** (9), 724–729.

10 Kapanidis, A.N., *et al.* (2004) Fluorescence-aided molecule sorting: analysis of structure and interactions by

alternating-laser excitation of single molecules. *Proc. Natl Acad. Sci. USA*, **101** (24), 8936–8941.
11 Muller, B.K., *et al.* (2005) Pulsed interleaved excitation. *Biophys. J.*, **89** (5), 3508–3522.
12 Margeat, E., *et al.* (2006) Direct observation of abortive initiation and promoter escape within single immobilized transcription complexes. *Biophys. J.*, **90** (4), 1419–1431.
13 Kapanidis, A.N., *et al.* (2005) Alternating-laser excitation of single molecules. *Acc. Chem. Res.*, **38** (7), 523–533.
14 Santoso, Y., *et al.* (2008) Red light, green light: probing single molecules using alternating-laser excitation. *Biochem. Soc. Trans.*, **36** (Pt 4), 738–744.
15 Kapanidis, A.N., *et al.* (2001) Mean DNA bend angle and distribution of DNA bend angles in the CAP-DNA complex in solution. *J. Mol. Biol.*, **312** (3), 453–468.
16 Ha, T., *et al.* (1999) Single-molecule fluorescence spectroscopy of enzyme conformational dynamics and cleavage mechanism. *Proc. Natl Acad. Sci. USA*, **96** (3), 893–898.
17 Lee, N.K., *et al.* (2005) Accurate FRET measurements within single diffusing biomolecules using alternating-laser excitation. *Biophys. J.*, **88** (4), 2939–2953.
18 Ha, T. (2001) Single-molecule fluorescence resonance energy transfer. *Methods*, **25**, 78–86.
19 Roy, R., Hohng, S., and Ha, T. (2008) A practical guide to single-molecule FRET. *Nat. Methods*, **5** (6), 507–516.
20 Kapanidis, A.N., *et al.* (2008) Alternating laser excitation of single molecules, in *Single-Molecule Techniques: A Laboratory Manual* (eds P.R. Selvin and T. Ha), Cold Spring Harbor Laboratory Press, Cold Spring Harbor, NY, pp. 85–119.
21 Nie, S., Chiu, D.T., and Zare, R.N. (1994) Probing individual molecules with confocal fluorescence microscopy. *Science*, **266** (5187), 1018–1021.
22 Eigen, M. and Rigler, R. (1994) Sorting single molecules – application to diagnostics and evolutionary biotechnology. *Proc. Natl Acad. Sci. USA*, **91** (13), 5740–5747.
23 Axelrod, D. (2001) Total internal reflection fluorescence microscopy in cell biology. *Traffic*, **2** (11), 764–774.
24 Axelrod, D. (2001) Selective imaging of surface fluorescence with very high aperture microscope objectives. *J. Biomed. Opt.*, **6** (1), 6–13.
25 Eggeling, C., *et al.* (2001) Data registration and selective single-molecule analysis using multi-parameter fluorescence detection. *J. Biotechnol.*, **86** (3), 163–180.
26 Nir, E., *et al.* (2006) Shot-noise limited single-molecule FRET histograms: comparison between theory and experiments. *J. Phys. Chem. B Condens. Matter Mater. Surf. Interfaces Biophys.*, **110** (44), 22103–22124.
27 Dahan, M., *et al.* (1999) Ratiometric measurement and identification of single diffusing molecules. *Chem. Phys.*, **247** (1), 85–106.
28 Antonik, M., *et al.* (2006) Separating structural heterogeneities from stochastic variations in fluorescence resonance energy transfer distributions via photon distribution analysis. *J. Phys. Chem. B*, **110** (13), 6970–6978.
29 Eggeling, C., *et al.* (2006) Analysis of photobleaching in single-molecule multicolor excitation and Forster resonance energy transfer measurements. *J. Phys. Chem. A*, **110** (9), 2979–2995.
30 Dale, R.E., Eisinger, J., and Blumberg, W.E. (1979) The orientational freedom of molecular probes. The orientation factor in intramolecular energy transfer. *Biophys. J.*, **26** (2), 161–193.
31 Murakami, K.S. and Darst, S.A. (2003) Bacterial RNA polymerases: the whole story. *Curr. Opin. Struct. Biol.*, **13** (1), 31–39.
32 Record, M.T., Jr., *et al.* (1996) Escherichia coli RNA polymerase (Esigma70), promoters, and the kinetics of the steps of transcription initiation, in *Escherichia coli and Salmonella* (ed. F. Neidhart), ASM Press, Washington, DC, pp. 792–820.
33 Young, B.A., Gruber, T.M., and Gross, C.A. (2002) Views of transcription initiation. *Cell*, **109** (4), 417–420.
34 Hsu, L.M. (2002) Promoter clearance and escape in prokaryotes. *Biochim. Biophys. Acta*, **1577** (2), 191–207.
35 Kapanidis, A.N., *et al.* (2005) Retention of transcription initiation factor sigma(70) in transcription elongation: single-molecule analysis. *Mol. Cell*, **20** (3), 347–356.

36 Straney, D.C. and Crothers, D.M. (1985) Intermediates in transcription initiation from the *E. coli lac* UV5 promoter. *Cell*, **43** (2 Pt 1), 449–459.
37 Travers, A.A. and Burgess, R.R. (1969) Cyclic re-use of the RNA polymerase sigma factor. *Nature*, **222** (193), 537–540.
38 Hansen, U.M. and McClure, W.R. (1980) Role of the sigma subunit of *Escherichia coli* RNA polymerase in initiation. II. Release of sigma from ternary complexes. *J. Biol. Chem.*, **255** (20), 9564–9570.
39 Metzger, W., *et al.* (1993) Nucleation of RNA chain formation by *Escherichia coli* DNA-dependent RNA polymerase. *J. Mol. Biol.*, **232** (1), 35–49.
40 Krummel, B. and Chamberlin, M.J. (1989) RNA chain initiation by *Escherichia coli* RNA polymerase. Structural transitions of the enzyme in early ternary complexes. *Biochemistry*, **28** (19), 7829–7842.
41 Mukhopadhyay, J., *et al.* (2001) Translocation of sigma(70) with RNA polymerase during transcription: fluorescence resonance energy transfer assay for movement relative to DNA. *Cell*, **106** (4), 453–463.
42 Bar-Nahum, G. and Nudler, E. (2001) Isolation and characterization of sigma(70)-retaining transcription elongation complexes from *Escherichia coli*. *Cell*, **106** (4), 443–451.
43 Ring, B.Z., Yarnell, W.S., and Roberts, J.W. (1996) Function of *E. coli* RNA polymerase sigma factor sigma 70 in promoter-proximal pausing. *Cell*, **86** (3), 485–493.
44 Osumi-Davis, P.A., Woody, A.Y., and Woody, R.W. (1987) Transcription initiation by *Escherichia coli* RNA polymerase at the gene II promoter of M13 phage: stability of ternary complex, direct photocrosslinking to nascent RNA, and retention of sigma subunit. *Biochim. Biophys. Acta*, **910** (2), 130–141.
45 Shimamoto, N., Kamigochi, T., and Utiyama, H. (1986) Release of the sigma subunit of *Escherichia coli* DNA-dependent RNA polymerase depends mainly on time elapsed after the start of initiation, not on length of product RNA. *J. Biol. Chem.*, **261** (25), 11859–11865.
46 Mukhopadhyay, J., *et al.* (2003) Fluorescence resonance energy transfer (FRET) in analysis of transcription-complex structure and function. *Methods Enzymol.*, **371**, 144–159.
47 Mekler, V., *et al.* (2002) Structural organization of bacterial RNA polymerase holoenzyme and the RNA polymerase-promoter open complex. *Cell*, **108** (5), 599–614.
48 Lawson, C.L., *et al.* (2004) Catabolite activator protein: DNA binding and transcription activation. *Curr. Opin. Struct. Biol.*, **14** (1), 10–20.
49 Kapanidis, A.N., *et al.* (2006) Initial transcription by RNA polymerase proceeds through a DNA-scrunching mechanism. *Science*, **314** (5802), 1144–1147.
50 Revyakin, A., *et al.* (2006) Abortive initiation and productive initiation by RNA polymerase involve DNA scrunching. *Science*, **314** (5802), 1139–1143.
51 Carpousis, A.J. and Gralla, J.D. (1985) Interaction of RNA polymerase with lacUV5 promoter DNA during mRNA initiation and elongation. Footprinting, methylation, and rifampicin-sensitivity changes accompanying transcription initiation. *J. Mol. Biol.*, **183** (2), 165–177.
52 Pal, M., Ponticelli, A.S., and Luse, D.S. (2005) The role of the transcription bubble and TFIIB in promoter clearance by RNA polymerase II. *Mol. Cell*, **19** (1), 101–110.
53 Straney, D.C. and Crothers, D.M. (1987) A stressed intermediate in the formation of stably initiated RNA chains at the *Escherichia coli* lac UV5 promoter. *J. Mol. Biol.*, **193** (2), 267–278.
54 Craighead, H.G. (2000) Nanoelectromechanical systems. *Science*, **290** (5496), 1532–1536.
55 Erben, C.M., Goodman, R.P., and Turberfield, A.J. (2006) Single molecule protein encapsulation in a rigid DNA cage. *Angew. Chem., Int. Ed.*, **45** (44), 7414–7417.
56 Rothemund, P.W. (2006) Folding DNA to create nanoscale shapes and patterns. *Nature*, **440** (7082), 297–302.
57 Goodman, R.P., Berry, R.M., and Turberfield, A.J. (2004) The single-step synthesis of a DNA tetrahedron. *Chem. Commun. (Camb)*, (12), 1372–1373.

58 Goodman, R.P., *et al.* (2005) Rapid chiral assembly of rigid DNA building blocks for molecular nanofabrication. *Science*, **310** (5754), 1661–1665.

59 Goodman, R.P., *et al.* (2008) Reconfigurable, braced, three-dimensional DNA nanostructures. *Nature Nanotechnol.*, **3** (2), 93–96.

60 Majumdar, D.S., *et al.* (2007) Single-molecule FRET reveals sugar-induced conformational dynamics in LacY. *Proc. Natl Acad. Sci. USA*, **104** (31), 12640–12645.

61 Jager, M., Michalet, X., and Weiss, S. (2005) Protein-protein interactions as a tool for site-specific labeling of proteins. *Protein Sci.*, **14** (8), 2059–2068.

62 Jager, M., Nir, E., and Weiss, S. (2006) Site-specific labeling of proteins for single-molecule FRET by combining chemical and enzymatic modification. *Protein Sci.*, **15** (3), 640–646.

63 Laurence, T.A., *et al.* (2005) Probing structural heterogeneities and fluctuations of nucleic acids and denatured proteins. *Proc. Natl Acad. Sci. USA*, **102** (48), 17348–17353.

64 Elson, E.L. and Magde, D. (1974) Fluorescence correlation spectroscopy. I. Conceptual basis and theory. *Biopolymers*, **13**, 1–27.

65 Haustein, E. and Schwille, P. (2004) Single-molecule spectroscopic methods. *Curr. Opin. Struct. Biol.*, **14** (5), 531–540.

66 Schwille, P. and Heinze, K.G. (2001) Two-photon fluorescence cross-correlation spectroscopy. *ChemPhysChem*, **V2** (N5), 269–272.

67 Deniz, A.A., *et al.* (2001) Ratiometric single-molecule studies of freely diffusing biomolecules. *Annu. Rev. Phys. Chem.*, **52**, 233–253.

68 Laurence, T.A., *et al.* (2007) Correlation spectroscopy of minor fluorescent species: signal purification and distribution analysis. *Biophys. J.*, **92** (6), 2184–2198.

69 Doi, M. and Edwards, S.F. (1986) *The Theory of Polymer Dynamics*, Oxford University Press, Oxford.

70 Hagerman, P.J. and Zimm, B.H. (1981) Monte-Carlo approach to the analysis of the rotational diffusion of wormlike chains. *Biopolymers*, **20** (7), 1481–1502.

71 Lapidus, L.J., *et al.* (2002) Effects of chain stiffness on the dynamics of loop formation in polypeptides. Appendix: testing a 1-dimensional diffusion model for peptide dynamics. *J. Phys. Chem. B*, **106** (44), 11628–11640.

72 Sharma, S., *et al.* (2008) Monitoring protein conformation along the pathway of chaperonin-assisted folding. *Cell*, **133** (1), 142–153.

73 Frydman, J. (2001) Folding of newly translated *in vivo*: the role of molecular chaperones. *Annu. Rev. Biochem.*, **70**, 603–647.

74 Muller, B.K., *et al.* (2006) Single-pair FRET characterization of DNA tweezers. *Nano Lett.*, **6** (12), 2814–2820.

75 Yurke, B., *et al.* (2000) A DNA-fuelled molecular machine made of DNA. *Nature*, **406** (6796), 605–608.

76 Myong, S., *et al.* (2007) Spring-loaded mechanism of DNA unwinding by hepatitis C virus NS3 helicase. *Science*, **317** (5837), 513–516.

77 Myong, S., *et al.* (2005) Repetitive shuttling of a motor protein on DNA. *Nature*, **437** (7063), 1321–1325.

78 Blanchard, S.C., *et al.* (2004) tRNA selection and kinetic proofreading in translation. *Nat. Struct. Mol. Biol.*, **11** (10), 1008–1014.

79 Blanchard, S.C., *et al.* (2004) tRNA dynamics on the ribosome during translation. *Proc. Natl Acad. Sci. USA*, **101** (35), 12893–12898.

80 Sabanayagam, C.R., Eid, J.S., and Meller, A. (2004) High-throughput scanning confocal microscopy for single molecule analysis. *Appl. Phys. Lett.*, **84** (7), 1216–1218.

81 Ha, T., *et al.* (2002) Initiation and re-initiation of DNA unwinding by the *Escherichia coli* Rep helicase. *Nature*, **419** (6907), 638–641.

82 Schluesche, P., *et al.* (2007) NC2 mobilizes TBP on core promoter TATA boxes. *Nat. Struct. Mol. Biol.*, **14** (12), 1196–1201.

83 Abbondanzieri, E.A., *et al.* (2008) Dynamic binding orientations direct activity of HIV reverse transcriptase. *Nature*, **453** (7192), 184-U2.

84 Andrecka, J., *et al.* (2008) Single-molecule tracking of mRNA exiting from RNA

polymerase II. *Proc. Natl Acad. Sci. USA*, **105** (1), 135–140.

85 Rasnik, I., *et al.* (2004) DNA-binding orientation and domain conformation of the *E. coli rep* helicase monomer bound to a partial duplex junction: single-molecule studies of fluorescently labeled enzymes. *J. Mol. Biol.*, **336** (2), 395–408.

86 Lee, N.K., *et al.* (2007) Three-color alternating-laser excitation of single molecules: monitoring multiple interactions and distances. *Biophys. J.*, **92** (1), 303–312.

87 Heinze, K.G., Jahnz, M., and Schwille, P. (2004) Triple-color coincidence analysis: one step further in following higher order molecular complex formation. *Biophys. J.*, **86** (1 Pt 1), 506–516.

88 Clamme, J.P. and Deniz, A.A. (2005) Three-color single-molecule fluorescence resonance energy transfer. *ChemPhysChem*, **6** (1), 74–77.

89 Hohng, S., Joo, C., Biophys, and Ha, T. (2004) Single-molecule three-color FRET. *Biophys. J.*, **87** (2), 1328–1337.

90 Ross, J., *et al.* (2007) Multicolor single-molecule spectroscopy with alternating laser excitation for the investigation of interactions and dynamics. *J. Phys. Chem. B*, **111** (2), 321–326.

91 Levene, M.J., *et al.* (2003) Zero-mode waveguides for single-molecule analysis at high concentrations. *Science*, **299** (5607), 682–686.

92 Vogelsang, J., *et al.* (2007) Single-molecule fluorescence resonance energy transfer in nanopipets: improving distance resolution and concentration range. *Anal. Chem.*, **79** (19), 7367–7375.

93 Mannion, J.T. and Craighead, H.G. (2007) Nanofluidic structures for single biomolecule fluorescent detection. *Biopolymers*, **85** (2), 131–143.

94 Elf, J., Li, G.W., and Xie, X.S. (2007) Probing transcription factor dynamics at the single-molecule level in a living cell. *Science*, **316** (5828), 1191–1194.

7 Unraveling the Dynamics Bridging Protein Structure and Function One Molecule at a Time

Jeffery A. Hanson, Yan-Wen Tan, and Haw Yang

7.1 Introduction

Proteins are inherently flexible molecules, and their constant bombardment by solvent molecules drives the spontaneous conformational fluctuations that occur at physiological temperatures. Accordingly, a folded protein may be viewed more appropriately as an ensemble of related low-energy conformations rather than as a single, unique structure. Although seemingly paradoxical, proteins have evolved to achieve many diverse functions, including the conversion of chemical energy to mechanical work, signaling, and catalysis, with high efficiency and specificity in the presence of constant conformational and thermal noise. The ubiquitous nature of conformational fluctuations suggests that they might play an integral part in the protein's structure–function relationship, providing evolutionary pressure to harness these motions in a productive manner. In order to fully understand the way in which these complex molecules operate, it will be beneficial to fully integrate the role of conformational dynamics in proteins with respect to the existing knowledge of their structure and function.

The problem of how protein conformational dynamics contributes to function can be formulated by several basic, yet poorly understood, questions:

- What range of motions can a protein sample, and what are their timescales?
- What roles do these motions play in protein function?
- How are these motions modulated by ligand binding or interactions with other proteins?
- How are these motions dictated by the protein's structure?

The fundamental challenge complicating the study of the functional consequences of conformational dynamics in proteins is the wide range of time and length scales on which these motions occur, namely timescales of 10^{-9} to 10^{2} s, and length scales of 10^{-11} to 10^{-9} m [1]. Indeed, while structural techniques are capable of revealing the atomistic details of proteins, and have contributed the bulk of the present understanding of the way in which proteins function, these methods typically

Single Particle Tracking and Single Molecule Energy Transfer
Edited by Christoph Bräuchle, Don C. Lamb, and Jens Michaelis

ISBN: 978-3-527-32296-1

resolve snapshots of low-energy conformations of proteins, revealing little about the range of motion, the timescales of conformational dynamics, or the energetics related to these structural fluctuations. Single-molecule spectroscopy can be used directly to investigate conformational fluctuations in proteins, and can be integrated with structural information to provide a more complete description of the motions of these molecules in solution. To appreciate how single-molecule spectroscopy contributes to addressing these important questions, it would be helpful briefly to discuss a few ensemble-averaged techniques that have been instrumental in advancing our understanding to the current state.

In the past, time-resolved methods such as neutron scattering [2, 3], crystallography [4] and optical spectroscopy [5–7], have been used to study conformational dynamics. Whilst, in general, these require the conformational changes of many molecules to be more-or-less synchronized in order to obtain sufficient signal, this limits the experiments that may be conducted to fast processes on the order of nanoseconds. Structural changes on the functionally important millisecond to second timescales tend to be averaged out for such ensemble-averaged experiments, and thus are not readily detectable by these methods.

Nuclear magnetic resonance (NMR) spectroscopy is arguably the most informative ensemble-averaged experimental technique currently available for studying conformational dynamics in proteins, due to its ability to measure time-dependent changes within molecules in solution. NMR can be used to measure protein dynamics with many structural details, since almost every atom in the molecule is capable of serving as a specific probe [8, 9]. While it is most sensitive to conformational dynamics on fast timescales (ns to ps), NMR can observe slower relaxation dynamics (μs to ms) with the help of spectral analysis and modeling [10]. However, NMR cannot identify the number of conformational states without *a priori* assumptions and models because it measures the *mean* conformation averaged over an ensemble of molecules. Due to this constraint, it is often not possible to extract kinetic parameters beyond a two-state model [11, 12]. Furthermore, the requirement for a high sample concentration makes NMR impractical for the study of intermolecular interactions; neither is it amenable for studying the mechanistic roles of slow conformational changes in various molecular processes. As with most ensemble-averaged experimental approaches, it is difficult to simultaneously resolve the timescale and range of motion of protein conformational changes – the two essential parameters that define dynamics – by using NMR.

Recently, optical single-molecule techniques have been applied to understanding conformational dynamics in proteins, and offer many unique advantages [13, 14]. As they do not rely on population averaging, single-molecule methods can detect rare events and can be used to measure the distributions of conformations accessible to proteins, as well as their rates of interconversion. Moreover, single-molecule fluorescence techniques are most powerful in the millisecond to minute (ms–min) time regime, and thus complement many of the more conventional techniques mentioned above. Characterizing conformational dynamics in the ms–min time regime is of particular interest, since it is a time regime that is poorly explored but covers many biologically important processes. Fluorescence methods are well suited to the study of conformational motions in individual proteins, as they have a relatively high signal-to-background ratio. Modern instrumentation can detect a single fluorophore-labeled protein, despite the fact that these molecules are typically two orders of magnitude smaller than the size of the diffraction limit for visible radiation [15]. In addition, the required equipment is relatively inexpensive and the techniques do not demand an extensive knowledge of optics or quantum mechanics [16, 17]. On the other hand, an unbiased interpretation of single-molecule data relies heavily on advanced statistics [18–20]. When combined with distance-sensitive probe mechanisms, such as Förster resonance energy transfer (FRET) [21], single-molecule experiments can report time-dependent protein structural information [22, 23] that is ideal for the study of conformational dynamics.

In this chapter we will explore the nature of conformational dynamics in proteins, and outline the insights gained from single-molecule fluorescence experiments, with emphasis placed on the relationship between dynamics and function. Although single-molecule studies of protein conformational dynamics are intimately tied to methodology developments, technical advances will not be described in detail at this point, as many recent excellent reviews have covered this topic [16, 17, 24]; neither will the chapter include recent single-molecule advances in the fields of protein folding [25], molecular motors [26–28], RNA folding and dynamics or ribozymes [29, 30], for which excellent reviews already exist. First, we will review the energy landscape model that will serve as the conceptual underpinning for the experiments to be described in the chapter.

7.1.1 Rough Energy Landscape

Although commonly associated with protein folding, the energy landscape governing protein conformational dynamics can be viewed as a low-energy basin at the center of the folding funnel [31]. This energy-landscape view of protein conformational dynamics was first proposed by Frauenfelder and colleagues over three decades ago [32]. It was developed primarily to describe the results of flash photolysis experiments measuring CO rebinding to the protein myoglobin.

Myoglobin (Mb), an oxygen-binding, heme iron-containing protein that is readily isolated from muscle tissue, has been studied extensively by many groups as a model system for flexibility in proteins. In the initial flash photolysis experiments, a laser pulse was used to disassociate CO from the heme into an internal cavity within the protein, while CO rebinding was monitored by changes in the Soret spectrum [32, 33]. Although this process exhibits single-exponential kinetics at physiological temperatures, it was shown that the rebinding of CO to Mb has a nonexponential decay at temperatures less then 200 K. This observation was attributed to the presence of multiple conformational states, with differential rates of CO rebinding, the interconversion of which is halted at low temperatures due to insufficient energy for barrier-crossing events. An infrared (IR) spectral analysis

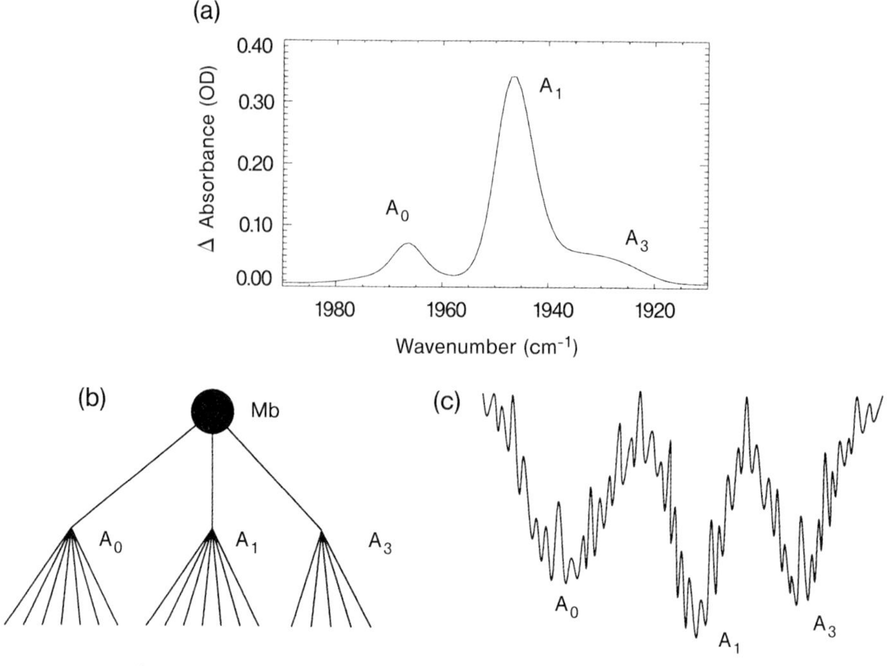

Figure 7.1 Conformational substates of myoglobin (Mb). (a) Infrared absorbance spectrum showing CO stretch frequencies of Mb at low temperature. (b) Schematic representation of hierarchy of conformational substates, with the highest tiers labeled A_0, A_1, and A_3. (c) Free-energy landscape, showing the hierarchy of conformational substates. Reproduced with permission from Ref. [35]; © 2001, National Academy of Sciences, USA.

revealed that three distinct CO stretch bands could be resolved for the Mb–CO complex [34] (Figure 7.1a). Further experiments revealed nonexponential rebinding kinetics within each conformational substate, that was attributed to the existence of a second tier of motions [36]: a hierarchy of conformational substates within each of the three spectrally resolvable states (Figure 7.1b,c). Hole-burning experiments at temperatures less then 1 K suggested the existence of an even lower tier of conformational states, the motions of which could only be detected at extremely low temperatures due to their small barriers for interconversion [37]. These experiments thus led to the modern view of the rough energy landscape consisting of a hierarchy of substates characterized by a wide range of free-energy barrier heights [38].

Neutron scattering experiments, capable of probing the time-dependent displacement of hydrogen atoms in proteins, showed that for temperatures less than ~200 K, the Mb hydrogen atom motions could be treated as harmonic, and indicated that the motions were due primarily to vibrations within a single energy well [3]. Above 200 K, however, the motions became anharmonic, which was interpreted as the onset of conformational dynamics when the system had sufficient thermal energy to enable the protein to overcome barriers between conformational states. Intriguingly, in many of the systems studied, this "dynamical transition" coincided with the onset of activity. Together with the conformation dynamics on-set interpretation, this observation led to the intuitive and reasonable hypothesis that conformational dynamics were necessary for protein function [3].

Although these studies discussed probed fast dynamics of crystalline Mb at low temperatures, the types of motion would be expected to form a ubiquitous part of the dynamics of a protein in solution at physiological temperatures. The origin of the rough energy landscape was suggested by analogy to a frustrated glass system [38]. Every motion by one part of the molecule in order to relieve an unfavorable interaction results in an unfavorable interaction being created in another region; thus, there is no true singular low-energy state accessible to the folded protein. These pioneering studies on Mb laid the framework for the current understanding of the existence of a hierarchy of states in proteins – the rough energy landscape – where conformational dynamics exist over a wide range of time and length scales, and are critical to protein function. Whilst the energy landscape picture is generally applicable to other proteins [39], and is a powerful tool for describing proteins, the generality of the model renders it ineffective in formulating a predictive understanding for individual proteins, without quantitatively characterizing the energy surfaces and further examining the extent to which the model is valid. Experimental approaches, such as single-molecule spectroscopy, that deliver this much-needed characterization of the energy landscape would therefore be expected to contribute greatly to an understanding of protein function.

7.1.2 Functional Roles of Conformational Dynamics

The presence of ubiquitous conformational dynamics in proteins leads to an obvious hypothesis: Conformational fluctuations in proteins play an integral role to their structure and function. Although attractive and intuitive, this hypothesis remains difficult to verify experimentally because of the challenges in characterizing protein dynamics, particularly in biologically relevant time regimes, which are not readily accessible with conventional techniques.

Although early views of proteins regarded them as relatively rigid (e.g., the "lock-and-key" model for enzyme catalysis [40]), Koshland's "induced-fit" hypothesis postulated a means by which structural rearrangements could play an important role in ligand binding and enzyme function [41]. Originally put forward to explain enzyme specificity between similar substrates, it was postulated that binding of the correct substrate would induce conformational changes that would lead to a complete reformation of the enzyme's active site, and result in the catalytic selectivity seen in many systems. X-ray crystallography studies performed on a great many systems showed this picture to be largely correct [42] (for a comprehensive list, see the Database of Molecular Motions, www.molmovdb.org [43]). Indeed, many functional roles have been postulated for conformational dynamics in enzymes, including a direction contribution to catalysis [44], allosteric regulation [45], ligand binding and recognition [46], and functional structural motions [47].

Whilst many of the proposed functional roles for conformational dynamics involve small-scale, relatively fast motions that ideally are studied with NMR experiments or molecular dynamics (MD) simulations, large-amplitude functional structural motions – such as the reorientation of domains of an enzyme during its catalytic cycle – are best investigated with single-molecule experiments. These motions, which frequently were implied by crystallography through the trapping of reaction intermediates, are commonly interpreted as being important to protein function, and typically involve large-amplitude conformational rearrangements predicted to occur on the ms–min timescale.

In this chapter, specific cases will be used to illustrate how optical single-molecule experiments can help in further advancing our understanding of proteins by connecting a protein's structure to its function through conformational dynamics. In Section 7.2, carefully designed and executed single-molecule experiments on molecular motors directly demonstrate the role of large-scale, functional conformational motions in these systems, testing previously inaccessible hypotheses regarding their function. Conformational motions represent the main means by which molecular motors achieve their functions, and are often not subtle since large transitions between discrete states are often required. These large, discrete changes in the molecular configuration produces huge signals in single-molecule experiments, and allow the conformational changes to be measured unambiguously. The same is also largely true of single-molecule experiments detecting protein or RNA folding [25, 29, 30]. On the other hand, conformational dynamics in proteins are generally expected to be much more subtle. Large, discrete transitions in conformation (resulting from dynamics with timescales that are separated by at least one decade, or large free energy barriers) rarely exist in proteins due to the complexity of their energy landscape. Thus, the visual assignment of switching between conformational substates is challenging, if not impossible. For this reason, many single-molecule experimentalists choose systems with clear behavior where hypotheses concerning function can be answered definitively. If these challenges can be overcome, then single-molecule fluorescence experiments may offer a unique perspective to the study of protein conformational dynamics, due to their ability to directly access the energy landscape on a biochemically relevant timescale. An example related to protein signaling is provided in Section 7.3, while examples focusing on enzyme catalysis are provided in Section 7.4.

7.2
Converting Chemical Energy to Mechanical Work: Molecular Motors

Molecular motors – proteins which do work or produce force, commonly by consuming ATP – are ideally studied using single-molecule techniques. Many of the earliest and most elegant single-molecule experiments have involved these systems. Since this class of proteins carries out their function almost exclusively through conformational changes, they are challenging to study with conventional ensemble techniques, whereas single-molecule techniques can directly investigate previously untestable hypothesis regarding their function. Although these types of systems are not the focus of this chapter, they clearly demonstrate the power of single-molecule techniques and their applications to biological systems. Therefore, two of the earliest systems studied will be covered briefly, namely F_0F_1 ATPase and kinesin.

7.2.1 F_0F_1 ATP Synthase

The F_0F_1 ATP synthase is a rotary motor from mitochondria which uses a pH gradient across the membrane to drive ATP synthesis [28, 48]. This protein consists of two main components: the F_1 complex, a hexameric ring that is responsible for ATP synthesis/hydrolysis; and the F_0 complex, a transmembrane proton pump that is responsible for coupling the proton gradient to rotation (Figure 7.2a). The two are linked by the γ subunit which is connected to F_0 and extends through the central pore of F_1. Although a rotary model was proposed on the basis of crystallographic and kinetic data [50], it was not until the advent of single-molecule techniques that this hypothesis could be demonstrated directly. In a famous set of single-molecule experiments, Noji *et al.* attached a fluorescently labeled actin to the γ subunit in the immobilized F_1 component and observed counterclockwise rotation in discrete 120° steps under ATP hydrolysis [51]. Since that pioneering experiment, a subsequent study in which a gold bead was attached to the γ subunit has revealed that the 120° step can be further dissected into 90° and 30° substeps [52]. The 90° step was assigned to ATP binding, since its lifetime was dependent on ATP concentration, while the 30° step–the lifetime of which was independent of ATP concentration–was assumed to be due to ATP hydrolysis or phosphate release [52]. In the first single-molecule experiment involving the full F_0F_1 complex, the full machine was shown to undergo rotation between three distinct states during ATP hydrolysis as well as ATP synthesis. By attaching FRET probes to the F_0 and F_1 components (Figure 7.2a), and observing the entire complex embedded in liposomes diffusing through the laser focus of a confocal microscope, it was possible to observe three FRET states under ATP synthesis, the order of which was reversed during ATP hydrolysis (Figure 7.2b,c) [49], demonstrating that the direction of rotation was coupled to the protein's function.

7.2.2 Kinesin

Another molecular motor where single-molecule experiments have provided invaluable insights into its function is the protein kinesin. This processive motor protein moves along the microtubule network in the cell, transporting large cargo from the interior of the cell to its periphery at the cost of one ATP hydrolysis per step [26, 53, 54]. The crystal structure of kinesin reveals two head domains, responsible for ATP hydrolysis and microtubule binding, attached to two cargo binding

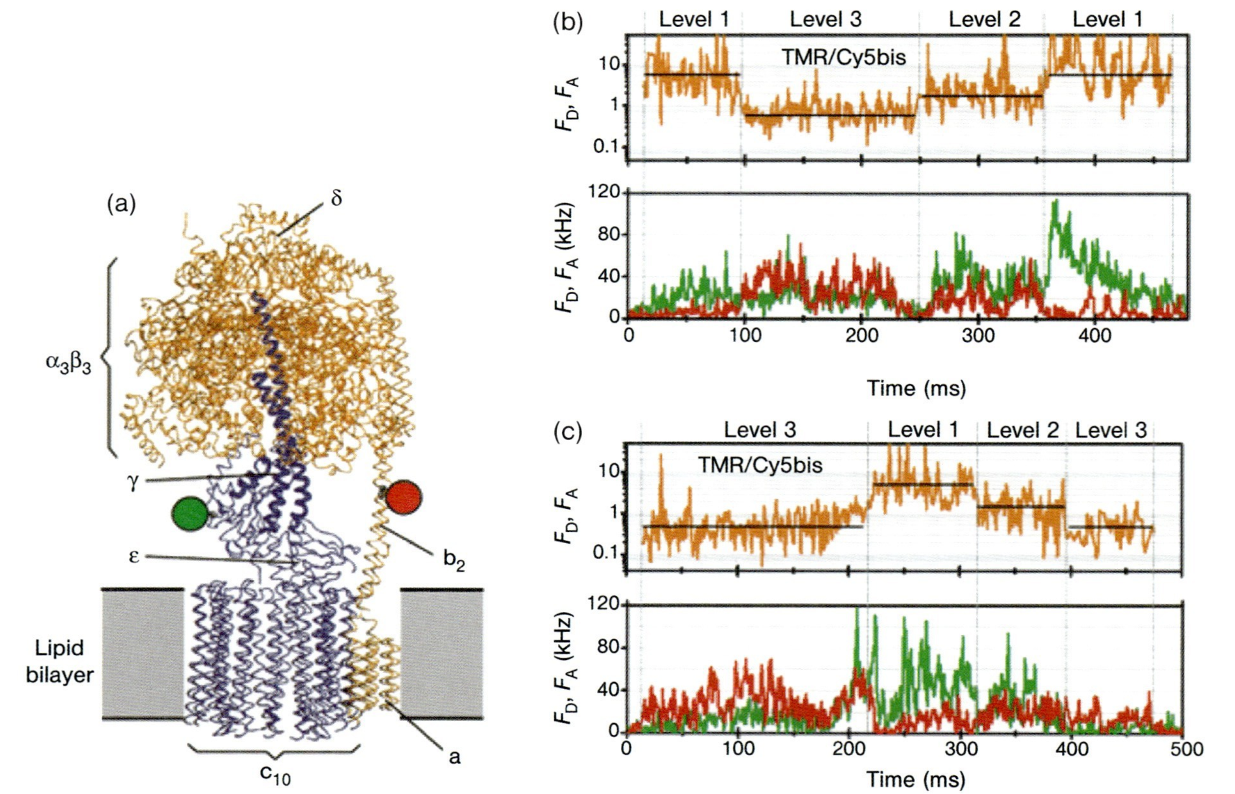

Figure 7.2 F_0F_1 ATPase structure and single-molecule experiments. (a) Structure of the F_0F_1 ATPase with the subunits labeled. The F_0 component is shown in blue, while the F_1 component is gold. The green and red spheres represent the placement of fluorophores of single-molecule experiments to observe rotation. (b) Single-molecule FRET data collected under ATP hydrolysis. The lower panel is a single burst of fluorescence emission observed in diffusion experiments. The donor emission is shown in green, and acceptor emission in red. The upper panel is the ratio of donor to acceptor emission. Three states are observed in the emission ratio. Horizontal black lines have been drawn to represent the average intensity. (c) Single-molecule FRET data collected during ATP synthesis. Note the opposite order of the three states indicating compared to (b). This indicates that rotation is occurring in the opposite direction. Reprinted with permission from Ref. [49]; © 2004, Macmillan Publishers Ltd.

domains via a long coiled-coil linker. Single-molecule experiments revealed kinesin to be processive, and to move with a step size of 8 nm, corresponding to a single ATP hydrolysis per step [55–57]. A key question regarding the mechanism of this movement was the manner in which these steps were achieved. Thus, two hypotheses were proposed [53]: (i) an *inchworm mechanism*, in which one head is always in the lead and the other always lags; and (ii) a *hand-over-hand mechanism*, similar to a bipedal movement in which the leading and lagging heads are interchanged with each step. Single-molecule fluorescence experiments, in which one head was labeled with an organic dye, showed that a single kinesin head took steps of 17.3 ± 3 nm, consistent with a step size of two 8 nm tubulin dimers, as predicted by the hand-over-hand model [58]. Further analysis of the dwell time between each step revealed it to be the convolution of two exponential processes – implying a hidden step from the unlabeled head – rather than the single exponential process (which would be expected for a single stochastic event). This elegant single-molecule experiment directly supported the hand-over-hand model for kinesin movement, as smaller step sizes were predicted by the inchworm model.

Kinesin represents an interesting system for understanding the molecular basis for processivity. There must be communication between the two head domains: if ATP hydrolysis and microtubule release were stochastic, then both heads of kinesin could simultaneously release the microtubule and processivity would be lost. Although the mechanism of this communication between heads is currently the subject of much debate [54], it is widely agreed that interactions involving strain, primarily through the neck linker, are critical for gating the motors. ATP can only be hydrolyzed in the trailing head after the leading head has bound to microtubulin. When the leading head has bound to ATP, an isomerization in the neck linker is produced, which results in the lagging head releasing the microtubulin and stimulating its forward movement and ADP release. This allows an interchange of leading and lagging heads, and the start of a new cycle. One question regarding this general mechanism is whether or not the lagging head is bound to the microtubule as the leading heads waits for ATP to bind [54]. In order to test this hypothesis, Mori *et al.* created two doubly labeled kinesin mutants which could be used to test whether kinesin has significant populations with one head unbound from the microtubule [59]. In one mutant, 215-43, either a high or low FRET state is given when kinesin is fully bound to the microtubule, depending on which head leads (Figure 7.3a). In the other mutant, 324-324, both orientations of the heads are expected to give the same FRET value (Figure 7.3a). By imaging static kinesin molecules bound to microtubules under various nucleotide states, in the absence of nucleotides (achieved by adding Apayrase), with ADP + inorganic phosphate or with AMP-PNP, FRET values consistent with a two-head-bound structure are observed (Figure 7.3b). However, at low ADP concentrations the FRET efficiencies were consistent with the trailing head unbound from the microtubule, yet still behind the leading head (Figure 7.3b) [59]. Single-molecule measurements of the same constructs moving under limiting ATP concentrations, revealed FRET efficiency histograms that were consistent with the intermediate waiting state. This suggested that a lagging head release precedes ATP binding to the leading head,

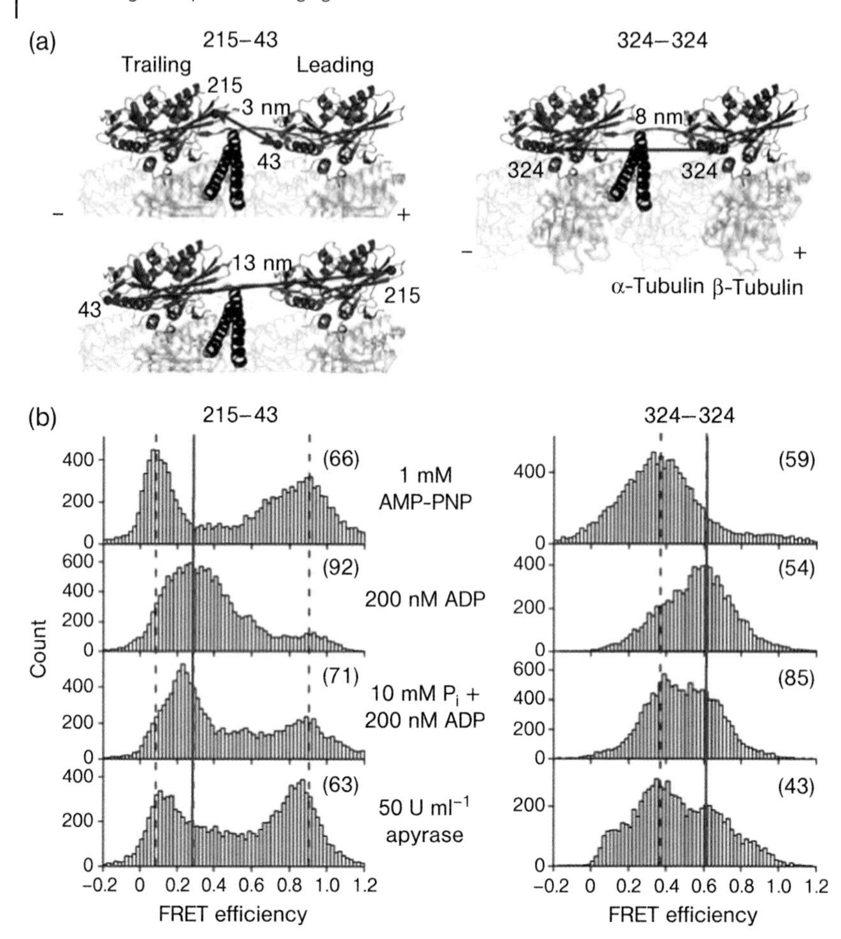

Figure 7.3 Single-molecule FRET experiments on kinesin. (a) Two FRET constructs for measuring distances in kinesin heads bound to microtubules. (b) Single-molecule probability distributions of FRET efficiency collected on labeled kinesin bound to microtubules under various nucleotide conditions. The dashed lines represent distances expected from kinesin with both heads bound to microtubules, while solid lines represent the new state which appears primarily at low ADP concentrations. Reprinted with permission from Ref. [59]; © 2007, Macmillan Publishers Ltd.

but that ATP binding is necessary to fully isomerize the neck linker on the leading head in order to propel the lagging head forward to bind the next tubulin dimer.

The intramolecular interactions within a motor protein complex are indispensable for a fundamental understanding of how chemical energies are converted to mechanical movement. The key concepts include cooperativity and allostery, both of which topics are anticipated to be an important future direction for single-molecule studies. Some recent examples of preliminary investigations into allostery in protein signaling are provided in the next section.

7.3 Allostery in Proteins

7.3.1 Shift in Conformational Population

Cooperative behavior in allosteric systems was proposed as an early functional role for conformational changes in proteins. Originally, allosteric proteins were defined as multimeric complexes consisting of multiple identical subunits where ligand binding to one subunit changed the affinity for the other subunits. On the basis of multiple hemoglobin structures, Monod *et al.* proposed that each subunit in an allosteric protein had multiple conformations; this became known as the Monod–Wyman–Changeux (MWC) or "concerted" model [60]. Here, one conformation with low-affinity ("R", relaxed) for substrates and another with high-affinity ("T", tense) could interconvert, but all subunits must have the same conformation; thus, subsequent ligand-binding events would shift the equilibrium in favor of the high-affinity state. Koshland *et al.* proposed that ligand binding to a single subunit caused a conformational change, and that this changed the affinity of neighboring subunits for the ligand; this became known as the Koshland–Nemethy–Filmer (KNF) or "sequential" model [41].

A more general model was subsequently advanced by Weber, in which each subunit of the protein was in equilibrium between conformational states such that ligand binding shifted the conformational equilibrium of the remainder of the subunits in the molecule [61]. This model formed a basic framework from which the modern understanding of functional conformational motions in proteins can be related to the energy landscape of conformational dynamics.

Recent evidence has revealed that a population shift upon ligand binding between pre-existing conformations of a protein (as suggested by Weber) may be a general feature of all proteins, and not just of multimeric allosteric proteins. For instance, NMR experiments have demonstrated a shift in equilibrium between interconverting populations upon inhibitor binding in Ribonuclease A [62], phosphorylation in NtrC [63], and galactose chemosensory receptor [64], to name a few. This has led to a generalized and simplified definition of allosteric behavior, whereby any ligand-binding event which changes the relative populations of conformational states of a protein may be considered "allosteric" [45, 65]. Nussinov and coworkers have proposed that, since the binding of a ligand to a protein necessarily changes the free energy landscape governing its conformational dynamics, all soluble proteins should be considered allosteric [45]. It can be argued that conformations which promote ligand binding or protein function must already exist as a part of the energy landscape of proteins; thus, proteins are expected to be able to sample these states, even though they were not readily observed in crystallographic studies [31, 45].

Single-molecule experiments can be used to measure the response of proteins to events such as ligand binding, since they can directly observe the conformational distributions and rates, which can then be related to the average free energy landscape on the millisecond timescale. This aspect is illustrated in the following section, and also in the context of enzyme catalysis (see Section 7.4.3).

7.3.2
Ligand Binding and Recognition: Calmodulin

Calmodulin (CaM), a small calcium-binding protein found in all eukaryotic cells, can bind to calcium and regulate the activity of a wide variety of proteins in the cell in response to changes in intracellular Ca^{2+} concentrations [66]. CaM consists of two pairs of Ca^{2+}-binding helix-turn-helix motifs (or EF-hands), separated by a flexible linker (Figure 7.4a). The binding of Ca^{2+} to the EF hands results in a conformational change which exposes hydrophobic surfaces that are responsible for binding to peptides on other proteins. Crystal structures have revealed Ca^{2+}-CaM as existing in two distinct states: an extended state, where the central linker forms a helix; and a more compact state, in which the central linker is sharply bent [68, 69] (Figure 7.4a). Extensive studies conducted in the Johnson laboratory have helped to characterize the conformational states of CaM in solution [67]. By introducing two FRET probes into the N- and C-terminal EF hand domains of CaM, the fluorescence intensity of the protein was monitored as it diffused through a laser focused in solution (Figure 7.4a). By binning each burst at 75 μs and calculating the resulting efficiency, three distinct subpopulations were identified in the distance histogram (Figure 7.4b) [70]. By comparison with the crystal structures of CaM, the longest distance was shown to arise from a conformation resembling the fully extended structure, while the shortest distance was expected to arise from conformations resembling the compact structure. Interestingly, single-molecule FRET experiments showed the existence of a third conformation which most of the CaM populated. This subpopulation was assumed to arise from a partially bent conformation of the central linker which was previously implied for Ca^{2+}-CaM [71] and observed in apo-CaM solution structures [72], although this single-molecule FRET experiment represents its first direct observation.

Although diffusion based on single-molecule experiments revealed on average only a millisecond's worth of data for each molecule being studied, it was shown that, by analyzing the longest fluorescent bursts, some of the molecules could be seen to interchange between the three observed populations [73]. This implied that the conformational states of CaM in solution were interchanging in the millisecond time regime. FRET changes in conformational fluctuations of CaM were also measured using fluorescence correlation spectroscopy (FCS). The experimental set-up for FCS is similar to that for single-molecule diffusion experiments, albeit the former is really an ensemble-averaged experimental technique. When combined with FRET measurements, FCS proved to sensitive to anticorrelated intensity fluctuations in fluorescence resulting from time-dependent energy transfer. In these experiments, a concentration of protein was used such that there were, on average, only a few molecules diffusing through the laser focal volume at any one time. Intensity autocorrelation functions for donor and acceptor emission were most sensitive to diffusion of the molecule in and out of the laser spot, but

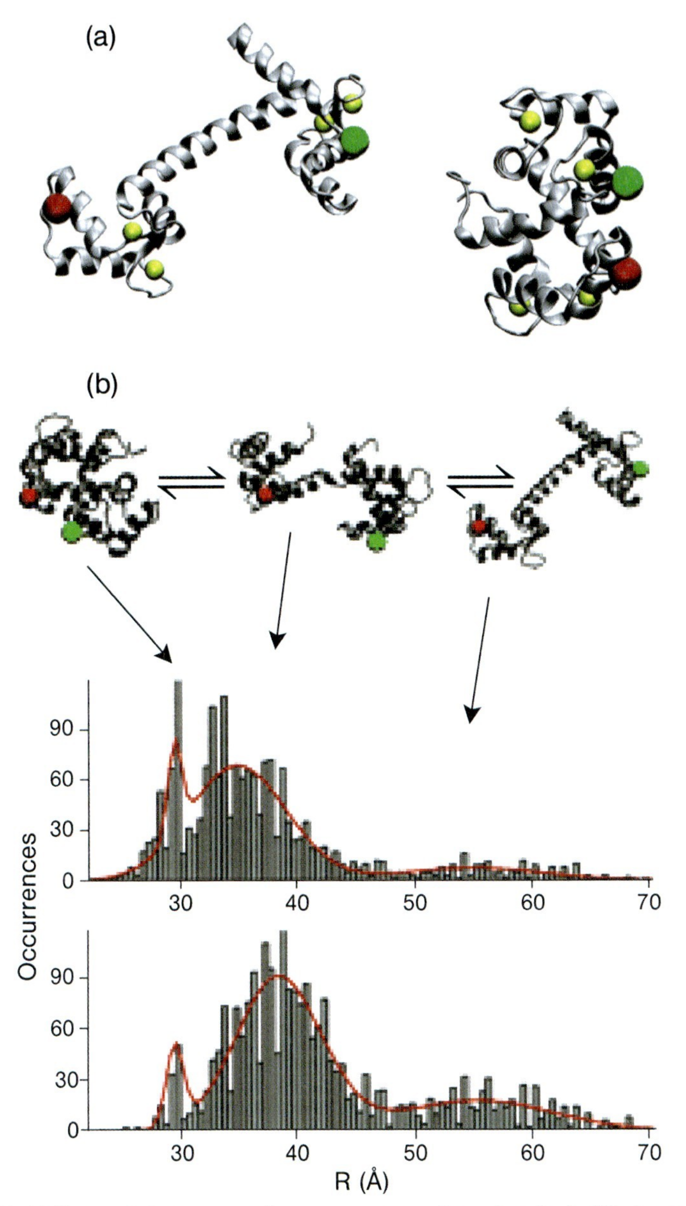

Figure 7.4 (a) Two crystal structures of calmodulin (CaM), depicting an extended and compact linker between the Ca^{2+}-binding EF hand motifs. Small spheres represent Ca^{2+} ions, while larger spheres represent fluorophore labeling positions for single-molecule experiments. (b) Distance histogram from single-molecule FRET experiments performed on freely diffusing CaM. The upper panel was performed at saturating Ca^{2+} concentrations (25 μm), and the lower panel at typical physiological concentrations (150 nm). Three substrates were identified in each experiment, and fitted to three Gaussian distributions (shown as a line). Reproduced with permission from Ref. [67].

were also able to reveal correlated intensity fluctuations occurring on faster timescales. FCS experiments performed on CaM revealed two timescales for conformational fluctuations at 100 μs and 1.5 ms, which were consistent with conformational dynamics estimated from single-molecule diffusion experiments [74]. The results of these experiments revealed that the distance distribution seen in CaM diffusion experiments was most likely derived from interconverting conformers, and not from static heterogeneity in the sample.

Taken together, these experiments demonstrated that Ca^{2+}-CaM was capable of sampling a wide range of conformations on the millisecond timescale in solution. Especially interesting was the previously uncharacterized intermediate state which was highly populated by Ca^{2+}-CaM conformers in solution. CaM is known to recognize and bind 300 different target peptides [75]; thus, it was suggested that the recognition of this diverse array of binding partners would be mediated by CaM's conformational heterogeneity, particularly from the partially closed state observed in single-molecule experiments. As each peptide requires unique structural interactions in the bound state, it would be expected that these interactions, or conformations of the protein favoring each of them, should also be a part of the energy landscape of CaM in the absence of its ligands.

Single-molecule experiments have been performed on fluorescently labeled CaM bound to physiological substrates [76–78]. Plasma-membrane Ca^{2+} ATPase (PMCA) is an ATP-dependent ion transporter that is ubiquitous to eukaryotic cells, and is responsible for exporting excess Ca^{2+} from the cell [79]. The C terminus of the pump contains an autoinhibitory peptide that binds to the ATPase active site; the binding of Ca^{2+}-CaM to this peptide relieves inhibition and activates the pump up to 10-fold. Interactions between Ca^{2+}-CaM and PMCA were observed on the single-molecule level by monitoring polarization modulation of the emission from a single dye attached to Ca^{2+}-CaM bound to PMCA [74]. Since the absorbance of a photon depends on the relative angle between the polarized excitation source and the dipole of the fluorophore, a rotationally restricted fluorophore would display a polarization-dependent emission, whereas a fluorophore freely rotating in solution would display little dependence [80]. Single-molecule polarization experiments performed on Ca^{2+}-CaM bound to PMCA revealed that with undersaturating Ca^{2+} concentrations, a relatively unrestricted motion of the probe was observed (Figure 7.5b). This was believed to arise from CaM binding the autoinhibitory peptide, which was in a disassociated state, from the ATPase active site. Under physiological Ca^{2+} concentrations, approximately 30% of the polarization histogram showed a more restricted species (Figure 7.5a) that was attributed to a configuration in which CaM had partially disassociated from the autoinhibitory peptide; this allowed it to rebind weakly to PMCA (see the model in Figure 7.5a). A subsequent experiment which extended to longer timescales by implementing a mechanical shutter to prevent photobleaching, showed that the on and off rates of the inhibitory peptide at normal Ca^{2+} concentrations were on the order of tens of seconds [76]. The existence of two modes of Ca^{2+}-CaM binding to the autoinhibitory peptide from PMCA was consistent with single-molecule FRET and polarization modulation experiments performed previously on the isolated C-terminal

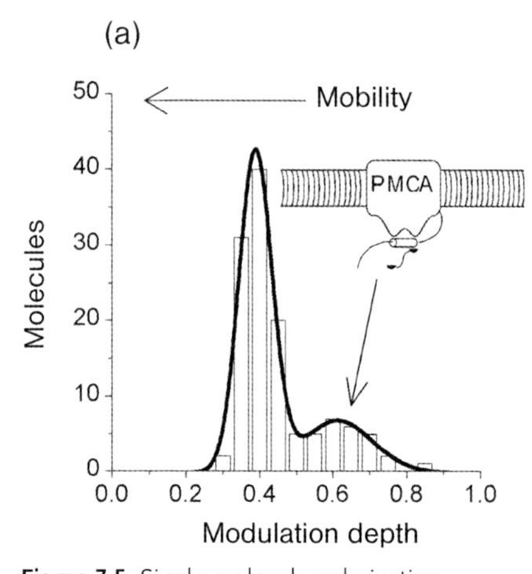

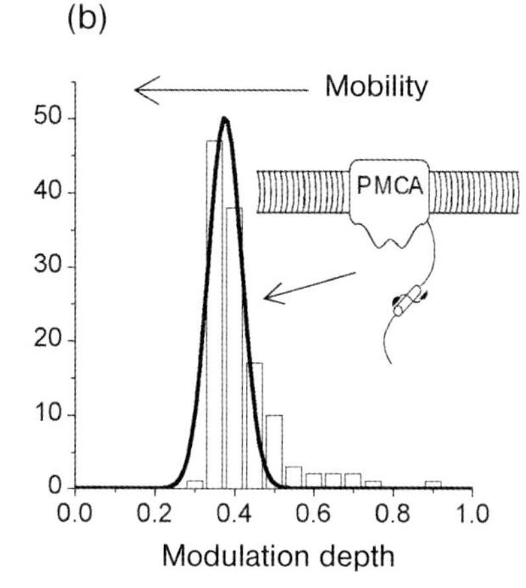

Figure 7.5 Single-molecule polarization experiments of Ca^{2+}-CaM bound to the plasma-membrane Ca^{2+}-ATPase (PMCA). Experiments were performed under (a) physiological (150 nm) Ca^{2+} and (b) saturating (25 μm) Ca^{2+}. A smaller modulation depth represents a freely rotating fluorophore, while a higher modulation depth represents a relatively restricted fluorophore. Reproduced with permission from Ref. [78]; © 2004, Biophysical Society.

PMCA peptide [81]. It was demonstrated that this peptide could bind to two configurations of Ca^{2+}-CaM: (i) a relatively open structure, where the peptide only bound to CaM's C-terminal EF-hand motif; and (ii) a compact structure, where both EF hand motifs bound the peptide [81]. This previously uncharacterized hybrid structure was proposed to act as a low-pass filter to prevent the pump from responding transient fluctuations of Ca^{2+} concentrations in the cell, but rather to the average concentration over several seconds [76]. Clearly, molecular signaling is an important and intriguing problem that warrants fruitful future research.

7.4 Enzyme Catalysis

Enzyme catalysis, which represents another important aspect of how conformational dynamics contribute to chemistry, is discussed in the following sections.

7.4.1 Slowly-Varying Catalytic Rate: Single-Molecule Enzymology

Many single-molecule experiments have used the direct observation of catalytic turnover to relate function to conformational dynamics. Although not direct observations of protein conformational dynamics, these studies have shed fascinating light on the possible role of slow conformational fluctuations in relation to enzyme function (for reviews, see Chapter 11 and Refs [24, 82]). By observing single turnover events, the behavior of individual molecules can be compared to that expected from the conventional Michaelis–Menten kinetic treatment. Any deviation from an ideal behavior would be expected to arise from a coupling of the catalysis to other events in the system, notably slow conformational transitions between substates. In these experiments, a fluorescent reporter–such as a cofactor [83–84], labeled enzyme [85] or fluorescent product [86, 87]–is used to monitor the time-dependent turnover of a single enzyme molecule and to characterize the statistics of individual events. The most important concepts to have arisen from this characterization of single-enzyme molecules are the presence of both static heterogeneity (molecule-to-molecule variation of catalytic rates) and dynamic heterogeneity (variation within an individual molecule as a function of time). These two phenomena are almost impossible to differentiate or characterize in conventional ensemble-averaged experiments. Interestingly, many experiments investigating the catalytic turnover of enzymes identified dynamic heterogeneity [83, 85–88] at the single-molecule level. Although these phenomena are almost always attributed to slow fluctuations in enzyme conformation, these experiments do not prove directly the presence of conformational changes in proteins. Therefore, a key question here is whether such slow conformational fluctuations exist in proteins.

7.4.2
Direct Measurement of Enzyme Conformational Dynamics

Single-molecule experiments have also been used to directly observe and characterize slow conformational fluctuations in proteins. Most often, this is achieved through single-molecule FRET, as it can be used to detect motions on the millisecond timescale. The study of molecules one at a time allows the direct observation of structural fluctuations of proteins, as well as measurement of the rates of interconversion between states; it also provides a direct access to the energy landscape governing conformational fluctuation on the millisecond timescale.

Single-molecule FRET was first used to study protein conformational dynamics of the enzyme staphylococcal nuclease (SNase) [21, 89]. This enzyme was specifically labeled with a donor fluorophore, tetramethylrhodamine (TMR) attached to a cysteine residue and labeled nonspecifically with the acceptor fluorophore (Cy5) on lysine residues at substoichiometric concentrations. By immobilizing individual molecules on a confocal microscope, it was possible to identify molecules with the desired 1 : 1 ratio of donor to acceptor. Fluorescence emission was collected from both dyes and transformed into a time-dependent energy-transfer-efficiency profile. In order to characterize the conformational dynamics present in SNase, an autocorrelation function was constructed and analyzed for the time traces obtained for each molecule. Interestingly, even though SNase was not expected to have significant conformational dynamics on slow timescales, fitting autocorrelation functions from 200 individual molecules to a single exponential model revealed dynamics on a range of timescales from 20 ms to 1 s [89].

In another experiment, Yang *et al.* used single-molecule electron transfer (ET) to directly observe the conformational dynamics in the FAD-containing enzyme flavin oxidoreductase (Fre) [90]. Electron transfer is a complimentary technique to FRET, in which an electron is transferred from the excited state of a fluorescent donor to an acceptor in a distance-dependent manner, resulting in a quenching of the observed fluorescence. Due to its exponential dependence on distance, ET is sensitive to distances of a few Angstroms. Here, the time-dependent quenching of a fluorescent FAD cofactor by a nearby tyrosine residue in Fre was measured on individual immobilized molecules. Photon-by-photon data collection allowed the simultaneous observation of both fluorescent lifetimes and intensities, enabling the construction of a lifetime autocorrelation function spanning timescales from 10^{-6} to 10^{2} s (Figure 7.6a) [90]. This correlation function demonstrated conformational fluctuations in the protein over a wide range of timescales, from microseconds to seconds. By calculating the mean lifetime from every 100 photons, a lifetime histogram could be created from an individual molecule (Figure 7.6b). This histogram was converted to a distance histogram (Figure 7.6c), which subsequently was transformed into a potential of mean force (PMF) (Figure 7.6d). A harmonic potential was used to model the PMF (the dashed lines in Figures 7.6b–d). From this potential, two models were created and compared to the lifetime correlation function in Figure 7.6a. The first, simple Brownian diffusion within the potential, clearly failed to capture the range of lifetime fluctuations seen in the system (Figure 7.6a, lower dashed line), whereas the second model, anomalous diffusion, was found to satisfactorily describe the observed lifetime fluctuations (Figure 7.6a). In these experiments, the highly multidimensional potential governing conformational dynamics in Fre was collapsed onto a single observable distance coordinate, $R(t)$. This effectively created a transient potential with fluctuating barrier heights as the system explored its conformational landscape, yet on average it could be described by a harmonic potential (inset in Figure 7.6d).

These studies demonstrated the existence of conformational dynamics in proteins spanning relatively long timescales due to conformational substates separated by a distribution of high barriers in the energy landscape. Although conformational motions in proteins in the micro- to millisecond regime have been demonstrated using NMR [63, 91–95], these single-molecule studies showed that motion in proteins could extend into seconds and possibly beyond, even in proteins where the functions did not obviously involve large-amplitude motions. These results implied that the dynamic heterogeneity seen in single-molecule catalysis experiments might indeed arise from slow conformational fluctuations. These conformational substates would represent an even higher degree of complexity in the protein energy landscape than had been observed previously by Frauenfelder, as the slow rates of interconversion implied high barriers between the states. Although these conformational states have been postulated to modulate enzyme activity, it is unclear whether such motions play an important mechanistic role in enzymatic turnover, or are merely a consequence of Nature using soft protein molecules for catalysis which requires a specific orientation of key residues within the protein's active site.

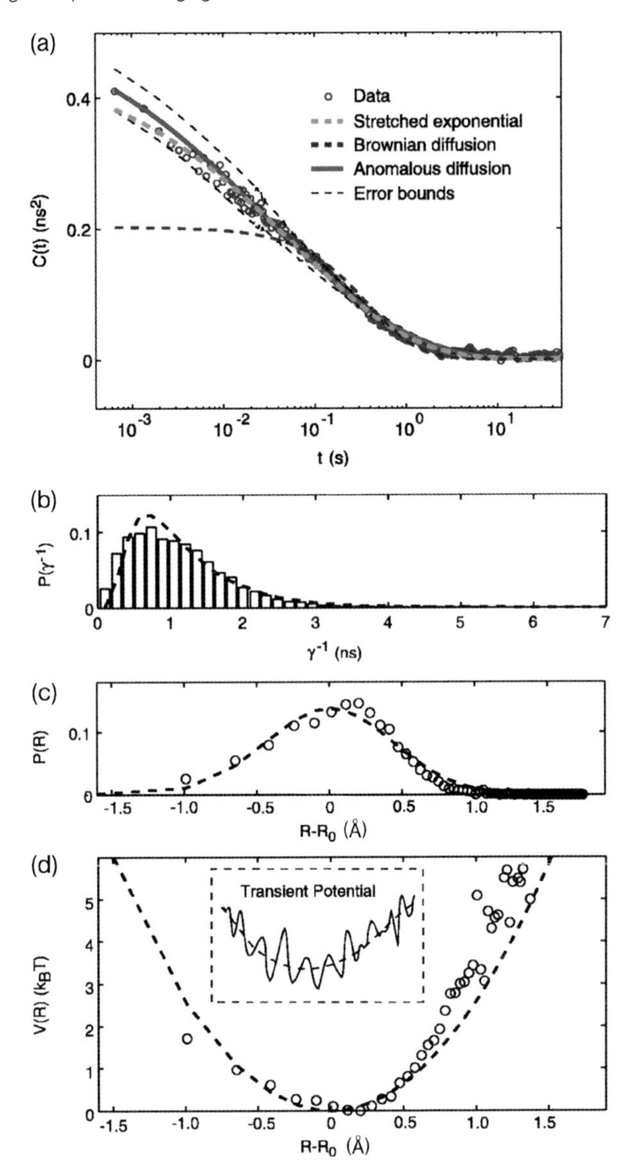

Figure 7.6 Analysis of flavin oxidoreductase (Fre) electron transfer trajectory. (a) Lifetime autocorrelation function from an electron-transfer experiment on a single Fre molecule. Lifetime fluctuations arise from distance fluctuations between the fluorescent FAD cofactor and a nearby tyrosine residue in the protein (Y35). This autocorrelation function demonstrates that structural fluctuations in Fre span a wide range of timescales. Fits with a stretched exponential model and anomalous diffusion model are shown, as well as a poor fit with a Brownian diffusion model. (b) Histogram of mean lifetime fluctuations calculated using a maximum likelihood estimator (MLE, bars). (c) Histogram from part (b) transformed into edge-to-edge distance between fluorescent FAD and the quenching tyrosine (Y35). (d) Potential of mean force calculated from histogram in (c). The dashed lines in (b), (c), and (d) are a fit of the potential of mean force to a harmonic potential. Adapted with permission of the American Association for the Advancement of Science from Ref. [90].

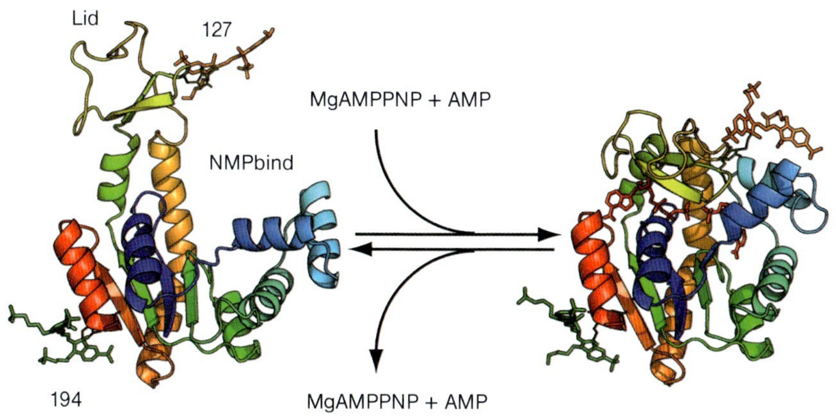

Figure 7.7 Adenylate kinase (AK) crystal structures. Left: Substrate-free structure with lids in an open state. Dyes have been modeled in at the labeling positions used in single-molecule experiments (A127C and A194C); Right: AMP-PNP and AMP-bound AK crystal structure with both lids in the closed state. Reproduced with permission from Ref. [100]; © 2007, National Academy of Sciences, USA.

7.4.3
Mechanistic Roles of Enzyme Conformational Dynamics: Adenylate Kinase

The enzyme adenylate kinase (AK) represents a model system for studying the relationship between functional conformational dynamics and enzyme activity [96–99]. This ubiquitous enzyme is responsible of catalyzing the reversible phosphate transfer between two nucleotide substrates:

$$Mg^{2+} \cdot ATP + AMP \underset{\text{reverse}}{\overset{\text{forward}}{\rightleftharpoons}} Mg^{2+} \cdot ADP + ADP$$

Escherichia coli AK consists of a central core domain which has two binding sites for its nucleotide substrates, each of which is covered by a lid, termed the ATP_{lid} and NMP_{bind} domains (Figure 7.7). From more than 20 available crystal structures, AK is observed to undergo a large-scale domain movement upon substrate binding, in which both the ATP_{lid} and NMP_{bind} domains close over their respective substrate-binding pockets. Conventionally, this motion is believed to protect the active site from water molecules during phosphate transfer [101]. The presence of a large-amplitude conformational change which is believed to be important for catalysis makes AK an ideal platform for studying the role of functional conformational dynamics to the general catalytic mechanism of this enzyme.

Hanson *et al.* have used advanced high-resolution, single-molecule methods to monitor the conformational dynamics of AK [100]. By labeling a dual-cysteine mutant on the ATP_{lid} and core domains of AK (Figure 7.7), the conformational fluctuations of AK's lid could be followed on the millisecond timescale. This was of particular interest as AK's catalytic rate predicted that turnover would occur within this regime. A sample single-molecule trajectory is shown in Figure 7.8a. In order to quantitatively follow, the millisecond movements in single molecules, advanced statistical methodologies that go beyond simple visualization are necessary. Conceptually, the new methods utilize the information carried by each and every single photon, in a photon-by-photon analysis. A photon-registered data collection with a single-photon detector, such as an avalanche photodiode (APD) rather then a CCD camera, is essential to achieve the maximum possible time resolution. This is because the frame rate of a modern CCD is ~10 ms, whereas single-photon arrival events can be detected with nanosecond accuracy. A maximum-likelihood analysis method was developed in order to achieve the optimum time resolution supported by the intrinsically noisy, low signal-to-background, single-molecule FRET data, without relying on mechanistic or kinetic modeling [102]. An analysis of the trajectory in Figure 7.8a with this method is presented in Figure 7.8b. By utilizing the Fisher information available from the photon-arrival data, the error for each distance measurement can be quantified. The gray boxes in Figure 7.8b represent the error in the time and distance measurements to 15% accuracy. Rather than employing a conventional constant time binning analysis, where the distance is calculated in equal time intervals, the maximum-information method employs constant information binning. Accordingly, the length of each single-molecule distance measurement is increased until the measurement accuracy meets a predetermined threshold (15% in Figure 7.8b). This method is superior to constant time binning, where the calculated distances often have wildly different (and often unknown) measurement errors which can lead to bias and misinterpretation of results. When applied to AK, this analysis revealed unambiguously that AK's lid spontaneously fluctuated over a wide range of distances, encompassing the distance expected from both the open and closed crystal structures, even in the absence of any substrate.

In order to characterize the energy landscape governing lid motions in AK, a probability density function (PDF) was constructed. Assuming a Gaussian kernel density estimator and summing over the distance measurements from more than 400 individual trajectories, resulted in a broad, relatively featureless distribution (the dotted line in Figure 7.8c). Rather than choosing a predetermined model – for instance, two Gaussians – and fitting this distribution, further statistical methods were developed in order to remove broadening caused by photon-counting noise in an unbiased, model-free manner (the solid line in Figure 7.8c) [103]. The broadening of the raw distribution in Figure 7.8c is assumed to arise from both experimental measurement error as well as the use of the Gaussian kernel in building the PDF. Since the maximum-information analysis has a well-characterized and constant measurement error [102], the raw PDF is assumed to be a convolution of the true underlying PDF with a Gaussian of known width. Maximum-entropy deconvolution is then applied to the raw PDF to recover the true PDF by minimizing the information entropy [103, 104]. This deconvolution method will yield the simplest solution – generally a smooth curve with the least number of states – which optimally satisfies the requirements of the raw PDF and the known measurement

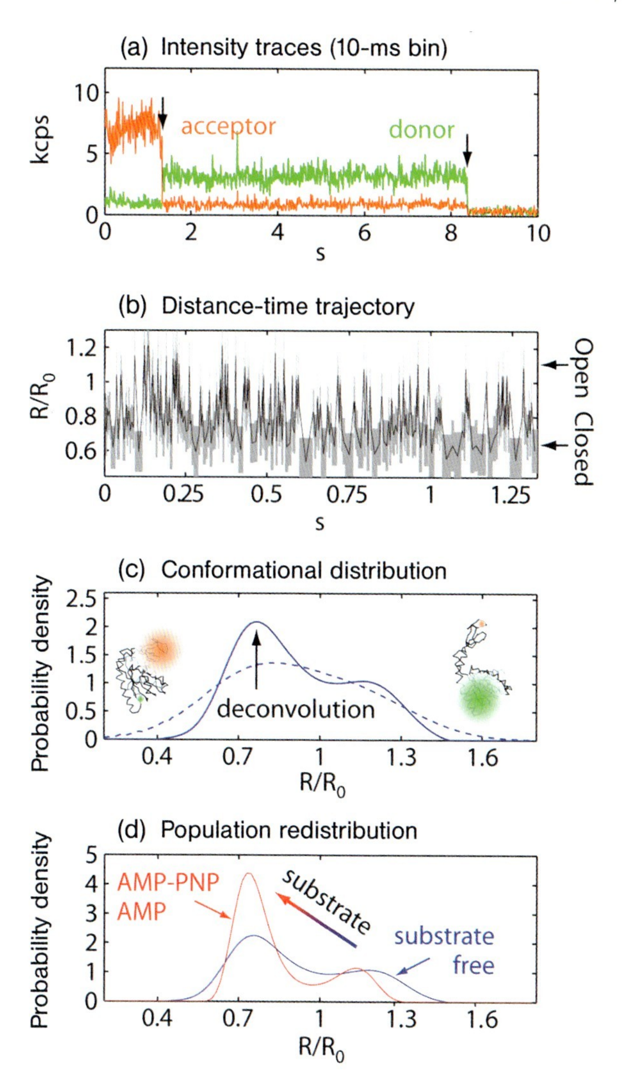

Figure 7.8 Adenylate kinase (AK) single-molecule data. (a) Intensity-versus-time trajectory for a single substrate-free AK molecule. The arrows indicate the time at which each fluorophore is irreversibly photobleached. (b) Distance-versus-time trajectory from (a), constructed using the maximum-likelihood method [102]. The gray boxes represent the uncertainty in distance and time for each measurement. (c) Probability density function constructed from >400 substrate-free AK molecules. The dotted line shows the raw probability density function, while the solid line has had the photon-counting error removed by entropy-regularized deconvolution [103]. (d) Comparison of deconvoluted probability density functions for substrate-free AK (blue) and AK in the presence of substrate analogues AMPPNP and AMP (red). All distances are normalized, R/R_0, where R_0 is the Forster radius (51 Å in the present study). Reproduced with permission from Ref. [100]; © 2007, National Academy of Sciences, USA.

error. In this way, it is possible to achieve high-resolution, single-molecule PDFs containing detailed information about the number of conformational states present in the system and their distances, without assuming a model and relying on fitting procedures.

The maximum-entropy deconvolution analysis reveals that AK's underlying PDF is distinctly bimodal (Figure 7.8c). By comparison with the crystal structures, it was concluded that the relatively closed state resulted from conformations similar to the closed, substrate-bound crystal structure, whereas the relatively open state was similar to the open crystal structure. Interestingly, even in the absence of substrate, AK can spontaneously sample both its open and closed conformations, but prefers the closed state. The finite width of each mode in the probability distribution is an indication that there is heterogeneity within each well. This broadening can be attributed to conformational substates with transitions that are too fast to be observed by the current time resolution, as might be expected due to the complex nature of conformational energy landscape in proteins.

Upon addition of the nonhydrolyzable substrate analogue AMP-PNP and AMP, the deconvoluted probability distribution still displays a bimodal distribution which favors the closed state (the red line in Figure 7.8d). The main differences between the substrate-bound and substrate-free distributions were a reweighting of populations to strongly favor the closed state, as well as a relative shortening of the distances measured in the open state. This redistribution of the populations reflected a modulation of AK's free energy landscape in response to substrate binding, due most likely to electrostatic interactions between the negatively charged nucleotide and positively charged residues in the lid.

In order to relate the conformational dynamics observed in AK's lid to its catalytic function, it is important to characterize the rate of interconversion between the two substates. By changing the time resolution at which the deconvoluted histograms are constructed over the range of 2 to 10 ms, the bimodal distribution seen at the fastest time resolutions coalesces into a single mode at the slowest time resolutions (Figure 7.9). By analogy to motional narrowing in NMR and optical spectroscopy [105, 106], this is an indication that the process being observed is occurring on a timescale similar to the experimental time resolution. By assuming a two-state model – an assumption justified by the bimodal behavior observed in the model-free, deconvoluted probability distributions – the time-dependent, single-molecule distributions can be fitted to a two-state motional narrowing model [107, 108]. This model determines not only the rates of lid opening and closing but also the mean position of each population. These results indicated that the primary means by which population redistribution occurred in AK upon substrate binding was by doubling the closing rate ($k_{\text{close, substrate-free}} = 220 \pm 40\,\text{s}^{-1}$; $k_{\text{close, AMPPNP}} = 440 \pm 110\,\text{s}^{-1}$), while the opening rate remained relatively unchanged. This implied that substrate binding stabilized the closed state relative to the open state, while the transition state for closing closely resembled the closed state.

It was found that the lid opening rate in the presence of substrates ($k_{\text{open}} = 160 \pm 40\,\text{s}^{-1}$) was almost identical to AK's bulk catalytic rate in the reverse direction ($k_{\text{cat,rev}} = 170 \pm 20\,\text{s}^{-1}$). This led to the possibility that lid opening com-

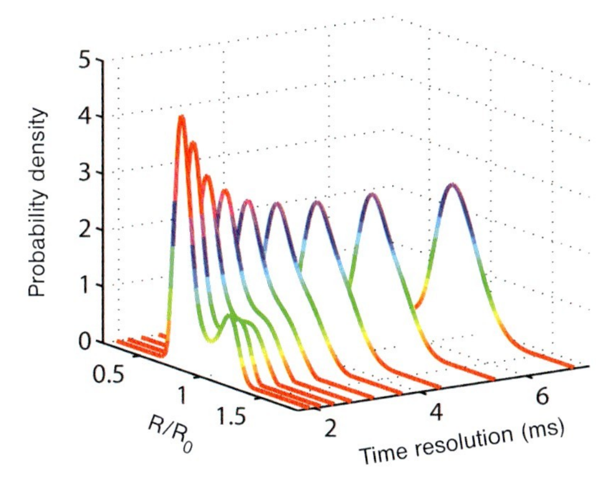

Figure 7.9 Time-dependent deconvoluted probability density functions for adenylate kinase (AK) with AMPNP and AMP, showing motional narrowing. Reproduced with permission from Ref. [100]; © 2007, National Academy of Sciences, USA.

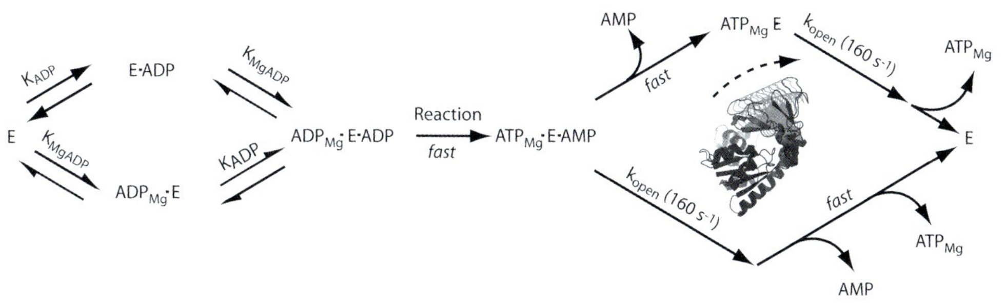

Figure 7.10 Mechanistic model of the adenylate kinase (AK) reverse reaction, integrating lid opening as the rate-limiting step of the reaction. Reproduced with permission from Ref. [100]; © 2007, National Academy of Sciences, USA.

prised the rate-limiting step in AK's reverse reaction. As the products of this reaction were ATP and AMP, the substrate-analogues AMP-PNP and AMP studied in single-molecule experiments should mimic AK's conformational dynamics in the reverse direction after catalysis and before product release. To test this hypothesis, a kinetic model was constructed in which phosphate transfer was fast once both substrates had been bound, and the rate-limiting step was the lid-opening gated product release, the rate of which was determined in single-molecule experiments (Figure 7.10). This model was fit to bulk kinetic data for AK's reverse reaction over a range of ADP and Mg^{2+} concentrations, and was found to be consistent with the measured data.

High-resolution, single-molecule experiments on AK are capable of directly detecting the conformational redistribution of the energy landscape upon substrate binding on a timescale which is relevant to the enzyme's biological function. Moreover, this process represents a new concept – "*dynamical induced fit*" – since both the open conformation (which is necessary for substrate binding and product release) and the closed conformation (required for catalysis) exist in the enzyme, even in the absence of any substrate. A reconfiguration of the energy landscape upon substrate binding may then be achieved through changes in the rates of transition between the pre-existing conformations. These experiments should also allow the construction of a quantitative model that would account for functional structural motions as the rate-limiting step of the enzyme's catalytic mechanism.

7.5 Conclusions

Single-molecule experiments can offer a unique view of protein conformational dynamics within a time regime significant for protein function. These experiments have allowed observations of crucial structural changes, and permitted the direct measurement of a time-averaged energy landscape which ultimately governs protein function. Here, some examples have been presented of how single-molecule experiments have shed light on the relationship between ubiquitous conformational fluctuations in proteins with regards to their structure and function. Yet, there are also instances in which new concepts are uncovered through single-molecule experiments, an excellent example being that of dynamical induced fit. In combination with other experimental and simulation methods, single-molecule methods continue to show great promise for advancing the present understanding of the way in which these molecules function. It appears that the future of single-molecule research is intimately linked with advances in single-molecule methodology, such as single-particle tracking [109], *in vivo* experiments, multicolor FRET [110], new optical probes, as well as a close interaction with theoretical developments such as improved data analysis procedures. The experiments discussed in this chapter clearly demonstrate the power of single-molecule experiments, whose contributions to the field of protein dynamics are guaranteed for years to come.

References

1 McCammon, J.A., and Harvey, S.C. (1987) *Dynamics of Proteins and Nucleic Acids*, Cambridge University Press.
2 Daniel, R.M., Dunn, R.V., Finney, J.L., and Smith, J.C. (2003) *Annu. Rev. Biophys. Biomol. Struct.*, **32**, 69–92.
3 Zaccai, G. (2000) *Science*, **288**, 1604–1607.
4 Bourgeois, D., Schotte, F., Brunori, M., and Vallone, B. (2007) *Photochem. Photobiol. Sci.*, **6**, 1047–1056.
5 Callender, R. and Dyer, R.B. (2006) *Chem. Rev.*, **106**, 3031–3042.
6 Carey, P.R. (2006) *Annu. Rev. Phys. Chem.*, **57**, 527–554.
7 Cho, M.H. (2008) *Chem. Rev.*, **108**, 1331–1418.

8 Boehr, D.D., Dyson, H.J., and Wright, P.E. (2006) *Chem. Rev.*, **106**, 3055–3079.
9 Kern, D., Eisenmesser, E.Z., and Wolf-Watz, M. (2005) *Methods Enzymol.*, **394**, 507–524.
10 Palmer, A.G., Kroenke, C.D., and Loria, J.P. (2001) Nuclear magnetic resonance methods for quantifying microsecond-to-millisecond motions in biological macromolecules. *Methods in Enzymology*, **339**, 204–238.
11 Korzhnev, D.M. and Kay, L.E. (2008) *Acc. Chem. Res.*, **41**, 442–451.
12 Mittermaier, A. and Kay, L.E. (2006) *Science*, **312**, 224–228.
13 Xie, X.S. and Trautman, J.K. (1998) *Annu. Rev. Phys. Chem.*, **49**, 441–480.
14 Weiss, S. (1999) *Science*, **283**, 1676–1683.
15 Moerner, W.E. and Fromm, D.P. (2003) *Rev. Sci. Instrum.*, **74**, 3597–3619.
16 Roy, R., Hohng, S., and Ha, T. (2008) *Nat. Methods*, **5**, 507–516.
17 Walter, N.G., Huang, C.Y., Manzo, A.J., and Sobhy, M.A. (2008) *Nat. Methods*, **5**, 475–489.
18 Barkai, E., Jung, Y.J., and Silbey, R. (2004) *Annu. Rev. Phys. Chem.*, **55**, 457–507.
19 Lippitz, M., Kulzer, F., and Orrit, M. (2005) *ChemPhysChem*, **6**, 770–789.
20 Barkai, E., Brown, F.L.H., Orrit, M., and Yang, H. (eds) (2008) *Theory and Evaluation of Single-Molecule Signals*, World Scientific, Singapore.
21 Ha, T.J., Ting, A.Y., Liang, J., Caldwell, W.B., Deniz, A.A., Chemla, D.S., Schultz, P.G., and Weiss, S. (1999) *Proc. Natl Acad. Sci. USA*, **96**, 893–898.
22 Jia, Y.W., Talaga, D.S., Lau, W.L., Lu, H.S.M., DeGrado, W.F., and Hochstrasser, R.M. (1999) *Chem. Phys.*, **247**, 69–83.
23 Talaga, D.S., Lau, W.L., Roder, H., Tang, J.Y., Jia, Y.W., DeGrado, W.F., and Hochstrasser, R.M. (2000) *Proc. Natl Acad. Sci. USA*, **97**, 13021–13026.
24 Michalet, X., Weiss, S., and Jager, M. (2006) *Chem. Rev.*, **106**, 1785–1813.
25 Schuler, B. and Eaton, W.A. (2008) *Curr. Opin. Struct. Biol.*, **18**, 16–26.
26 Park, H., Toprak, E., and Selvin, P.R. (2007) *Q. Rev. Biophys.*, **40**, 87–111.
27 Seidel, R. and Dekker, C. (2007) *Curr. Opin. Struct. Biol.*, **17**, 80–86.
28 Kinosita, K., Adachi, K., and Itoh, H. (2004) *Annu. Rev. Biophys. Biomol. Struct.*, **33**, 245–268.
29 Zhuang, X.W. (2005) *Annu. Rev. Biophys. Biomol. Struct.*, **34**, 399–414.
30 Ditzler, M.A., Aleman, E.A., Rueda, D., and Walter, N.G. (2007) *Biopolymers*, **87**, 302–316.
31 Kumar, S., Ma, B.Y., Tsai, C.J., Wolfson, H., and Nussinov, R. (1999) *Cell Biochem. Biophys.*, **31**, 141–164.
32 Austin, R.H., Beeson, K.W., Eisenstein, L., Frauenfelder, H., and Gunsalus, I.C. (1975) *Biochemistry*, **14**, 5355–5373.
33 Ansari, A., Berendzen, J., Braunstein, D., Cowen, B.R., Frauenfelder, H., Hong, M.K., Iben, I.E.T., Johnson, J.B., Ormos, P., Sauke, T.B., Scholl, R., Schulte, A., Steinbach, P.J., Vittitow, J., and Young, R.D. (1987) *Biophys. Chem.*, **26**, 337–355.
34 Makinen, M.W., Houtchens, R.A., and Caughey, W.S. (1979) *Proc. Natl Acad. Sci. USA*, **76**, 6042–6046.
35 Frauenfelder, H., McMahon, B.H., Austin, R.H., Chu, K., and Groves, J.T. (2001) *Proc. Natl Acad. Sci. USA*, **98**, 2370–2374.
36 Johnson, J.B., Lamb, D.C., Frauenfelder, H., Muller, J.D., McMahon, B., Nienhaus, G.U., and Young, R.D. (1996) *Biophys. J.*, **71**, 1563–1573.
37 Campbell, B.F., Chance, M.R., and Friedman, J.M. (1987) *Science*, **238**, 373–376.
38 Frauenfelder, H., Sligar, S.G., and Wolynes, P.G. (1991) *Science*, **254**, 1598–1603.
39 Tsai, C.J., Kumar, S., Ma, B.Y., and Nussinov, R. (1999) *Protein Sci.*, **8**, 1181–1190.
40 Fischer, E. (1894) *Ber. Deutsch. Chem. Ges.*, **27**, 2985–2993.
41 Koshland, D.E. (1958) *Proc. Natl Acad. Sci. USA*, **44**, 98–104.
42 Gerstein, M., Lesk, A.M., and Chothia, C. (1994) *Biochemistry*, **33**, 6739–6749.
43 Gerstein, M. and Krebs, W. (1998) *Nucleic Acids Res.*, **26**, 4280–4290.
44 Hammes-Schiffer, S. and Benkovic, S.J. (2006) *Annu. Rev. Biochem.*, **75**, 519–541.
45 Gunasekaran, K., Ma, B.Y., and Nussinov, R. (2004) *Proteins: Struct. Funct., Bioinf.*, **57**, 433–443.

46 Tsai, C.J., Ma, B.Y., and Nussinov, R. (1999) *Proc. Natl Acad. Sci. USA*, **96**, 9970–9972.
47 Miyashita, O., Wolynes, P.G., and Onuchic, J.N. (2005) *J. Phys. Chem. B*, **109**, 1959–1969.
48 Yoshida, M., Muneyuki, E., and Hisabori, T. (2001) *Nat. Rev. Mol. Cell. Biol.*, **2**, 669–677.
49 Diez, M., Zimmermann, B., Borsch, M., Konig, M., Schweinberger, E., Steigmiller, S., Reuter, R., Felekyan, S., Kudryavtsev, V., Seidel, C.A.M., and Graber, P. (2004) *Nat. Struct. Mol. Biol.*, **11**, 135–141.
50 Gresser, M.J., Myers, J.A., and Boyer, P.D. (1982) *J. Biol. Chem.*, **257**, 2030–2038.
51 Noji, H., Yasuda, R., Yoshida, M., and Kinosita, K. (1997) *Nature*, **386**, 299–302.
52 Yasuda, R., Noji, H., Yoshida, M., Kinosita, K., and Itoh, H. (2001) *Nature*, **410**, 898–904.
53 Asbury, C.L. (2005) *Curr. Opin. Cell Biol.*, **17**, 89–97.
54 Block, S.M. (2007) *Biophys. J.*, **92**, 2986–2995.
55 Svoboda, K., Schmidt, C.F., Schnapp, B.J., and Block, S.M. (1993) *Nature*, **365**, 721–727.
56 Schnitzer, M.J. and Block, S.M. (1997) *Nature*, **388**, 386–390.
57 Hua, W., Young, E.C., Fleming, M.L., and Gelles, J. (1997) *Nature*, **388**, 390–393.
58 Yildiz, A., Tomishige, M., Vale, R.D., and Selvin, P.R. (2004) *Science*, **303**, 676–678.
59 Mori, T., Vale, R.D., and Tomishige, M. (2007) *Nature*, **450**, 750–754.
60 Monod, J., Wyman, J., and Changeux, J.P. (1965) *J. Mol. Biol.*, **12**, 88–118.
61 Weber, G. (1972) *Biochemistry*, **11**, 864–878.
62 Beach, H., Cole, R., Gill, M.L., and Loria, J.P. (2005) *J. Am. Chem. Soc.*, **127**, 9167–9176.
63 Volkman, B.F., Lipson, D., Wemmer, D.E., and Kern, D. (2001) *Science*, **291**, 2429–2433.
64 Bornhorst, J.A. and Falke, J.J. (2001) *J. Gen. Physiol.*, **118**, 693–710.
65 Kern, D. and Zuiderweg, E.R.P. (2003) *Curr. Opin. Struct. Biol.*, **13**, 748–757.
66 Hoeflich, K.P. and Ikura, M. (2002) *Cell*, **108**, 739–742.
67 Johnson, C.K. (2006) *Biochemistry*, **45**, 14233–14246.
68 Babu, Y.S., Bugg, C.E., and Cook, W.J. (1988) *J. Mol. Biol.*, **204**, 191–204.
69 Fallon, J.L. and Quiocho, F.A. (2003) *Structure*, **11**, 1303–1307.
70 Slaughter, B.D., Unruh, J.R., Allen, M.W., Urbauer, R.J.B., and Johnson, C.K. (2005) *Biochemistry*, **44**, 3694–3707.
71 Barbato, G., Ikura, M., Kay, L.E., Pastor, R.W., and Bax, A. (1992) *Biochemistry*, **31**, 5269–5278.
72 Kuboniwa, H., Tjandra, N., Grzesiek, S., Ren, H., Klee, C.B., and Bax, A. (1995) *Nat. Struct. Biol.*, **2**, 768–776.
73 Slaughter, B.D., Bieber-Urbauer, R.J., and Johnson, C.K. (2005) *J. Phys. Chem. B*, **109**, 12658–12662.
74 Slaughter, B.D., Allen, M.W., Unruh, J.R., Urbauer, R.J.B., and Johnson, C.K. (2004) *J. Phys. Chem. B*, **108**, 10388–10397.
75 Yap, K.L., Kim, J., Truong, K., Sherman, M., Yuan, T., and Ikura, M. (2000) *J. Struct. Funct. Genomics*, **1**, 8–14.
76 Mandal, A., Liyanage, M.R., Zaidi, A., and Johnson, C.K. (2008) *Protein Sci.*, **17**, 555–562.
77 Priddy, T.S., Price, E.S., Johnson, C.K., and Carlson, G.M. (2007) *Protein Sci.*, **16**, 1017–1023.
78 Osborn, K.D., Zaidi, A., Urbauer, R.J.B., and Johnson, C.K. (2004) *Biophys. J.*, **87**, 1892–1899.
79 Strehler, E.E., Caride, A.J., Filoteo, A.G., Xiong, Y.N., Penniston, J.T., and Enyedi, A. (2007) *Ann. New York Acad. Sci.*, **1099**, 226–236.
80 Xie, X.S. and Dunn, R.C. (1994) *Science*, **265**, 361–364.
81 Liu, R.C., Hu, D.H., Tan, X., and Lu, H.P. (2006) *J. Am. Chem. Soc.*, **128**, 10034–10042.
82 Smiley, R.D. and Hammes, G.G. (2006) *Chem. Rev.*, **106**, 3080–3094.
83 Lu, H.P., Xun, L.Y., and Xie, X.S. (1998) *Science*, **282**, 1877–1882.
84 Shi, J., Palfey, B.A., Dertouzos, J., Jensen, K.F., Gafni, A., and Steel, D. (2004) *J. Am. Chem. Soc.*, **126**, 6914–6922.

85 Kuznetsova, S., Zauner, G., Aartsma, T.J., Engelkamp, H., Hatzakis, N., Rowan, A.E., Nolte, R.J.M., Christianen, P.C.M., and Canters, G.W. (2008) *Proc. Natl Acad. Sci. USA*, **105**, 3250–3255.

86 Edman, L., Foldes-Papp, Z., Wennmalm, S., and Rigler, R. (1999) *Chem. Phys.*, **247**, 11–22.

87 Flomenbom, O., Velonia, K., Loos, D., Masuo, S., Cotlet, M., Engelborghs, Y., Hofkens, J., Rowan, A.E., Nolte, R.J.M., Van der Auweraer, M., de Schryver, F.C., and Klafter, J. (2005) *Proc. Natl Acad. Sci. USA*, **102**, 2368–2372.

88 van Oijen, A.M., Blainey, P.C., Crampton, D.J., Richardson, C.C., Ellenberger, T., and Xie, X.S. (2003) *Science*, **301**, 1235–1238.

89 Ha, T.J., Ting, A.Y., Liang, J., Deniz, A.A., Chemla, D.S., Schultz, P.G., and Weiss, S. (1999) *Chem. Phys.*, **247**, 107–118.

90 Yang, H., Luo, G.B., Karnchanaphanurach, P., Louie, T.M., Rech, I., Cova, S., Xun, L.Y., and Xie, X.S. (2003) *Science*, **302**, 262–266.

91 Williams, J.C., and Mcdermott, A.E. (1995) *Biochemistry*, **34**, 8309–8319.

92 Nicholson, L.K., Yamazaki, T., Torchia, D.A., Grzesiek, S., Bax, A., Stahl, S.J., Kaufman, J.D., Wingfield, P.T., Lam, P.Y.S., Jadhav, P.K., Hodge, C.N., Domaille, P.J., and Chang, C.H. (1995) *Nat. Struct. Biol.*, **2**, 274–280.

93 Ishima, R., Freedberg, D.I., Wang, Y.X., Louis, J.M., and Torchia, D.A. (1999) *Structure*, **7**, 1047–1055.

94 Cole, R. and Loria, J.P. (2002) *Biochemistry*, **41**, 6072–6081.

95 Eisenmesser, E.Z., Millet, O., Labeikovsky, W., Korzhnev, D.M., Wolf-Watz, M., Bosco, D.A., Skalicky, J.J., Kay, L.E., and Kern, D. (2005) *Nature*, **438**, 117–121.

96 Muller, C.W., Schlauderer, G.J., Reinstein, J., and Schulz, G.E. (1996) *Structure*, **4**, 147–156.

97 Miyashita, O., Onuchic, J.N., and Wolynes, P.G. (2003) *Proc. Natl Acad. Sci. USA*, **100**, 12570–12575.

98 Maragakis, P. and Karplus, M. (2005) *J. Mol. Biol.*, **352**, 807–822.

99 Henzler-Wildman, K.A., Thai, V., Lei, M., Ott, M., Wolf-Watz, M., Fenn, T., Pozharski, E., Wilson, M.A., Petsko, G.A., Karplus, M., Hubner, C.G., and Kern, D. (2007) *Nature*, **450**, 838–844.

100 Hanson, J.A., Duclerstacit, K., Watkins, L.P., Bhattacharyya, S., Brokaw, J., Chu, J.W., and Yang, H. (2007) *Proc. Natl Acad. Sci. USA*, **104**, 18055–18060.

101 Jencks, W.P. (1975) *Adv. Enzymol. Relat. Areas Mol. Biol.*, **43**, 219–410.

102 Watkins, L.P. and Yang, H. (2004) *Biophys. J.*, **86**, 4015–4029.

103 Watkins, L.P., Chang, H., and Yang, H. (2006) *J. Phys. Chem. A*, **110**, 5191–5203.

104 Jaynes, E.T. (1982) *Proc. IEEE*, **70**, 939–952.

105 Anderson, P.W. (1954) *J. Phys. Soc. Jpn*, **9**, 316–339.

106 Kubo, R. (1954) *J. Phys. Soc. Jpn*, **9**, 935–944.

107 Geva, E. and Skinner, J.L. (1998) *Chem. Phys. Lett.*, **288**, 225–229.

108 Gopich, I.V. and Szabo, A. (2003) *J. Phys. Chem. B*, **107**, 5058–5063.

109 Cang, H., Xu, C.S., and Yang, H. (2008) *Chem. Phys. Lett.*, **457**, 285–291.

110 Hohng, S., Joo, C., and Ha, T. (2004) *Biophys. J.*, **87**, 1328–1337.

8
Quantitative Distance and Position Measurement Using Single-Molecule FRET

Jens Michaelis

8.1
Introduction

Since its first quantitative description [1], fluorescence resonance energy transfer (FRET) between two dye molecules has received much attention [2–6]. At a very early stage, the distance-dependence of FRET, and particularly its sensitivity to distances between 10–80 Å, suggested that it could be used to precisely measure distances in the nanometer range. In fact, on this basis FRET quickly became known as the "molecular ruler" [7]. However, due to the dependences of the critical parameter of FRET, the so called Förster distance R_0, knowing the precise length scale of the ruler often provides an experimental challenge. Notably, variations in not only the fluorescence quantum yield of the donor molecule but also the orientational freedom of the donor and acceptor, leads to variations in R_0. The restrictions are even more severe when FRET is used to probe the dynamics of molecules. In these measurements variations in quantum yield and/or the orientational freedom of the dye molecules, may cause the length scale of the ruler to be altered during the course of the experiment [8]. If such dynamic changes occur it is important that the experimental measurement is capable to resolve these changes. The consequence of this was the development of a single-molecule method, in which FRET between a single donor and a single acceptor molecule is recorded, thus opening the door to dynamics measurements [9]. This technique, which is referred to as either single-molecule FRET (smFRET) or single-pair FRET (spFRET), has been used in a wide variety of applications [10–14]. Since the experiments require the use of highly sensitive fluorescence microscopes, practical guidelines also exist that will facilitate the use of smFRET by a much larger group of scientists in future [15–18].

Single-molecule FRET experiments are typically conducted either in a confocal microscope, where molecules diffuse through the excitation volume, or on molecules that are immobilized on surfaces where confocal microscopy, wide-field microscopy or total internal reflection (TIRF) microscopy can be employed. Typically, hundreds of single molecules are investigated, with the resultant FRET

Single Particle Tracking and Single Molecule Energy Transfer
Edited by Christoph Bräuchle, Don C. Lamb, and Jens Michaelis

ISBN: 978-3-527-32296-1

efficiencies being analyzed in histograms. Due to the low light levels that arise being due to the signal deriving only from the electronic transitions of a single-molecule, single-photon counting techniques are used for data collection. Depending on the signal level in any particular experiment, the "*shot-noise*"–which is inherent to the detection counting process–causes a broadening of the observed histograms. In order to differentiate broadening due to shot-noise from that due to heterogeneities (or dynamics) in the sample, new methods for data analysis have been developed [19, 20] (for details, see Chapter 5). At this point, the discussion will be focused on the FRET efficiencies extracted from the peak of the observed histograms, with any inhomogeneities that might have occurred being neglected (albeit some noted exceptions).

Within this chapter, attention will be focused on the different approaches aimed at extracting quantitative information regarding distances and positions, using data obtained from smFRET measurements. Following a brief review of the fundamentals of FRET, some classic experiments are described in which FRET was used as a molecular ruler. This leads on to details of the two main error sources in quantitative distance and/or position determination using smFRET, namely quantum yield variations and orientation of the fluorophores. Experiments aimed at quantitative distance measurements and quantitative position determination are then discussed. The aim of the chapter is not to provide a complete overview of the experiments performed, nor to compare the existing techniques in minute detail, but rather to raise awareness of the possibilities and problems associated with quantitative distance and position measurements when using smFRET.

8.2
Fundamentals of FRET

Fluorescence resonance energy transfer usually involves a resonance between two fluorophores, the fluorescence *donor*, the emission dipole moment of which overlaps spectrally with the absorption dipole moment of the *acceptor* [1]. As a result, energy is transferred from the donor to the acceptor, with the amount of energy transfer following a sharp R^6 distance dependence, as it originates from the dipole–dipole coupling of the two chromophores:

$$E_{\text{FRET}} = \frac{R_0^6}{R_0^6 + R^6} \tag{8.1}$$

The characteristic distance at which $E_{\text{FRET}} = 0.5$ is called the Förster distance, R_0 (Figure 8.1a). The value of R_0 can be derived both classically and quantum mechanically [1, 3], and is given by:

$$R_0 = \left(8.79\times10^{23}\kappa^2\Phi_D n^{-4} J(\lambda)\right)^{1/6} \text{ (in units of Å)} \tag{8.2}$$

where Φ_D is the donor fluorescence quantum yield, n is the index of refraction of the medium, and $J(\lambda)$ is the spectral overlap of the donor emission and acceptor

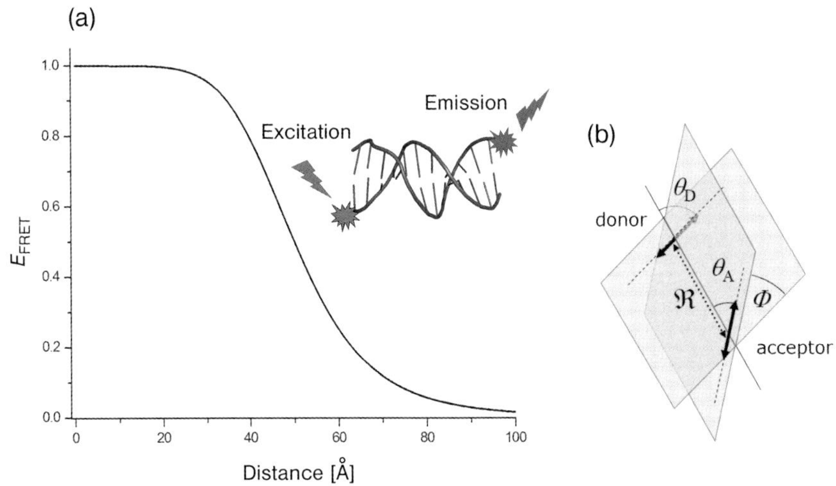

Figure 8.1 The basic principle of fluorescence resonance energy transfer (FRET). (a) The energy transfer between a donor and an acceptor dye molecule depends critically on the distance. The $1/R^6$ distance behavior of E_{FRET} provides an extremely sensitive tool for distance measurements around the 50% level. The displayed curve shows the distance dependence for a donor–acceptor pair with $R_0 = 50$ Å, a value typically observed in smFRET. Therefore, the distance range accessible to FRET is well-suited to measurements on proteins and small DNA molecules. (b) The orientations of dye molecules. The relative orientation of the donor and acceptor molecule is important for determining κ^2, and thus to calibrate the molecular ruler by determining R_0.

excitation (in units of M^{-1} cm^3). κ^2 is a parameter that depends on the relative orientation of the donor and acceptor transition moments, and is given by [21]:

$$\kappa^2 = (\cos\theta_T - 3\cos\theta_D \cos\theta_A)^2 \tag{8.3}$$

where θ_T is the angle between the donor and acceptor transition dipole moments $\boldsymbol{D}$ and $\boldsymbol{A}$, which can be expressed as: $\cos\theta_T = \sin\theta_D \sin\theta_A \cos\phi + \cos\theta_D \cos\theta_A$. Here, θ_D and θ_A are the angles between these dipoles and the vector $\Re$ which connects their position, and ϕ is the angle between the planes $(\boldsymbol{D},\Re)$ and $(\boldsymbol{A},\Re)$ (Figure 8.1b). Depending on the relative orientation of the donor and acceptor, κ^2 can take values from 0 to 4. Therefore, if the relative orientation of donor and acceptor molecules in a FRET experiment is unknown, a large uncertainty arises. This situation is often referred to as the κ^2 or "orientational problem" [22]. If, however, both the donor and acceptor are completely free to rotate on the timescale of the fluorescence lifetime, averaging about all possible relative orientations results in $\kappa^2 = 2/3$. In order to reach this regime of free rotation, the dye molecules are typically attached via long flexible linkers of between three and 12 C atoms. However, due to interactions of the dye molecules with the protein or nucleic acid to which they are attached (e.g., stacking of the chromophore onto bases in case of nucleic acids, or binding to hydrophobic patches on protein surfaces), the rotations are frequently hindered. Furthermore, the local environment often provides a confined geometry such that the rotational movement is typically hindered or spatially restricted. Therefore, a fairly large uncertainty in R_0 arises from the imprecise knowledge of the relative orientations of the dye molecules. One means of accounting for such an uncertainty will be discussed in Section 8.5.

8.3
FRET as a Spectroscopic Ruler: Initial Experiments and Limitations

In their classic experiments, Stryer and Haugland [5, 7] used proline polymers of various lengths with terminally attached dyes to experimentally verify the R^6 dependence of FRET. Polyproline forms relatively stiff helices, and thus can be thought of as a "molecular yard-stick." When Stryer and Haugland measured the FRET efficiency between a naphthyl and a dansyl dye as a function of polymer length, and plotted $log(E^{-1} - 1)$ against log (R), the data were well described by a linear fit, which had a slope of 5.9 ± 0.3, in excellent agreement with Förster theory. The determined R_0 in these experiments was ~35 Å. Similar experiments were also performed on double-stranded DNA (ds-DNA) molecules by Lilley and coworkers [23]. Due to the helical arrangement of the bases in a DNA molecule, a correct analysis of the length-dependent FRET data must include distance changes due to the relative orientations of the dye molecules. This leads to an oscillatory behavior in FRET efficiency upon changing the length of the DNA molecule. With these early experiments having provided excellent proof of Förster theory, FRET became widely used as a molecular ruler during the following years.

In more recent studies, Eaton and coworkers repeated the experiments on polyproline using Alexa488 as a donor molecule and Alexa594 as an acceptor, which resulted in a fairly long R_0 of 54 Å [24]. In these studies, deviations were found from the ideal theory, both for long and short distances. Whilst the long-distance behavior was shown to result from chain bending (probably caused by *cis–trans* isomerizations within the poly-proline chain [25]), the short-distance behavior was of interest from a general perspective. Two possible explanations were proposed: (i) that the shortened donor lifetimes would cause orientational effects to become more important; and (ii) that the distance could already be so short, that the point approximation in the Förster theory would break down. In order to avoid such complications, FRET is typically used to measure distances in excess of 20 Å. In fact, a general "rule of thumb" is that, for accurate measurements, the distance should fall within the range of $0.5 \leq R/R_0 \leq 1.75$ [26].

One important application of smFRET as a spectroscopic ruler is in the investigation of protein folding [27]. By using a rapid exchange of GdCl, small proteins can deliberately be unfolded or refolded in a spontaneous fashion [28]. By using smFRET, it is possible to study the distances changes between dye molecules attached at different positions along the polypeptide chain, and then to use these changes to unravel folding pathways. Since, at least in the unfolded state, the chain is very flexible, the free dye rotation assumption is probably well justified. However, during folding the local environment of the chromophores will change continuously, and therefore variations in Φ_D (and probably also in κ^2) are to be expected.

While it is important to realize that experimental uncertainties will ultimately limit the accuracy with which FRET can be used as a spectroscopic ruler, it is important to raise the awareness about what uncertainties have the dominant contributions. By comparing the data of some 240 donor–acceptor pairs attached to bacterial RNA polymerase complexes [29, 30], Ebright and coworkers were able to estimate errors in typical ensemble FRET experiments of macromolecular complexes [26]. The observed relative errors $\Delta E_i/E_i$ from multiple independent measurements were found to vary between 0% and 50% (Figure 8.2a). However, if a test is conducted to determine how much distance uncertainty is caused by these experimental errors, the relative distance uncertainty is found to range between 0% and 10% (Figure 8.2). In contrast, if a 25%, 50%, or 100% error is added to this error estimate to account for uncertainties in R_0, this will result in a distance error of ~5%, 10%, and 15%, respectively. Such large uncertainties are frequently encountered, mainly due to the uncertainty arising from the κ^2 problem. For a given donor–acceptor pair, the particular uncertainty can be calculated directly (as will be described in Section 8.5). This computed uncertainty is much larger than the uncertainty of the index of refraction, or of the calculation of the overlap integral [2]. Another important source of uncertainty is the *quantum yield* of the donor, although a careful experimental design will overcome this uncertainty. This will be discussed for the case of immobilized molecules in the following section, and for freely diffusing molecules in Section 8.6.

8.4 Measuring the Quantum Yield

As mentioned above, in order to perform accurate FRET measurements, the donor quantum yield must be known precisely, since it is needed to calculate R_0. The quantum yield of the donor and acceptor molecules are typically measured on the ensemble level [31]. To this end, samples labeled with only the donor as well as the acceptor are used at high concentration. After determining the absorbance of the sample, the fluorescence is compared to a known fluorescence standard (e.g., rhodamine). This method, however, has several limitations:

- The preparation of samples with a sufficiently high concentration such that the absorbance can be determined accurately may be difficult, as one dye molecule is attached to a large macromolecule.
- In some instances, not all of the dye molecules are necessarily fluorescent, even though they still might be absorbing. Such sample heterogeneity causes an underestimation of the fluorescence quantum yield.
- The fluorescence quantum yield of the dye molecules can fluctuate with time. Often, a small variation in the local environment of the dye molecule leads to a strong change in its fluorescence quantum yield.

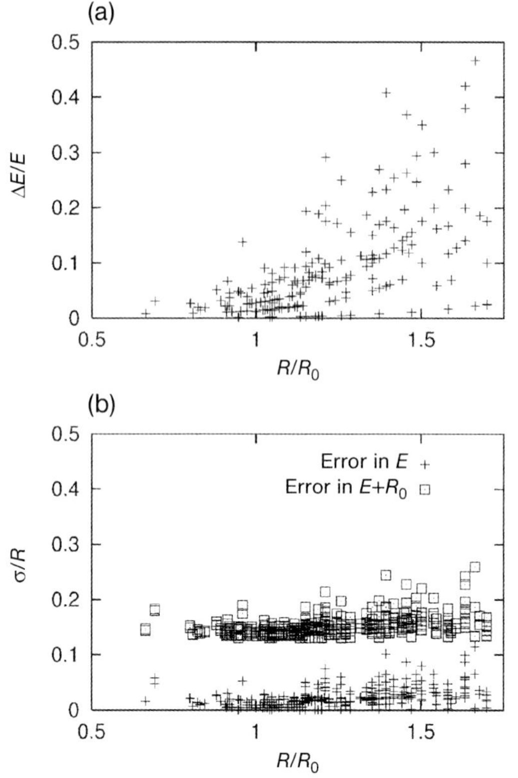

Figure 8.2 Uncertainties affecting the distance computation from FRET measurements. (a) Errors from 242 different donor–acceptor pairs in RNA polymerase (RNAP) complexes. The observed uncertainties vary between 0% and 50%. However, for measurements around the Förster radius, the observed errors are much smaller, and all lie within 0–10%. (b) The simulated relative uncertainty in R for data in panel (a) (shown as +). The computed uncertainty is independent of the absolute FRET efficiency, which means that the higher observed relative uncertainty in FRET efficiency does not carry over in a higher uncertainty in the relative distance determination. The primary source of error in FRET-based distance measurements arises from errors in determining R_0. A 10% error in R_0, which is a realistic estimate for typical measurements, results in a much larger uncertainty in distance determination. © Biophysical Journal [26].

Another possible method of determining the quantum yield with enhanced sensitivity is to measure the lifetime of a dye molecule attached to the protein, and to compare this to the lifetime of the free dye molecule. The quantum yield of the free dye molecule can be accurately determined using the standard method; hence, only a change in quantum yield from free to bound dye needs to be determined. The fluorescence lifetime can be determined using time-correlated, single-photon counting. While lifetime measurements can be carried out with very high sensitivity (down to the single-molecule level), the experimentally observed decay often does not follow a single exponential. Thus, by changing from a free to a bound dye, both the amplitudes and the decay times of the different components will be altered, adding a clear challenge to the analysis.

As the quantum yield varies from molecule to molecule, it is best to correct for this experimentally on the single-molecule level. For this, the data may be collected until the acceptor photobleaches. At the same time, the intensity of the donor will increase, as it is no longer quenched by the presence of the acceptor [32, 33]. To calculate the FRET efficiency of the individual FRET pairs, the following formula can be used [34]:

$$E = \frac{I_A - \beta \cdot I_D}{I_A + \gamma \cdot I_D} \tag{8.4}$$

where $\gamma = \frac{I_A - I'_A}{I'_D - I_D}$ and $\beta = \frac{I'_A}{I'_D}$. Here, I_A and I_D are background-corrected intensities from the acceptor and donor channels, and $I_{A,D}$ and $I'_{A,D}$ are the intensities before and after acceptor photobleaching, respectively. β and γ are experimental correction factors; β accounts for the leakage of the donor emission into the acceptor detection channel, while γ is a factor that includes the quantum yields of the fluorophores and the detection efficiencies of the two channels, such that $\gamma + \beta = \phi\eta$.[1] Here, ϕ is the ratio of acceptor to donor quantum yield, and η is the ratio of detection efficiencies in the acceptor versus donor channel. It should be noted that Equation 8.4 neglects the effects of direct excitation of the acceptor with the donor excitation wavelength, a condition which is typically met if donor–acceptor pairs are spectrally well separated. In single-molecule experiments, the normal approach is to record a rather large distribution of γ-values [34]. Since, typically, the spectral fluctuations are small compared to the bandwidth of the emission filters used in the experiments (and therefore the detection efficiencies remain constant), variations in γ can be caused either by variations in donor or acceptor quantum yield. The latter can be investigated directly using a scheme where the acceptor is directly excited alternately with FRET excitation for every time point; this technique is referred to as alternating laser excitation (ALEX; see Chapter 6) [35] or pulsed interleaved excitation (PIE) [36].

Since there is a distribution of γ-values, it is important to determine the correction factors for all FRET pairs individually by time averaging the intensities I and I'. FRET pairs where no acceptor bleaching was observed are discarded from the analysis, since for these FRET pairs γ cannot be determined. The variations in quantum yield within an ensemble of molecules can be quite large; the effect depends not only on the biological sample but also on the choice of dye molecules [34]. The described methodology allows for the correction of donor quantum yield from molecule to molecule, as well as during the experiment.

1) The presented treatment uses the actual measured intensity in donor and acceptor channel for defining the γ factor. Occasionally, an alternative approach is used, where the leakthrough and direct excitation are subtracted first from the measured signals, in which case γ is defined by: $\gamma = \phi\eta$.

Such a correction is required for accurate FRET measurements; however, the arising distance uncertainties are typically smaller than those arising from orientational variations, which will now be discussed.

8.5
The Orientation of Donor and Acceptor Molecules

As mentioned above, one of the most critical parameters in quantitative FRET measurements is the relative orientation of the donor and acceptor molecules [21, 22]. In most applications, the anisotropy of the dye molecules is measured with typical values of r between 0.05 and 0.3. Most authors then go on to state that there is enough rotational freedom to assume free rotation, and thus $\kappa^2 = 2/3$ and R_0 can be determined. However, as will now be shown, this assumption is often an oversimplification that is simply not valid when determining distances.

First, it is important to determine the time-resolved anisotropy, as both fast and slow rotational components are typically present [37]. Even if the molecule samples all possible orientations, it is important to know the timescale of the rotations. Rotations which are faster than the fluorescence lifetime lead to an averaged value of $\kappa^2 = 2/3$ (dynamic averaging), while slower rotations lead to a value which depends on the FRET value (static averaging) [21, 38]. Typically, the statically averaged value leads to a value somewhat lower than $\kappa^2 = 2/3$, and thus to shorter Förster radii [39].

Second, if there is neither complete static nor complete dynamic averaging, there will be a distribution of possible Förster radii, which can be computed. If beyond the dynamic averaging, there is no additional reorientation as is often the case, then the Förster radius will depend upon the relative orientation of the rotational cones of the donor and acceptor molecules. The distribution of possible Förster radii can then be determined by using Monte Carlo simulations, assuming that there is no *a priori* information on the angle [40]. Figure 8.3 shows the results for different values of donor and acceptor anisotropies. Even for dye molecules which are relatively free to rotate, for example, with a determined anisotropy of $r = 0.05$, the uncertainty in R_0 is quite large with a full-width half-maximum (FWHM) of about 10%. For less-mobile chromophores, the distribution broadens even further and becomes increasingly asymmetric, with the majority of orientations yielding a shorter than expected R_0, but some orientations yielding a much higher R_0. The calculated distribution of possible Förster radii also remains quite broad if only one molecule is rotationally restricted (Figure 8.3d). Thus, instead of assuming that $\kappa^2 = 2/3$, the fluorescence anisotropies of the donor and acceptor should be determined, and the histogram of possible Förster radii computed; this may then be used to estimate the experimental uncertainties. If an experiment requires a somewhat smaller uncertainty, then a different single-molecule fluorescence polarization technique may be used, such as polarization modulation [41, 42], annular illumination [43], or defocused imaging [44], to determine the orientations of the dye molecules during the experiment.

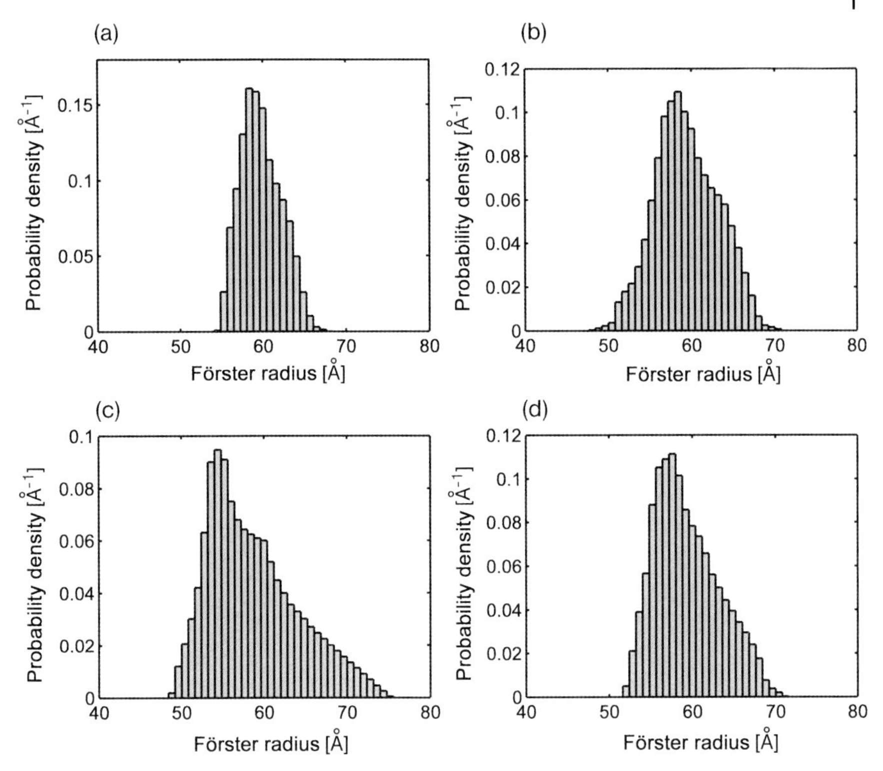

Figure 8.3 Distribution of possible Förster radii in the dynamic averaging regime. For this computation, it was assumed that there was no orientational freedom of the dye molecules beyond the rotations on the time scale of the fluorescence lifetime. Then, if there is no *a priori* knowledge of the relative orientation, all orientations are treated with equal probability. The resulting distribution of Förster radii can then be calculated using Monte Carlo simulations. Shown are the computed distribution of Förster radii for $R_0^{iso} = 60$ Å (this is the isotropic Förster radius; i.e., calculated using an estimate for the index of refraction, the measured donor quantum yield, the measured spectral overlap and assuming $\kappa^2 = 2/3$) and (a) $r_D = r_A = 0.05$; (b) $r_D = r_A = 0.1$; (c) $r_D = r_A = 0.2$; (d) $r_D = 0.05$, $r_A = 0.2$. The distributions are quite broad, even for rather low anisotropies, and cannot be described by a single Gaussian.

Recently, the effect of dye molecule orientation on the result of FRET distance measurements was investigated systematically by Lilley, Ha and coworkers for dye molecules attached terminally to the end of short DNA molecules [45]. In these experiments, Cy3 and Cy5 molecules were attached via C3 linkers to the 5′-end of complementary single-stranded DNA (ss-DNA) oligomers of various lengths. By hybridization, ds-DNA oligomers were obtained with dye molecules attached terminally to both ends. When using nuclear magnetic resonance (NMR), Lilley and coworkers had shown previously that terminally attached Cy5 and Cy3 molecules could be found predominantly in a stacked conformation [46, 47]. By measuring the FRET efficiency as a function of length of the DNA, a modulation of FRET

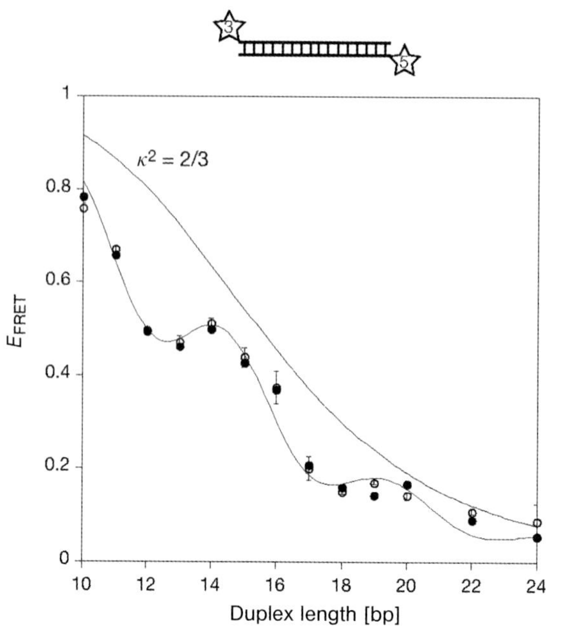

Figure 8.4 Experimental observations of orientational dependence of FRET efficiency. FRET efficiencies between donor and acceptor molecules attached terminally to both ends of ds-DNA oligonucleotides of varying length were measured (open circles = steady-state ensemble measurements; closed circles = single-molecule measurements). The FRET efficiency decreases with increasing distance. However, a modulation with periodicity of half the helicity of the DNA was observed, a clear indication of an orientational effect. The data can be fitted by a model in which the dye molecules are stacked and make transient excursions to a freely rotating conformation (solid line). In contrast, neither the description for completely free molecules (solid line) nor for perfectly stacked molecules (oscillations down to $E_{FRET} = 0$, model not shown) describe the data. © Proceedings of the National Academy of Sciences [45].

efficiency with a periodicity of approximately five bases was observed (Figure 8.4). This modulation period corresponded to half the helicity of the DNA, and was therefore a clear manifestation of an orientational modulation. These results were in contrast to those of earlier experiments conducted by Lilley and coworkers, who used dye molecules attached via more flexible linkers, where only a modulation with a periodicity of the DNA helix (i.e., 10 bases) was reported. A modulation with the periodicity equal to that of the DNA can be explained by pure geometric effects; in other words, it is the vector distance between the dye molecules is important, and not just the mere number of bases between them [23]. In the experiments displayed in Figure 8.4, the modulation of FRET efficiency with number of base pairs was twice as fast, and thus the Förster radius in this measurement was not constant but rather a function of the number of bases separating the donor and acceptor. In order to observe such orientational effects, in these experiments the dye molecules were attached by unusually short C3 linkers, and

NMR structures showed the stacking of the dye molecules onto the last base. Nonetheless, still only a slight modulation of the FRET efficiency was observed. In contrast, if the stacking were perfect, and thus the orientation of the dye molecules completely fixed, for some distances the energy transfer should vanish completely. The deviations from this ideal behavior can be explained by a model in which transient excursions of the molecules from the stacked conformation occur.

In summary, orientational effects can cause drastic deviations if compared to the $\kappa^2 = 2/3$ situation. Orientational effects will most likely play a role: (i) in confined geometries, where the dye has very little freedom to move and reorient; (ii) when strong adsorption sites are present; or (iii) when the rotational movement is highly constrained by multiple covalent bonds [48]. In fact, it had been demonstrated earlier that when a stilbene dicarboxoamide was used as part of a DNA hairpin capped with perylenedicarboxoamide, κ^2-values ranging between 0 and 1 were observed for different hairpin lengths [49].

8.6
Accurate FRET Measurements Using Fluorescence Correlation Spectroscopy

In solution experiments, when single molecules are diffusing freely through the excitation volume of a confocal microscope, there is – in addition to the ratiometric method of determining the FRET efficiency (which was described until now) – a second method which uses the measurement of fluorescence lifetimes. The method, which was developed by Seidel and coworkers, and referred to as multiparameter fluorescence detection (MFD) [50], is based on the simultaneous measurement of fluorescence intensity, lifetime, and anisotropy of the donor–acceptor pairs. From the fluorescence intensity of the donor and acceptor, it is possible to calculate the distance between the donor and acceptor, and to plot these data in a two-dimensional (2-D) histogram over the simultaneously recorded donor fluorescence lifetime. One of the first applications of MFD to biology was the investigation of conformations and conformational changes of the HIV-1 reverse transcriptase (RT) [51]. The computed 2-D histograms for the labeled RT–DNA complexes are shown in Figure 8.5. For the cases of the donor-labeled RT and acceptor-labeled DNA primer/DNA template, three different populations – termed Ia, Ib, and II – were observed. In order to obtain a quantitative understanding of the nature of these states, it is necessary to determine whether these populations differ only by distance, or also by FRET-independent variations of the donor quantum yield (which can be determined by measuring donor-only samples; see Figure 8.5b), or by variations in the orientational behavior of donor and acceptor. Interestingly, Seidel *et al.* showed that all three populations fell on a theoretical line calculated by assuming that the FRET-independent quenching of donor fluorescence remained unchanged. Therefore, the observed variations were caused by changes in the effective donor quantum yield due to quenching by the acceptor. In addition, information about orientational effects can be obtained by examining

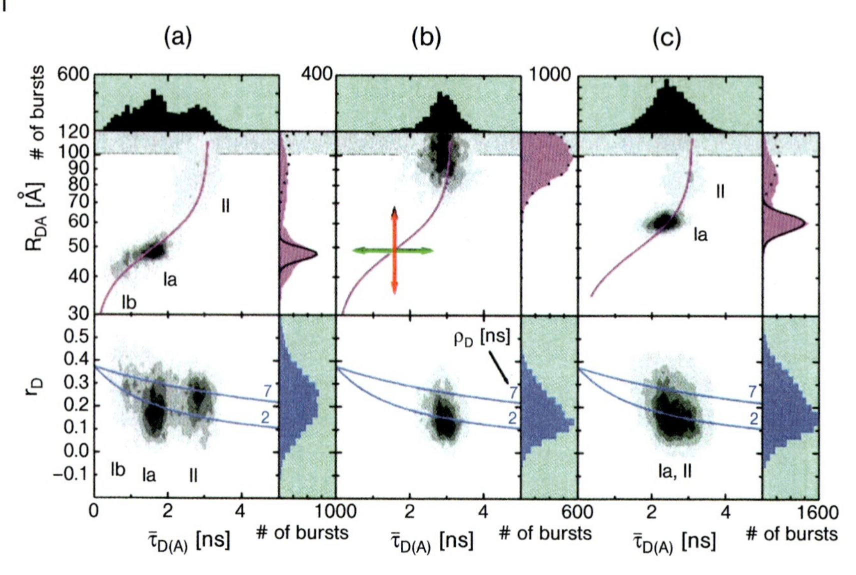

Figure 8.5 Multiparameter FRET analysis. The graphs shows 2-D histograms of average donor lifetimes versus computed donor–acceptor distances (assuming $\kappa^2 = 2/3$) (upper panels) as well as versus donor anisotropy (lower panels). Each panel includes the 1-D projections of the 2-D histograms. (a) Data from donor-labeled reverse transcriptase (RT) and acceptor-labeled DNA molecule are shown. The histogram displays three populations termed Ia, Ib, and II. The red line in the upper graph is a theoretical calculation assuming that the three observed states differ only in donor acceptor distance and not in quantum yield or κ^2. Local donor quenching would lead to a horizontal shift, and local acceptor quenching to a vertical shift, with respect to the red line. The blue lines in the lower graph are computed from the Perrin equation using 2 ns and 7 ns as the rotational correlation times. (b) Data shown for the donor only sample. (c) 2-D histograms after incubation with four nucleotides, leading to a shift of populations Ia and Ib, and thus differentiating between active (Ia and Ib) and inactive (II) subpopulations. © Proceedings of the National Academy of Sciences [51].

the 2-D histogram of donor anisotropy over donor lifetime. The different observed states were seen to have different rotational correlation times (which can be calculated using the Perrin equation). If it is assumed that the acceptor is free to rotate completely, then these variations would cause distance uncertainties of ±16 Å for state II, and ±3 Å for states Ia and Ib.

The nature of these three states was revealed by comparing these FRET data, to those obtained from experiments after the addition of nucleotides. The nucleotides were incorporated by the RT, and thus the position of the RT on the nucleic acid scaffold was changed. In these experiments, a change in the position of peaks Ia and Ib was observed which resulted in only a single peak at larger distances in the 2-D histogram. In contrast, the population II remained unchanged. These histograms, as well as other compelling data, allowed the building of a model, in which in conformation Ia the RT complex was in the educt state, while conformation Ib corresponded to the product state of the RT. State II was attributed to a nonproductive state in which RT was bound, for example, in a different orientation. In summary, by using MFD it is possible to determine whether observed FRET variations are caused by FRET-dependent or -independent donor quenching. In addition, if the rotation of the dye molecules is not severely restricted, then accurate FRET measurements become possible. If, however, the local environment limits the rotational mobility of the dye molecule, the obtainable accuracy decreases rapidly such that other methods are required.

An alternative method for measuring accurate FRET in solution is the application of ALEX (see Chapter 6) [52] or PIE [36]. With ALEX/PIE, it is possible to perform background-, crosstalk and γ-corrected ratiometric measurements of the FRET efficiency, E. However, again accurate distances can only be computed for rotationally unrestricted donor–acceptor pairs and thus, the κ^2 problem persists with this method.

8.7 FRET-Based Triangulation and the Nanopositioning System

Although, during the past centuries, navigation was based on the measurement of star positions in the night sky when, by measuring the distances to several known points in the sky, the navigator could infer his or her position. Yet, today's global positioning system (GPS) modules that everybody uses are, intriguingly, based on a similar idea to navigation using the stars. For the GPS, the distance measurement is performed by converting the transit times of microwave signals emitted by satellites located at fixed positions in space into distances. Then, by measuring these distances from at least three different satellites, the navigator's position on Earth can be calculated by using triangulation.

The same basic idea can also be applied to the nanometer regime using FRET, as first demonstrated by Ebright and coworkers in their groundbreaking studies of the bacterial RNA polymerase (RNAP) [26, 29, 30]. In these studies, a large number of FRET measurements was used as well as distance-constrained docking (using Markov Chain Monte Carlo simulations) to determine the structure of the RNAP holoenzyme, as well as the RNAP–promoter open complex. The method allowed most likely positions to be calculated given the uncertainties (a uniform uncertainty of 15% was assumed in Förster radius for all measurements; cf. the discussion in Section 8.3), but no three-dimensional (3-D) confidence regions were computed. Similar approaches were also applied to the study of lambda-integrase Holliday junctions [53], regulatory DNA–protein complexes [54], and a quantum dot–protein bioconjugate assembly [55].

Often, the molecules under investigation are quite flexible or exist in different well-defined states, and thus ensemble averaging would lead to a false interpretation of the FRET data. Averaging can, however, be avoided by performing triangulation on the single-molecule level. This was first shown by Seidel and coworkers in studies of the conformational dynamics of syntaxin 1, using syntaxin molecules diffusing through the excitation volume of a confocal microscope [56]. Ha and coworkers demonstrated that the positional information could be obtained also for immobilized macromolecular complexes in studies investigating the conformation of *Escherichia coli* Rep helicase [57]. Recently, bends and kinks in DNA structures were analyzed using MFD, molecular dynamics simulations, and quantitative statistical (χ^2) triangulation analysis [38].

Recently, the triangulation of smFRET distances measurements was applied to determine the position of nascent RNA in RNA polymerase II (Pol II) transcription elongation complexes (Figure 8.6) [58]. In these experiments, mismatched nucleic acid scaffolds of template DNA and nontemplate DNA were used, as well as RNA molecules of different lengths, which were incubated with Pol II. These artificial elongation complexes are competent in elongating the RNA primer *in vitro* [59, 60]. By making use of X-ray crystallographic data [60], acceptor dye molecules were attached to known positions of the complex, namely different positions on the template DNA (marked by green circles in Figure 8.6), as well as to positions on the heterodimer Rpb4/7, which is part of the 12 polypeptide Pol II complex, using single cysteine mutants of Rpb4/7. The smFRET was then measured between a donor attached to the 5′-end of the RNA and an acceptor at one of the known positions. The experiments were then repeated for all other known positions, as well as for various RNA lengths. For each length of the RNA, namely a 17, 20, 23, 26 and 29 mer, triangulation of three distance measurements allowed determination of the position of the RNA 5′-end (Figure 8.6). In contrast, in the crystallographic studies (albeit also here long RNA molecules were used), only the first 10 RNA nucleotides were visible due to an increasing flexibility of the growing RNA chain, and therefore the RNA exit pathway had not been determined. The results of the smFRET triangulation confirmed that the RNA leaves the active cleft of the polymerase through the previously proposed RNA exit tunnel [61, 62]. Beyond the exit tunnel, the RNA follows a previously unexpected path, crossing over the Pol II dock domain, which is occupied during initiation by the transcription factor TFIIB [63]. This path was unexpected, as UV-crosslinking studies had indicated that the RNA is positioned much closer to the Pol II subunit Rpb7 [64].

While smFRET triangulation allows remarkable new insight in the structure and conformation of macromolecular complexes, the above discussion relating to uncertainties in smFRET distance measurements raises certain concerns. For example, How accurate are the results of the smFRET triangulation? How can the data analysis be improved so as to obtain information about not only the most likely position, but also the uncertainty of the position determination, given the experimental data? Early approaches used the information of additional distance measurements in order to test the triangulation results [58], or combined the results of many measurements into a most likely value [65]. However, a thorough treatment is needed which includes all uncertainties and computes the most likely position as well as the full 3-D positional uncertainty. In addition, once such a treatment has been obtained, it should be easily applicable to a large variety of triangulation problems.

In order to identify such a solution, it would help to recall the comparison of FRET-based triangulation with GPS, and introduce a nomenclature that will help

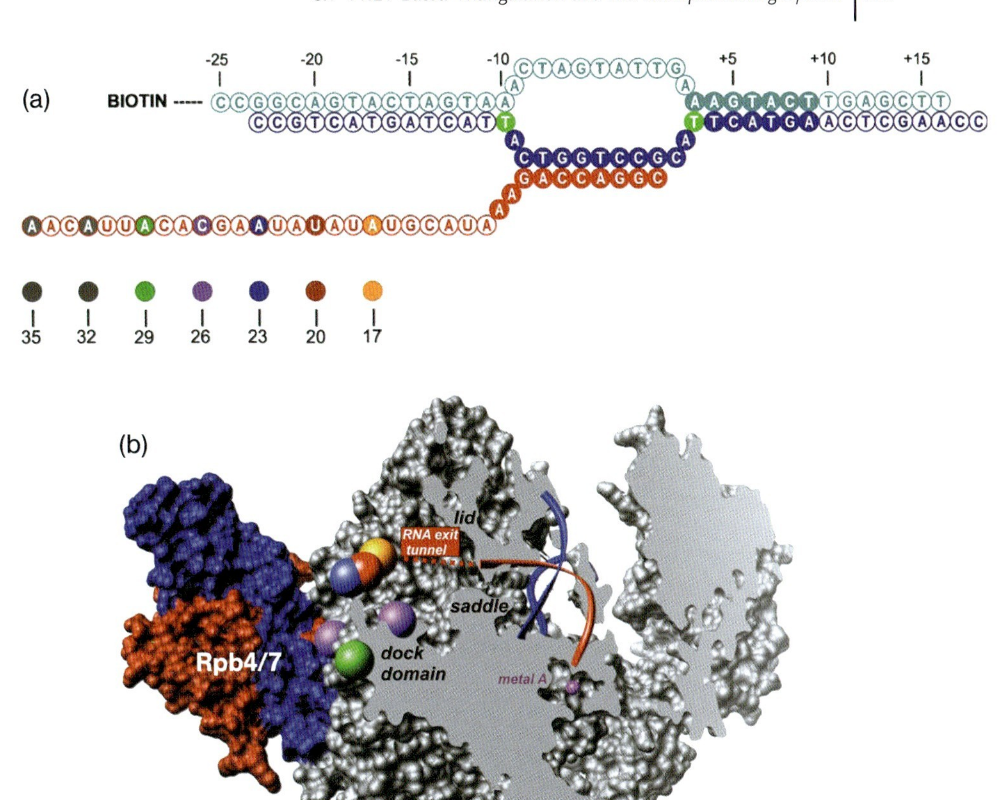

Figure 8.6 Position of nascent RNA in Pol II elongation complexes. (a) Nucleic acid scaffolds used for the formation of the elongation complexes. An 11 nt mismatched region between template and nontemplate DNA allows for the formation of elongation complexes in the absence of initiation factors. The solid circles indicate nucleotides, the positions of which were determined using X-ray crystallography [60], while open circles indicate positions that could not be determined. The labeling sites on the template DNA (known positions) as well as on the RNA (unknown positions) are marked by colored circles. (b) Position of the 5′-end of the RNA determined using single-molecule triangulation. The color code of the positions for different RNA lengths is as in panel (a). The circles have a radius of 5 Å, since test measurements yielded deviations ≤5 Å. At a length of 26 nt, two positions are shown. For this length, the resulting smFRET histograms showed two peaks of equal amplitude corresponding to two possible positions. The single-molecule data revealed dynamic switching between these two positions in this case. © Proceedings of the National Academy of Sciences [69].

to visualize the following discussion. In the FRET measurements, one dye molecule, the position of which is to be determined, acts as an "antenna dye molecule" (ADM), while several other molecules at known positions can be thought of as "satellite dye molecules" (SDMs). The outcome of the measurement of FRET efficiency between a pair of SDMs and ADMs is the knowledge of a distance between the two. Thus, it is known that the position of the ADM must be somewhere on a surface of a sphere, the radius of which is given by the determined distance. The same procedure can be repeated for an arbitrary number of SDMs. The ADM position can then be determined by the point at which all spheres intersect. In general, four measurements are necessary to yield a unique solution. However, by taking into account the fact that all measurements will have a certain experimental error, which is caused, for example, by the uncertainty in determining the correct R_0 for a given SDM–ADM pair, there will in general no longer be a common intersect (Figure 8.7a–c). One solution to the problem can be obtained if fuzzy spheres rather than infinitely sharp spheres are used, where the fuzziness accounts for the total uncertainty (Figure 8.7d–f). By overlaying several fuzzy spheres, there will no longer be a single intersection point in space but rather a fuzzy intersection volume (Figure 8.7g–i). The uncertainty in the triangulation process can then be determined directly from the fuzziness of the intersection volume. It is important to point out that, given a certain experimental uncertainty (e.g., in the Förster radius), the uncertainty of position determination is highly dependent on the relative positions of the SDMs. What is now required is a thorough mathematical treatment of the described situation, and such an approach has recently been developed to directly account for the experimental uncertainties and compute the intersection volume. In stressing the similarities with GPS, this approach has been termed the nanopositioning system (NPS) [40].

The main idea behind NPS is to use Bayesian Data Analysis [66] in order to account for all experimental data, as well as its uncertainties. By using a Bayesian analysis, it is possible to compute not only the most likely position of the ADM but also the full 3-D probability density of its positions that reflects the credibility of the knowledge, given the experimental uncertainties. In Bayes' theory, this is called the *posterior* and it is given by:

$$p(\mathbf{x},\{s_i\},\{R_i\}|\{E_i\},I) \propto p(\mathbf{x},\{s_i\},\{R_i\}|I)\,p(\{E_i\}|\mathbf{x},\{s_i\},\{R_i\},I) \tag{8.5}$$

where $\mathbf{x}$ and $\mathbf{s}_i$ are parameters describing the position of the ADM and the i-th SDM, respectively. R_i is a parameter that describes the Förster radius for the measurement of the FRET efficiency, E_i. The curly brackets denote the set of variables such as $\{E_i\} = E_1, \ldots, E_N$ for a set of N SDMs. The variables to the right of the vertical bar are the conditions and the "," denotes a logical "and". The right-hand site of Equation 8.5 consists of a product of $p(\mathbf{x},\{\mathbf{s}_i\},\{R_i\}|I)$, the so-called *prior* in which the information about the parameters is contained, and $p(\{E_i\}|\mathbf{x},\{\mathbf{s}_i\},\{R_i\},I)$, the so-called *likelihood*, which mathematically connects the measured data to the model used to describe it, that is, Förster theory [40].

Equation 8.5 describes the *posterior*, the outcome of the experiment as a function of the parameters $\mathbf{x},\{\mathbf{s}_i\},\{R_i\}$. In order to obtain the desired result, the probability

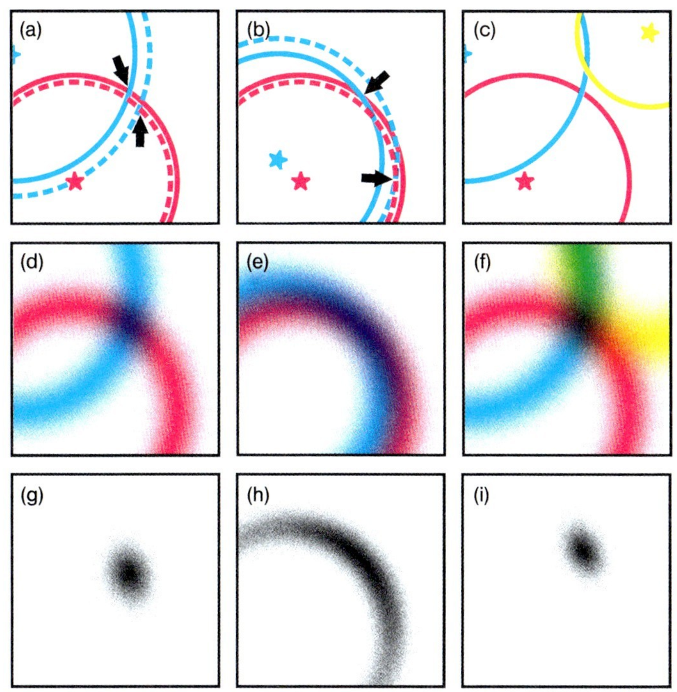

Figure 8.7 Geometric aspects in position determination and the concept of "fuzzy circles." For a given experimental uncertainty, the accuracy of the triangulation depends on the relative positions of the satellite dye molecules (SDMs), and must be described using probability densities, as illustrated here in two dimensions. (a–c) Infinitely sharp circles. As a result of a distance measurement, the position of the antenna dye molecule (ADM) is somewhere on a circle around the SDM. Experimental uncertainties result in circles with different radii (dashed line). In panel (a), the circles from two SDMs intersect perpendicularly, yielding minimal experimental uncertainty. (b) Here, the circles intersect at a quite sharp angle, resulting in an increase of triangulation uncertainty in one direction. (c) If the results from many SDMs are compared, due to experimental errors in general no common intersection point can be defined. (d–f) Experimental uncertainties can be represented geometrically by fuzzy spheres, or here as circles for the 2-D representation. Mathematically, this corresponds to probability density functions. (g–i) The result of the triangulation can be found directly in the geometric representation by overlaying the respective fuzzy spheres (circles in two dimensions). Again, mathematically this results in a probability density for the position of the ADM. Copyright Nature Methods [40].

density for the ADM position, the dependencies on the other parameters must be remove; this is achieved by integration over all possible values, a process called *marginalization* in Bayes' theory.

For most measurements, a Gaussian *likelihood* around the mean FRET efficiency can be used with an uncertainty determined by statistical errors (accuracy of fit to experimental data) and systematic errors. Computation of the *prior* is more complicated, as it requires that uncertainties about the Förster radius, the SDM position, as well as geometric constraints for both the ADM and SDM positions are all accounted for.

From the discussion of the previous sections it is clear that FRET measurements typically have substantial uncertainties in the determination of the Förster radius. Nevertheless, if the anisotropy of donor and acceptor is measured, then an estimate can be obtained for the uncertainty, for example, using a fully static model and performing a Monte Carlo simulation (as described in Section 8.5). Together with measurements of the overlap integral and the donor quantum yield, it is then possible to estimate the Förster radius and its uncertainty, and to use this for the prior calculation.

The second uncertainty arises from the fact that SDM molecules are attached via long flexible linkers in order to allow for a free rotation of the dye molecule, and thus smaller uncertainties in the κ^2 factor. Hence, the precise position is not known. One way to address this uncertainty is to use molecular dynamics simulations to determine the possible positions of the dye molecule [56, 67, 68]. However, this is currently not feasible if the SDM is attached to a large macromolecule, due to the enormous computation time and the limited accuracy of this approach. The alternative is simply to calculate the total volume accessible to the SDM, and to use a flat *prior* within this volume – that is, all accessible positions are equally likely [40].

The same approach, that is, a *prior* with equal probability within a restricted volume, can also be used in order to account for geometric constraints of the position of the ADM. Since the ADM is attached to a macromolecule with a known structure, it is obvious that the position cannot lie at an already occupied point in space. Moreover, sometimes it is possible to estimate the maximum distance from a point of the known structure, for example, if the ADM is attached to a nucleic acid of known length.

Thus, all parts of the *prior* and the *likelihood* can be determined, such that the *posterior* can be calculated for any set of FRET measurements. Since the mathematics of the computations are quite time-consuming, an easy-to-use graphical user interface software has been developed (this can be downloaded from cup.uni-muenchen.de/pc/michaelis/software) in order to facilitate the frequent use of NPS.

Given an easy-to-use NPS software, smFRET-based triangulation can in the future be applied to a large variety of biological systems. The question might then be asked as to how many measurements would be needed to obtain an accurate localization. This important point is, of course, quite difficult to answer, but most of all it is important to recall that the uncertainty of NPS-based position determination depends not only on the number of measurements, but also on the relative positions of the SDMs. Moreover, the local environment often severely restricts the accessible space and thus helps in the positioning process. In the present test measurements it could be shown that, for a good localization, typically five SDMs are necessary, while accuracies of up to 3 Å can be obtained when using six satellites [40].

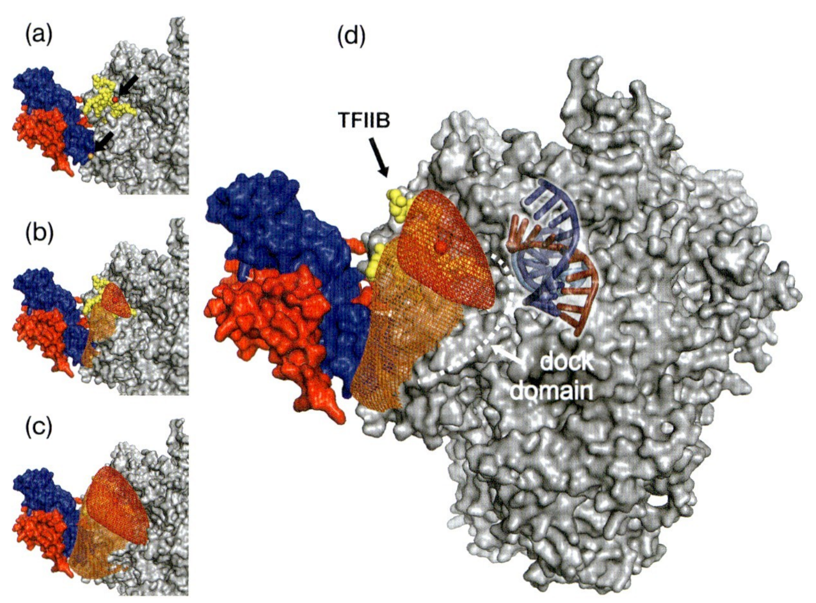

Figure 8.8 Application of the nanopositioning system (NPS) to Pol II transcription elongation. The position of an ADM attached to the 5′-end of a 29 nt RNA within a Pol II transcription elongation complex, both, in the presence (orange) and absence (red) of TFIIB is determined using NPS. Displayed are: (a) the most likely positions, as well as the probability density contoured at (b) 38%, (d) 68%, and (c) 95%. Copyright Nature Methods [40].

Recently, NPS has been applied to the investigation of Pol II transcription elongation [40]. For this, six different SDMs were used to determine the position of the RNA 5′-end of a 29 nt RNA, both in the presence and absence of transcription factor TFIIB (Figure 8.8). In the absence of TFIIB, the ADM position is again found in an area close to the dock domain, while TFIIB causes a reorganization of the RNA such that a position much closer to the Rpb4/7 heterodimer is now seen, in accordance with the UV-crosslinking data [64]. Therefore, TFIIB is likely to stay bound beyond the early stages of transcription elongation, thus guiding the nascent RNA to RNA binding folds at Rpb7. As seen in Figure 8.8, NPS provides not only the most *likely* position [as indicated by an orange (TFIIB) or yellow (no TFIIB) sphere in Figure 8.8a], but allows for the calculation of the 3-D probability density of the ADM position (meshed surfaces). The probability density is contoured at different credibility levels, for example, 38% (0.5 σ), 68% (1 σ), or 95% (2 σ), in analogy to the displayed density of X-ray crystallographic data (Figure 8.8b–d). Thus, NPS allows, for the first time, a quantitative analysis of the uncertainties in a FRET-based position triangulation approach. When examining Figure 8.8, it becomes evident that the experimental uncertainty is spatially quite heterogeneous. The factors that influence this spatial heterogeneity include the relative position of the SDMs, as well as the freely accessible volume of the ADM. In order to provide an idea of what is possible, the uncertainties for the localization of the ADM attached to the RNA 5′-end (in the absence of TFIIB) was estimated, and showed values between 3–9 Å (depending on the labeling locations) [40].

While the described method is static in nature, it can also be applied to investigate the dynamic switching between well-defined positions. For example, in the case of Pol II elongation complexes at an RNA length of 26 nt, single-molecule data showed the dynamic switching with dwell times of approximately 1 s between two well-defined FRET values [58]. The resulting histograms were well described by double Gaussians, and triangulation thus resulted in two possible positions. Thus, NPS provides a novel method for the accurate determination of the position of flexible domains, and provides much-sought help towards determining the structural conformations of larger dynamic macromolecular complexes.

8.8
Conclusions and Outlook

Performing accurate quantitative distance measurements using smFRET is a difficult task, since typically large experimental uncertainties exist with regards to the most important parameter, the Förster radius. Thus, while energy transfer can be thought of as a molecular ruler, extreme caution must be exercised as the length units of the ruler may be inaccurate, or even subject to change during the course of an experiment. One source of such uncertainty is the effect of variations in quantum yield, which could occur due to local environmental inhomogeneities. Such variations can be accounted for by using either multiparameter fluorescence detection, ALEX, or by the careful analysis of intensity changes upon acceptor photobleaching. The most important experimental uncertainty, however, arises from the relative orientation of the donor and acceptor. It is possible to calculate the respective experimental uncertainty when it is clear how the rotational motion of the chromophores is restricted. For the worst case scenario, where complete statical averaging is allowed, the possible distance distribution can be calculated using Monte Carlo simulations. Accurate results can be obtained only in rare cases, where either both the donor and acceptor are completely free to reorient, or their orientation is fixed and known.

In contrast, if there is a need to determine the position of a dye molecule, that position can be determined very accurately by using multiple distance measurements and performing a statistical analysis such as NPS, which includes all experimental uncertainties. Currently, NPS is limited to situations where labels can be placed at many known positions after which, by means of multiple smFRET measurements, the position of a dye attached to another domain at an unknown position can be inferred. However, in the future, it might become possible to expand NPS to a situation where a meshwork of labeling sites, all at unknown positions, are used for FRET measurements, with the relative 3-D positions of the dye molecules being obtained from the results. This will become especially important for understanding the conformation of large multicomponent complexes, the architecture of which remains elusive to current structural biology methods, due to their large flexibility.

Acknowledgments

The author acknowledges the members of his group, namely Joanna Andrecka, Volodymyr Kudryavtsev, Wolfgang Kügel, Robert Lewis, Adam Muschielok, Peter Schwaderer, and Barbara Treutlein for the hard work that enabled not only the writing of this chapter, but also encouraged many lively discussions. The author's thanks also extend to Patrick Cramer and the members of his laboratory for introducing us to Pol II and providing proteins, ideas, and biological assistance, and to colleagues Christoph Bräuchle and Don Lamb for many stimulating discussions throughout the years. These studies were supported by the Sonderforschungsbereich 486 and 646, the Center for Integrated Protein Science Munich (CIPS^{M}), the Nanosystems Initiative Munich (NIM), and the Center for Nanoscience (CeNS).

References

1 Förster, T. (1948) Zwischenmolekulare Energiewanderung Und Fluoreszenz. *Annalen Der Physik*, **437** (1–2), 55–75.

2 Clegg, R.M. (1992) Fluorescence resonance energy transfer and nucleic acids. *Methods Enzymol.*, **211**, 353–388.

3 Clegg, R.M. (1996) Fluorescence resonance energy transfer, in *Fluorescence Imaging Spectroscopy and Microscopy* (eds X.F. Wang and B. Herman), John Wiley & Sons, Inc., New York, pp. 179–525.

4 Selvin, P.R. (1995) Fluorescence resonance energy transfer. *Methods Enzymol.*, **246**, 300–334.

5 Stryer, L. (1978) Fluorescence energy transfer as a spectroscopic ruler. *Annu. Rev. Biochem.*, **47**, 819–846.

6 Hillisch, A., Lorenz, M., and Diekmann, S. (2001) Recent advances in FRET: distance determination in protein-DNA complexes. *Curr. Opin. Struct. Biol.*, **11** (2), 201–207.

7 Stryer, L. and Haugland, R.P. (1967) Energy transfer–a spectroscopic ruler. *Proc. Natl Acad. Sci. USA*, **58** (2), 719–726.

8 Selvin, P.R. (2000) The renaissance of fluorescence resonance energy transfer. *Nat. Struct. Biol.*, **7** (9), 730–734.

9 Ha, T., *et al.* (1996) Probing the interaction between two single molecules: fluorescence resonance energy transfer between a single donor and a single acceptor. *Proc. Natl Acad. Sci. USA*, **93** (13), 6264–6268.

10 Weiss, S. (1999) Fluorescence spectroscopy of single biomolecules. *Science*, **283** (5408), 1676–1683.

11 Joo, C., *et al.* (2008) Advances in single-molecule fluorescence methods for molecular biology. *Annu. Rev. Biochem.*, **77**, 51–76.

12 Zhuang, X. and Rief, M. (2003) Single-molecule folding. *Curr. Opin. Struct. Biol.*, **13** (1), 88–97.

13 Lilley, D.M. and Wilson, T.J. (2000) Fluorescence resonance energy transfer as a structural tool for nucleic acids. *Curr. Opin. Chem. Biol.*, **4** (5), 507–517.

14 Ha, T. (2004) Structural dynamics and processing of nucleic acids revealed by single-molecule spectroscopy. *Biochemistry*, **43** (14), 4055–4063.

15 Roy, R., Hohng, S., and Ha, T. (2008) A practical guide to single-molecule FRET. *Nat. Methods*, **5** (6), 507–516.

16 Joo, C. and Ha, T. (2007) Single-molecule FRET with total internal reflection

microscopy, in *Single Molecule Techniques: A Laboratory Manual* (eds P. Selvin and T. Ha), Cold Spring Harbor Laboratory Press, Cold Spring Harbor, New York, pp. 3–36.

17 Ha, T. (2001) Single-molecule fluorescence resonance energy transfer. *Methods*, **25** (1), 78–86.

18 Walter, N.G., *et al.* (2008) Do-it-yourself guide: how to use the modern single-molecule toolkit. *Nat. Methods*, **5** (6), 475–489.

19 Antonik, M., *et al.* (2006) Separating structural heterogeneities from stochastic variations in fluorescence resonance energy transfer distributions via photon distribution analysis. *J. Phys. Chem. B*, **110** (13), 6970–6978.

20 Nir, E., *et al.* (2006) Shot-noise limited single-molecule FRET histograms: comparison between theory and experiments. *J. Phys. Chem. B*, **110** (44), 22103–22124.

21 Dale, R.E., Eisinger, J., and Blumberg, W.E. (1979) The orientational freedom of molecular probes. The orientation factor in intramolecular energy transfer. *Biophys. J.*, **26** (2), 161–193.

22 van der Meer, B.W. (2002) Kappa-squared: from nuisance to new sense. *J. Biotechnol.*, **82** (3), 181–196.

23 Clegg, R.M., *et al.* (1993) Observing the helical geometry of double-stranded DNA in solution by fluorescence resonance energy transfer. *Proc. Natl Acad. Sci. USA*, **90** (7), 2994–2998.

24 Schuler, B., *et al.* (2005) Polyproline and the “spectroscopic ruler” revisited with single-molecule fluorescence. *Proc. Natl Acad. Sci. USA*, **102** (8), 2754–2759.

25 Best, R.B., *et al.* (2007) Effect of flexibility and cis residues in single-molecule FRET studies of polyproline. *Proc. Natl Acad. Sci. USA*, **104** (48), 18964–18969.

26 Knight, J.L., *et al.* (2005) Distance-restrained docking of rifampicin and rifamycin SV to RNA polymerase using systematic FRET measurements: developing benchmarks of model quality and reliability. *Biophys. J.*, **88** (2), 925–938.

27 Schuler, B. and Eaton, W.A. (2008) Protein folding studied by single-molecule FRET. *Curr. Opin. Struct. Biol.*, **18** (1), 16–26.

28 Lipman, E.A., *et al.* (2003) Single-molecule measurement of protein folding kinetics. *Science*, **301** (5637), 1233–1235.

29 Mekler, V., *et al.* (2002) Structural organization of bacterial RNA polymerase holoenzyme and the RNA polymerase-promoter open complex. *Cell*, **108** (5), 599–614.

30 Mukhopadhyay, J., *et al.* (2004) Antibacterial peptide microcin J25 inhibits transcription by binding within and obstructing the RNA polymerase secondary channel. *Mol. Cell*, **14** (6), 739–751.

31 Lakowicz, J.R. (2006) *Principles of Fluorescence Spectroscopy*, 3rd edn, Springer, New York.

32 Ha, T., *et al.* (1999) Single-molecule fluorescence spectroscopy of enzyme conformational dynamics and cleavage mechanism. *Proc. Natl Acad. Sci. USA*, **96** (3), 893–898.

33 Jia, Y.W., *et al.* (1999) Folding dynamics of single GCN4 peptides by fluorescence resonant energy transfer confocal microscopy. *Chem. Phys.*, **247** (1), 69–83.

34 Sabanayagam, C.R., Eid, J.S., and Meller, A. (2005) Using fluorescence resonance energy transfer to measure distances along individual DNA molecules: corrections due to nonideal transfer. *J. Chem. Phys.*, **122** (6), 061103.

35 Kapanidis, A.N., *et al.* (2004) Fluorescence-aided molecule sorting: analysis of structure and interactions by alternating-laser excitation of single molecules. *Proc. Natl Acad. Sci. USA*, **101** (24), 8936–8941.

36 Muller, B.K., *et al.* (2005) Pulsed interleaved excitation. *Biophys. J.*, **89** (5), 3508–3522.

37 Sanborn, M.E., *et al.* (2007) Fluorescence properties and photophysics of the sulfoindocyanine Cy3 linked covalently to DNA. *J. Phys. Chem. B*, **111** (37), 11064–11074.

38 Wozniak, A.K., *et al.* (2008) Single-molecule FRET measures bends and kinks in DNA. *Proc. Natl Acad. Sci. USA*, **105** (47), 18337–18342.

39 VanBeek, D.B., *et al.* (2007) Fretting about FRET: correlation between kappa and R. *Biophys. J.*, **92** (12), 4168–4178.

40 Muschielok, A., *et al.* (2008) A nano-positioning system for macromolecular structural analysis. *Nat. Methods*, **5** (11), 965–971.

41 Ha, T., *et al.* (1996) Single molecule dynamics studied by polarization modulation. *Phys. Rev. Lett.*, **77** (19), 3979–3982.

42 Jung, C., *et al.* (2007) Single-molecule traffic in mesoporous materials: translational, orientational, and spectral dynamics. *Adv. Mater.*, **19** (7), 956.

43 Hubner, C.G., *et al.* (2004) Three-dimensional orientational colocalization of individual donor–acceptor pairs. *J. Chem. Phys.*, **120** (23), 10867–10870.

44 Toprak, E., *et al.* (2006) Defocused orientation and position imaging (DOPI) of myosin V. *Proc. Natl Acad. Sci. USA*, **103** (17), 6495–6499.

45 Iqbal, A., *et al.* (2008) Orientation dependence in fluorescent energy transfer between Cy3 and Cy5 terminally attached to double-stranded nucleic acids. *Proc. Natl Acad. Sci. USA*, **105** (32), 11176–11181.

46 Norman, D.G., *et al.* (2000) Location of cyanine-3 on double-stranded DNA: importance for fluorescence resonance energy transfer studies. *Biochemistry*, **39** (21), 6317–6324.

47 Iqbal, A., *et al.* (2008) The structure of cyanine 5 terminally attached to double-stranded DNA: implications for FRET studies. *Biochemistry*, **47** (30), 7857–7862.

48 Forkey, J.N., *et al.* (2003) Three-dimensional structural dynamics of myosin V by single-molecule fluorescence polarization. *Nature*, **422** (6930), 399–404.

49 Lewis, F.D., Zhang, L., and Zuo, X. (2005) Orientation control of fluorescence resonance energy transfer using DNA as a helical scaffold. *J. Am. Chem. Soc.*, **127** (28), 10002–10003.

50 Widengren, J., *et al.* (2006) Single-molecule detection and identification of multiple species by multiparameter fluorescence detection. *Anal. Chem.*, **78** (6), 2039–2050.

51 Rothwell, P.J., *et al.* (2003) Multiparameter single-molecule fluorescence spectroscopy reveals heterogeneity of HIV-1 reverse transcriptase:primer/template complexes. *Proc. Natl Acad. Sci. USA*, **100** (4), 1655–1660.

52 Lee, N.K., *et al.* (2005) Accurate FRET measurements within single diffusing biomolecules using alternating-laser excitation. *Biophys. J.*, **88** (4), 2939–2953.

53 Radman-Livaja, M., *et al.* (2005) Architecture of recombination intermediates visualized by in-gel FRET of lambda integrase-Holliday junction-arm DNA complexes. *Proc. Natl Acad. Sci. USA*, **102** (11), 3913–3920.

54 Sun, X., *et al.* (2006) Architecture of the 99 bp DNA-six-protein regulatory complex of the lambda att site. *Mol. Cell*, **24** (4), 569–580.

55 Medintz, I.L., *et al.* (2004) A fluorescence resonance energy transfer-derived structure of a quantum dot-protein bioconjugate nanoassembly. *Proc. Natl Acad. Sci. USA*, **101** (26), 9612–9617.

56 Margittai, M., *et al.* (2003) Single-molecule fluorescence resonance energy transfer reveals a dynamic equilibrium between closed and open conformations of syntaxin 1. *Proc. Natl Acad. Sci. USA*, **100** (26), 15516–15521.

57 Rasnik, I., *et al.* (2004) DNA-binding orientation and domain conformation of the *E. coli* rep helicase monomer bound to a partial duplex junction: single-molecule studies of fluorescently labeled enzymes. *J. Mol. Biol.*, **336** (2), 395–408.

58 Andrecka, J., *et al.* (2008) Single-molecule tracking of mRNA exiting from RNA polymerase II. *Proc. Natl Acad. Sci. USA*, **105** (1), 135–140.

59 Gnatt, A.L., *et al.* (2001) Structural basis of transcription: an RNA polymerase II elongation complex at 3.3 Å resolution. *Science*, **292** (5523), 1876–1882.

60 Kettenberger, H., Armache, K.J., and Cramer, P. (2004) Complete RNA polymerase II elongation complex structure and its interactions with NTP and TFIIS. *Mol. Cell*, **16** (6), 955–965.

61 Cramer, P., *et al.* (2000) Architecture of RNA polymerase II and implications for the transcription mechanism. *Science*, **288** (5466), 640–649.

62 Vassylyev, D.G., *et al.* (2007) Structural basis for transcription elongation by bacterial RNA polymerase. *Nature*, **448** (7150), 157–162.

63 Bushnell, D.A., *et al.* (2004) Structural basis of transcription: an RNA polymerase II-TFIIB cocrystal at 4.5 angstroms. *Science*, **303** (5660), 983–988.

64 Ujvari, A. and Luse, D.S. (2006) RNA emerging from the active site of RNA polymerase II interacts with the Rpb7 subunit. *Nat. Struct. Mol. Biol.*, **13** (1), 49–54.

65 Ha, T., *et al.* (2002) Initiation and re-initiation of DNA unwinding by the *Escherichia coli* Rep helicase. *Nature*, **419** (6907), 638–641.

66 Sivia, D.S. (2006) *Data Analysis: A Bayesian Tutorial*, Oxford University Press, Oxford, pp. 1–128.

67 Schroder, G.F. and Grubmuller, H. (2004) FRETsg: biomolecular structure model building from multiple FRET experiments. *Comput. Phys. Commun.*, **158** (3), 150–157.

68 Dolghih, E., Roitberg, A.E., and Krause, J.L. (2007) Fluorescence resonance energy transfer in dye-labeled DNA. *J. Photochem. Photobiol. A: Chemistry*, **190**, 321–327.

69 Kettenberger, H., Armache, K.J., and Cramer, P. (2003) Architecture of the RNA polymerase II-TFIIS complex and implications for mRNA cleavage. *Cell*, **114** (3), 347–357.

Part III
Single Molecules in Nanosystems

Single Particle Tracking and Single Molecule Energy Transfer
Edited by Christoph Bräuchle, Don C. Lamb, and Jens Michaelis

ISBN: 978-3-527-32296-1

9
Coherent and Incoherent Coupling Between a Single Dipolar Emitter and Its Nanoenvironment

Vahid Sandoghdar

9.1
Introduction

Understanding and controlling the coupling of individual physical systems are important in technology and science, with two main applications in the transfer of energy and information. The commercialization of electric power during the nineteenth century made it possible to transfer energy via electric currents and to transmit information via electric signals, which in turn helped to fuel the second industrial revolution. The development of radio- and microwave communication during the first half of the twentieth century extended this mechanism to a wireless mode at higher frequencies. The advent of lasers brought about another paradigm shift to yet higher frequencies in the optical and near-infrared (NIR) regimes. Today, telecommunication channels are based almost exclusively on optical fibers, although copper wires still dominate the transmission of signals in electronics equipment.

The above-mentioned applications involve the one-way transfer of signals or energy. Yet, it is also possible to couple two physical systems coherently so that energy is exchanged between the two in a reversible manner. In fact, in parallel to the research on new integrated optical technologies, mechanisms for the coherent coupling of individual quantum systems have also been sought during the past two decades, within the context of quantum information processing and computation. Here, it is desirable to prepare well-defined superpositions of quantum states of a system, to perform various operations on them, and then to communicate the results to other quantum systems over large distances. Some of these approaches have involved the mechanical oscillations of ions in a trap, the coupling of spins, interaction between superconducting electron pairs, and the coupling of atom-like states via photons [1]. Photons show great promise in this endeavor as they can be transmitted with a small loss over long distances, and are also robust against decoherence [2]. Indeed, quantum cryptography using photons as information carriers has been already commercialized as the first product of this line of research.

Single Particle Tracking and Single Molecule Energy Transfer
Edited by Christoph Bräuchle, Don C. Lamb, and Jens Michaelis

ISBN: 978-3-527-32296-1

The optical coupling of two emitters requires an efficient exchange of photons. In the near field, this takes place via nonradiative exchange of virtual photons, whereas at large distances the connection between emitters uses traveling real photons. As will be seen, the efficiency of the former process depends on the distance between the two individual dipoles and their relative orientations. The success of the latter process, however, depends on how well the propagating photons can be coupled with emitters. It should be noted that typical spectroscopic measurements are not concerned with this issue, and are performed using nW-mW laser power, equivalent to 10^9–10^{15} photons per second. However, a simple estimate shows that it should be possible to excite atoms efficiently even with very few photons. As an example, consider the absorption cross-section $3\lambda^2/2\pi$ of a two-level atom with transition wavelength λ [3]. This expression immediately reveals that the interaction cross-section is larger than the area $(\lambda/2\mathrm{NA})^2$, given by the diffraction limit for a focusing system of numerical aperture *NA*. In fact, rigorous calculations reveal that a dipolar emitter can experience 100% coupling efficiency with a directional dipolar light beam [4].

The implementation of efficient coupling between photons and dipolar emitters poses a major challenge in experimental quantum optics. Progress in cavity quantum electrodynamics during the 1980s and 1990s has provided several proof of principle examples for the coherent coupling of a single photon with a single atom in high-finesse cavities [5]. Similar concepts have also been extended to dissipative energy transfer mediated via a microcavity [6]. Recently, it has been demonstrated that, in both near and far fields, it is possible to couple freely propagating laser light to a single molecule with a very high efficiency [7, 8]. The fascinating extension of such experiments would be to use a single photon source instead of a laser beam. In this scheme, one molecule would be used as a light source to excite a second molecule, achieving long-distance transfer of energy between two individual elementary sources of light.

This chapter is restricted to various experiments where a single dipolar radiator is coupled to its nanoscopic environment via its *near field*. In particular, discussions will center on the coupling of two single molecules, the plasmon oscillations of two gold nanoparticles, on a single molecule or gold nanoparticle to a mirror, and finally of a single molecule and a nanoparticle. Finally, a brief insight is provided into the prospects of strong modification of the molecular photophysics via coupling to nanoantennae. Given the limited space and scope of the chapter, attention will be focused on studies conducted in the author's laboratory. Thus, apologies are offered for any other relevant reports that have been omitted.

9.2 Systems

Of the two different types of optical radiator – fluorescent molecules and gold nanoparticles – the former is an intrinsically quantum mechanical system with strong transition dipole moments, whereas the latter is a fully classical system with induced dipoles under illumination. It is well known, however, that the great majority of the optical and spectroscopic phenomena known currently can be treated to a very good approximation in a classical or semi-classical formalism. In this chapter, this point is stressed by discussing several near-field coupling effects.

9.2.1 Single Molecules

Fluorescent emitters have been studied for about two decades at the single molecule level in a variety of contexts. Given the large number of books and reviews on this topic [9–11], it will not be discussed in any great detail here. It suffices to mention that single-molecule experiments can be grouped into two categories of *cryogenic* and *ambient* measurements. Although, in both cases, high-resolution optical microscopy has been used to obtain spatial information, low-temperature experiments have the added advantage that the dephasing processes can be suppressed, reducing the homogeneous linewidths down to the natural linewidth of the molecular transition. The inhomogeneous distribution of transition frequencies in the sample then helps to isolate one single molecule.

9.2.2 Plasmonic Nanoparticles as Classical Atoms

The response of a nanoparticle to optical illumination is entailed in its scattering properties. For a spherical particle, these can be calculated in terms of multipoles by using the theory developed by Mie [12]. In the especial case that the sphere diameter D is much smaller than the wavelength of the incident light in the medium surrounding the particle, it can be treated as a dipole with a polarizability [13]

$$\alpha(\lambda) = \frac{\pi D^3}{2} \frac{\varepsilon_{\mathrm{p}}(\lambda) - \varepsilon_{\mathrm{m}}(\lambda)}{\varepsilon_{\mathrm{p}}(\lambda) + 2\varepsilon_{\mathrm{m}}(\lambda)} \tag{9.1}$$

Here, $\varepsilon_p(\lambda)$ and $\varepsilon_m(\lambda)$ are the dielectric constants of the particle and its surrounding medium respectively, and λ is the vacuum wavelength. If the material conditions are such that the denominator has a minimum at a certain wavelength, the polarizability–and therefore the scattering cross-section–are enhanced. This can take place for metallic nanoparticles, leading to localized plasmon polariton resonances [13, 14]. The scattering cross-section of a subwavelength particle is then given by:

$$\sigma = \frac{(2\pi)^3 \varepsilon_{\mathrm{m}}^2 |\alpha|^2}{3\lambda^4} \tag{9.2}$$

It transpires out that a gold nanoparticle, when placed in air or in a low-index dielectric material such as glass, shows a resonance in the visible domain (see Figure 9.1a). However, plasmon spectra can be varied by changing the size, shape, and the dielectric functions of the medium surrounding the particle [13, 14].

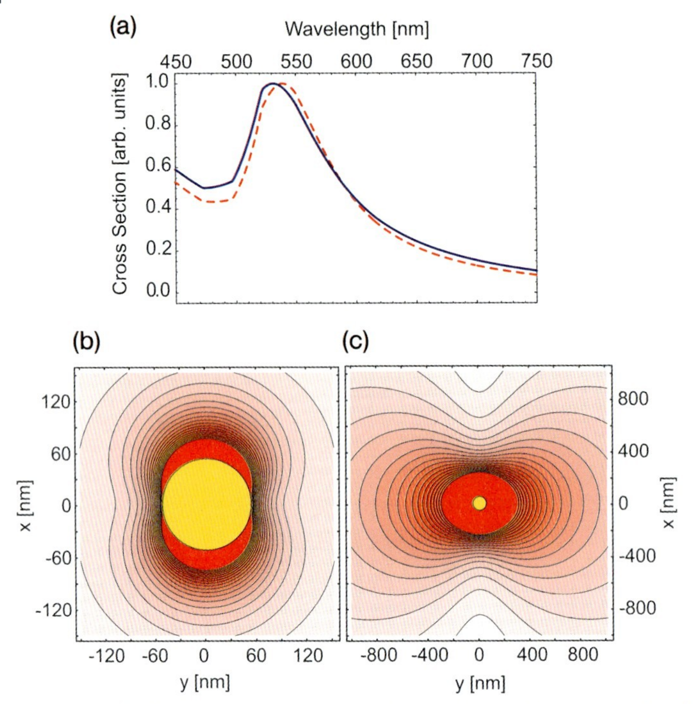

Figure 9.1 (a) Plasmon spectrum of a gold nanoparticle ($D = 100\,\text{nm}$) in air ($\varepsilon_m = 1$) computed using Mie theory (solid line) and an effective polarizability (dashed line). (b, c) Amplitudes of the scattered electric field calculated using Mie theory in the near and far fields, respectively. The illumination was a plane wave propagating along z and polarized along the x-direction. The yellow disks depict the nanoparticle to scale. Reproduced from Ref. [15].

Gold nanospheres of diameters ranging from 50 to approximately 100 nm are in the regime where the dynamical effects become important and the observed plasmon resonance begins to deviate from that predicted by the simple formula in Equation 9.1. Nevertheless, it has been shown that plasmon spectra of such particles can be reproduced very well by an effective dipolar polarizability α_{eff} if radiation damping and dynamic depolarization are taken into account [15, 16]. Figure 9.1a shows that the far-field scattering cross-section calculated according to Mie theory (solid line) agrees quite well with that evaluated by using α_{eff} (dashed line). Whilst the contribution of higher multipoles is negligible in the far field, it could amount to up to 10% of the dipolar one in the near field. It follows that the electric field lines of the light scattered by the nanoparticle trace a dipolar radiation pattern. Figure 9.1b,c display the strength of the scattered electric field in the near and far fields of the particle, respectively.

The above-mentioned features allows gold nanospheres to serve as subwavelength classical antennae [17, 18]. The dipolar character of a nanoparticle makes it an ideal approximation to a point-like oscillator, which has been a very useful conceptual construct for relating the classical and quantum mechanical features of an atom [19, 20]. Here, the plasmon resonance plays the role of the transition between the ground and the excited states. The Rayleigh-like scattering replaces spontaneous emission [21] whereby the particle polarizability, and hence its scattering cross-section σ, provide a measure for the strength of this process. Furthermore, absorption in a gold particle mimics a nonradiative decay channel in an atom [22]. Thus, many of the central features of the coupling between a molecule and its nanoenvironment can be revisited in the interaction of a metallic nanoparticle with its immediate surrounding.

9.3
Coupling of Two Oscillating Dipoles

9.3.1
Two Single Molecules

Optical transitions are predominantly of a dipolar nature, which means that the light emitted by a molecule follows a dipolar pattern. In reverse, the perfect absorption of light also requires the excitation light to be in a dipolar mode [4, 23]. Thus, the most ideal source of light for exciting an optical emitter is a second identical emitter. However, the realization of two single emitters placed at a large distance and efficient coupling of their dipolar modes via lenses, mirrors, and fibers, is a challenging task. One way of overcoming the problems of freely propagating dipolar fields with high fidelity is to bring the two emitters very close to each other. Indeed, it is easy to see that the modes of the two emitters are perfectly matched at a separation of $r = 0$, leading to a perfect coupling.

The coupling of two emitters can be treated by considering the interaction Hamiltonian $H_I = -E_1 \cdot D_2$, where E_1 is the electric field of molecule 1 at the position of molecule 2 with a transition dipole moment operator D_2. Now, in the far field, E_1 scales as $1/r$ so that the excitation probability drops rapidly as $1/r^2$. In the near field, however, E_1 can be substantial. In this limit, the coupling frequency δ can be written as:

$$\delta = \frac{3\sqrt{\gamma_1\gamma_2}}{8\pi(knr)^3}\left[\left(\hat{\mathbf{d}}_1\cdot\hat{\mathbf{d}}_2\right) - 3\left(\hat{\mathbf{d}}_1\cdot\hat{\mathbf{r}}\right)\left(\hat{\mathbf{d}}_2\cdot\hat{\mathbf{r}}\right)\right] \tag{9.3}$$

Here, we have introduced the transition dipole moments $\vec{\mathbf{d}}_i \equiv \langle g_i|\vec{\mathbf{D}}_i|e_i\rangle$, where $|\vec{\mathbf{D}}_i|$ and γ_i are the dipole operators, and the spontaneous emission rates of molecules i = 1, 2, respectively. The quantity $k = 2\pi/\lambda$ is the wavenumber, and λ is the wavelength of the transition. It is evident that E_1 and thus δ grow as $1/(kr)^3$ when the distance to the emitter is reduced.

As sketched in Figure 9.2a, two molecules coupled in this manner can be considered as a four-level system with the ground state

$$|G\rangle = |g_1g_2\rangle$$

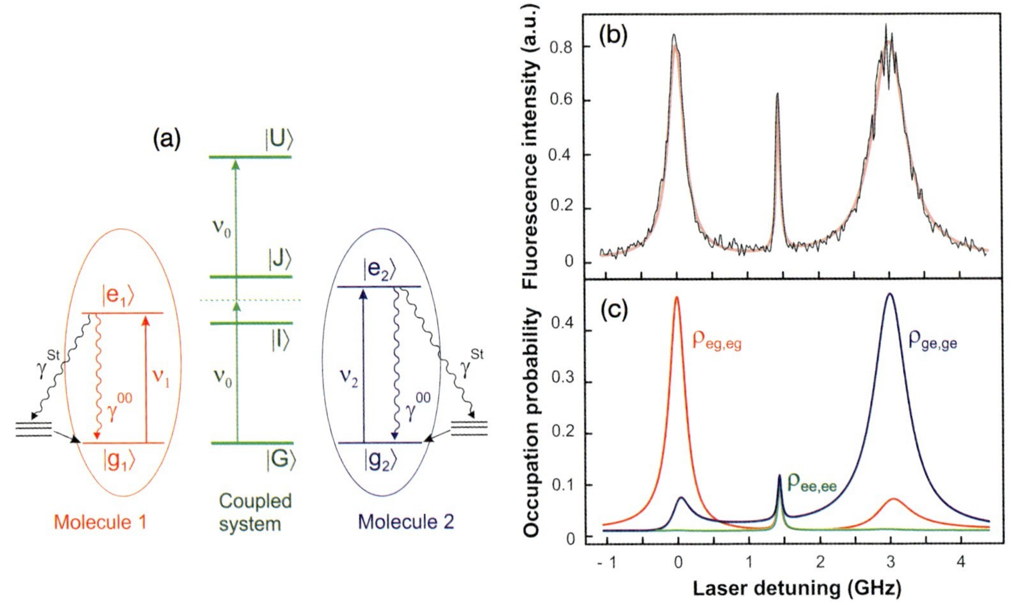

Figure 9.2 (a) Energy level scheme of the uncoupled (red and blue) molecules and of the coupled system (green). See text for details. (b) A fluorescence excitation experimental spectrum of two coupled molecules. (c) Calculations of the populations $\rho_{eg,eg}$, $\rho_{ge,ge}$, and $\rho_{ee,ee}$ for the first molecule being in the excited state and the second in the ground state, *vice versa*, and both being in the excited state. Modified from Ref. [24].

the upper state

$$|U\rangle = |e_1 e_2\rangle \tag{9.4}$$

and the intermediate superposition states

$$|I\rangle = \cos\theta\,|e_1 g_2\rangle - \sin\theta\,|g_1 e_2\rangle \tag{9.5}$$

$$|J\rangle = \sin\theta\,|e_1 g_2\rangle + \cos\theta\,|g_1 e_2\rangle \tag{9.6}$$

whereby $\tan(2\theta) = \delta/\Delta\nu$ and the coefficients $\sin\theta$ and $\cos\theta$ are determined by diagonalizing the new full Hamiltonian [25]. The frequencies of the coupled system then read $\nu_G = 0$,

$$\nu_{I,J} = \frac{\nu_2 + \nu_1}{2} \mp \sqrt{\left(\frac{\nu_2 - \nu_1}{2}\right)^2 + \delta^2}$$

and

$$\nu_U = \nu_1 + \nu_2$$

where ν_i are the transition frequencies of the unperturbed molecules $i = 1, 2$.

As pointed out by Dicke [26], the interaction between the two molecules is also expected to modify their spontaneous emission rates because the existence of one dipole in the near field of the other opens an additional decay channel. In other words, each dipole acts as an antenna for the other one. For very small separations, the single-photon emission rates of the new excited states can be readily obtained by calculating the Einstein A-coefficient [19] $\gamma = 8\pi^2\nu^3|\vec{\mathbf{d}}|^2/(3\varepsilon\hbar c^3)$ in conjunction with the matrix elements of the total dipole operator $\vec{\mathbf{D}}_S = \vec{\mathbf{D}}_1 + \vec{\mathbf{D}}_2$ for the system (where ε_0 is the vacuum permittivity). The linewidths read [24]

$$\begin{aligned}
\gamma_I &= \gamma_1 \cos^2\theta + \gamma_2 \sin^2\theta - 2\gamma_{12}\sin\theta\cos\theta \\
\gamma_J &= \gamma_1 \sin^2\theta + \gamma_2 \cos^2\theta + 2\gamma_{12}\sin\theta\cos\theta \qquad (9.7) \\
\gamma_U &= \gamma_1 + \gamma_2
\end{aligned}$$

Here, γ_{12} denotes the incoherent cross-damping rate [27], which takes the form $\gamma_{12} = \sqrt{\gamma_1\gamma_2}\,(\hat{\mathbf{d}}_1 \cdot \hat{\mathbf{d}}_2)$ for $k_0 r << 1$. The positive and negative contributions of $2\gamma_{12}$ in Equation 9.7 stem from the constructive and destructive interference of the decay channels of the two molecules, portraying the essence of Dicke's superradiance and subradiance. In the extreme case, where the two molecules have initially the same transition frequencies and orientations, and can be brought arbitrarily close to each other, the coupling leads to a super-radiant state with a lifetime that is twice shorter than the original excited state lifetimes of the individual uncoupled molecules plus a fully dark state, which does not radiate at all. In an intuitive picture, the super-radiant system has a large dipole moment that is the sum of the transition dipole moments of the two molecules whereas the dark state has a zero dipole moment resulting from an out-of-phase addition of the individual dipole moments.

The coherent coupling of emitters has been known to affect the radiative properties of highly doped glasses and crystals, and was studied in gaseous atomic systems in the microwave domain about three decades ago. Studying these effects in the optical regime and at the single emitter level is particularly challenging. In order to obtain a significant coupling, the separation between the two emitters must be kept much smaller than the wavelength of transition, and they must be spectrally as close as possible. Although the latter requirement is almost automatically met in cold gas systems, the first condition is very difficult to achieve, even in the case of tight traps [28]. Several years ago, the coherent coupling of two single emitters in the solid state was successfully demonstrated for the first time, by combining cryogenic single-molecule spectroscopy with ultrahigh-resolution optical microscopy [24].

Figure 9.3a shows the heart of the experimental arrangement in a cryostat at $T = 1.4\,\mathrm{K}$. A microelectrode was used to apply an inhomogeneous electric field, resulting in a position-dependent Stark shift of a single terrylene molecule located in the thin *p*-terphenyl (pT) crystal. Fluorescence excitation spectroscopy [29] was used to excite single molecules in the sample via their zero-phonon absorption spectra, with typical linewidths of approximately 40 MHz. By scanning the sample, the Stark shift of the molecular line was recorded as a function of position. Figure 9.3b displays an example of such measurements. The smooth surface of this two-

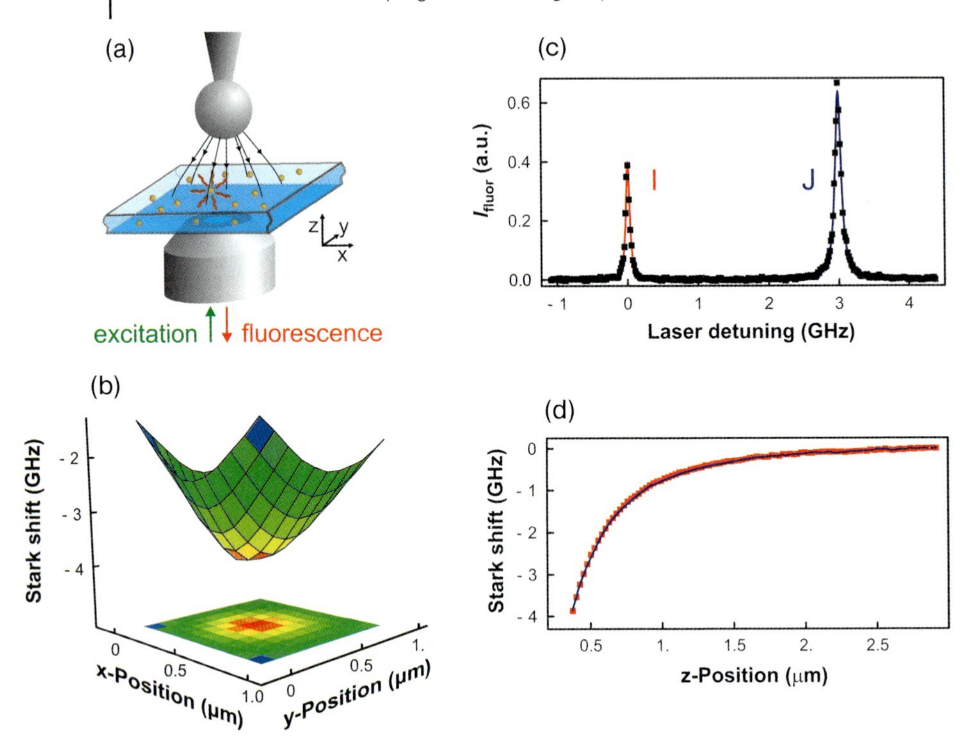

Figure 9.3 (a) Schematic of the microelectrode set-up for manometer localization of single molecules. (b) The Stark shift map of a single molecule transition frequency as a function of displacement under the microelectrode. (c) Fluorescence excitation spectrum of two coupled molecules. (d) Stark shifts of resonances I (left-hand peak) and J (right-hand peak) in panel (c) as a function of microelectrode displacement away from the sample. Modified from Ref. [24].

dimensional (2-D) plot resulting from the low noise in determining Stark shifts allowed the center of this profile to be determined with an accuracy superior to 2 nm [24]. Figure 9.3c shows the spectrum recorded from two close-lying molecules found in this fashion. The two curves in Figure 9.3d plot the Stark shifts of the two resonances I and J as a function of the vertical displacement between the molecule and the microelectrode. The small vertical separation of less than 10 nm between the two molecules is not noticeable on this scale, but a careful analysis of Stark shifts obtained from lateral and vertical scans yielded a distance of 12 ± 2 nm between the two molecules.

Detailed studies of the spectral lines in Figure 9.3c confirmed that they indeed correspond to transitions from the ground state G to states I and J (see Figure 9.2a). In particular, spectra recorded at high excitation power revealed a third resonance nearly mid-way between resonances I and J, as shown in Figure 9.2b. This resonance turns out to correspond to a two-photon absorption by the coupled system, as indicated in Figure 9.2a. Figure 9.2c shows the results of calculations for the occupation probabilities of the uncoupled states, yielding a very good fit to the measured spectrum in Figure 9.2b. These calculations allow a determination of the coupling frequency $\delta \simeq 1\,\text{GHz}$ and the detuning $\Delta\nu \simeq 2.3\,\text{GHz}$ between the two molecules. It should be possible to apply an electric field with a gradient (see Figure 9.3a) that is large enough to Stark shift the resonances of the two molecules into each other; that is, to obtain $\Delta\nu = 0$. This would yield perfectly sub-radiant and super-radiant coupled states. In fact, in this case even molecules with larger separations r, and thus a small coupling δ, would result in sub-radiant and super-radiant states.

The system discussed above consisted of two molecules with spectra as narrow as their natural linewidths; that is, with the maximum coherence. However, in most cases, the coherence of the system is lowered by phononic interactions with the matrix, leading to dephasing or quenching effects. As a result, the linewidth of a single molecule is broaden to up to several nanometers at room temperature. Thus, the coupling frequency δ is lower than the molecular decoherence rate, and the interaction turns into a simple one-way energy transfer. If the coherence of the system is low, then the two molecules must be placed closer to each other in order to experience a substantial energy transfer. This is the mechanism that is used for Förster-type fluorescence resonant energy transfer (FRET). It should be noted here that, although typical FRET applications use different species as donor and acceptor molecules, dissipative energy transfer can (and does) also take place if the donor and acceptor are both of the same type. Indeed, this is the process that leads to quenching of fluorescence if a host is doped too strongly. If the emitters do not have 100% quantum efficiency, the radiation of one hops to its neighbors and eventually is converted to heat before it can be radiated. It should also be noted that the FRET rate scales as $1/r^6$, whereas the coherent coupling frequency δ depends on $1/r^3$. This difference can be traced to the fact that the FRET process can be described by "Fermi's Golden Rule" in a second-order perturbation theory formalism, considering a finite spread in the available states for coupling of the donor to the acceptor. Coherent dipole–dipole coupling, on the other hand, involves a first-order perturbation process and interaction between well-defined resonances.

9.3.2 Two Plasmonic Nanoparticles

In the previous section, the quantum mechanical coupling of two molecules in the near field was discussed. In fact, much of the physics involved was quite analogous to the mechanical coupling of two classical oscillators, for example, pendula. In this section, the interaction of a gold nanoparticle with a second particle that is separated by less than one wavelength, will be described.

In order to investigate the coupling of two gold nanoparticles in a well-defined manner, scanning probe technology has been exploited to control their relative positions *in situ*. One gold particle (P_{tip}) is attached to the end of a glass fiber tip, following the preparation technique reported previously [30, 31]. The second

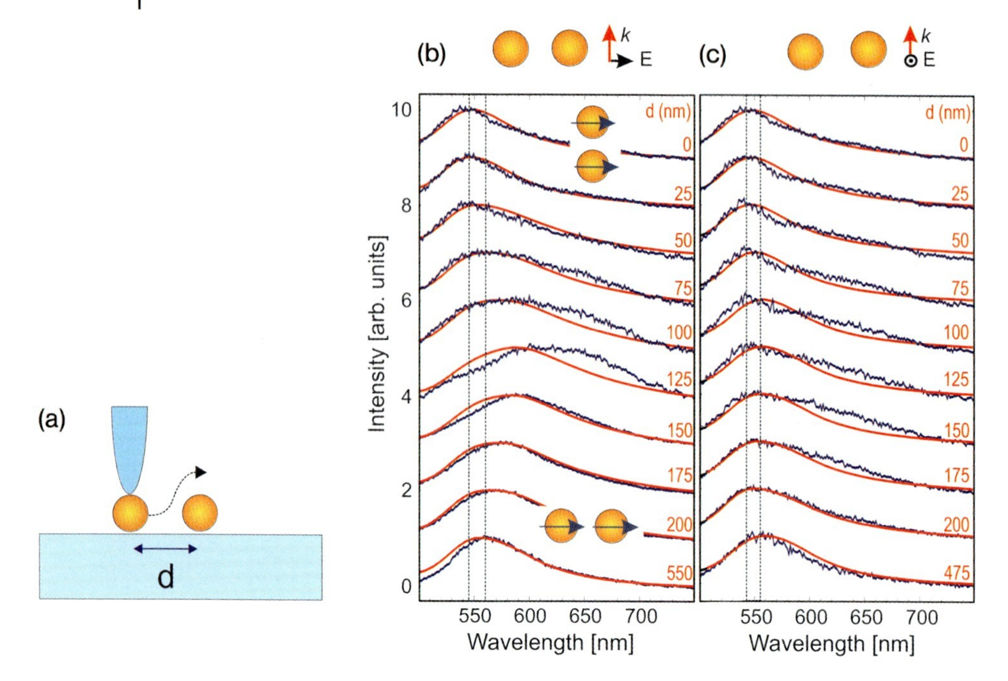

Figure 9.4 (a) Schematic of the experiment, and series of spectra acquired for the particle pair with polarization of the exciting field in parallel to (b) and orthogonal to (c) the particle pair axis. The spectra have been normalized and shifted in intensity for clarity. Calculated scattering spectra for the situations are shown as red lines. The parameter d denotes the projection of the center-to-center particle separation onto the substrate plane. The dashed vertical lines mark the plasmon resonance peaks for d large and $d = 0$. Modified from Ref. [15].

particle (P_{sub}) is selected from a very low concentration of nanoparticles that were spin-coated onto a glass substrate. Both particles have nominal diameters of 100 nm.

The plasmon resonances of the particles were recorded separately, and the two particles were chosen to have very similar plasmon spectra. Furthermore, by examining these spectra as a function of the polarization of the excitation light, it could be verified that they were spherical [31]. Then, by using scanning near-field optical microscopy (SNOM), both the tip and the sample were positioned with nanometer accuracy (see Figure 9.4a). At each scan pixel, the spectrometer was triggered and the scattering spectra of the two gold nanoparticles recorded; these were spaced within the confocal image spot of the detection system. Figure 9.4b,c show selections of spectra for two different incident polarizations. In the case of Figure 9.4b, as the two particles are brought closer from the large separation of $d = 550$ nm, the spectrum becomes broader and is shifted to longer wavelengths. Here, the parameter d denotes the projection of the center-to-center particle separation onto the substrate (see Figure 9.4a). At $d \sim 125$ nm, the spectral shift amounts to a maximum of approximately 60 nm. As the tip is scanned further in shear-force control, P_{tip} is lifted upwards, resulting in a narrower and blue-shifted spectrum. When P_{tip} is above P_{sub} (i.e., $d = 0$), the spectrum has become even *narrower* than the starting point, and its center wavelength is shifted to a lower wavelength by about 15 nm. In an intuitive picture, each particle acts as an induced electric dipole moment directed along the incident field polarization. Dipole–dipole coupling for the head-to-tail and side-by-side configurations (see the cartoons in Figure 9.4b) leads to attractive or repulsive interactions, respectively [32–34].

The agreement between the measured and calculated spectra in Figure 9.4a are generally very good, aside from the region at about $d \sim 125$ nm. At such small separations, the plasmon resonance is very sensitive to the exact distance between the particles, which was not regulated well in these experiments. Indeed, it is this sensitivity to displacements that constitutes the core idea of the recently proposed "plasmon rulers" [35, 36].

For the configuration of Figure 9.4c, the polarization is perpendicular to the axis joining the particles, so that the relative orientation of the dipoles in the two nanoparticles does not change during the entire scan. As the particles approach each other, there is a shift toward lower wavelengths, signifying a repulsive force. Again, the theoretical calculations show the same trend.

Although the coupling of two gold nanoparticles might seem different from the results of the previous section for molecules, both interactions involve coherent near-field dipole–dipole coupling. There are, however, differences, the most important being that the dipole moments induced in the gold particles are much larger than the transition dipoles of a molecular transition. This can be intuitively understood by the physical size of the particle, and thus the dipole. As a result, plasmon resonances are very broad. In other words, plasmonic nanoparticles scatter very strongly. Because the dipole–dipole coupling energy is proportional to the strengths of the two dipoles (note the $\sqrt{\gamma_1\gamma_2}$ term in Equation 9.3), the shifts in the interaction of two gold nanoparticles are easily resolvable by using a grating spectrometer. It must be borne in mind, however, that if the coupling frequency is normalized by the unperturbed radiative linewidths γ, the outcome (δ/γ) is actually much larger in the case of the two coupled molecules (25γ) than the observed effects on the coupled nanoparticles $(<\gamma)$. This is partly because the molecules studied were spaced by approximately 10 nm, whereas the gold particles reached a minimum separation of the order of 100 nm. On the other hand, the oscillating charges on a spherical geometry are distributed so that, at close distances, the two particles cannot be described as pure dipoles. This results in another evident difference between the spectra in Figure 9.3d and Figure 9.4, namely that no splitting is observed in the latter case. It transpires that, for both illumination configurations of Figure 9.4, only those symmetric modes analogous to the states of Equation 9.5 are detected. The excitation of the antisymmetric states is not strong enough for two spheres; however, such states do exist [37] and have been observed for ensembles of coupled-disk metallic nanostructures [38].

9.4
A Dipole Close to a Surface

In the following sections, one of the two dipoles discussed above will be replaced by a "macroscopic" simple object such as a plane or sphere, providing a confined geometry in the near-field of the other dipole. It is known that the radiative properties of an emitter, such as the angular distribution of its emitted power, its spectrum, and fluorescence lifetime can be strongly modified in the presence of boundary conditions for the electromagnetic field. One of the earliest indications of these phenomena was pointed out by Sommerfeld, who considered the radiation of a dipole close to a surface [39]. Another seminal proposal was made by Purcell on the enhancement of the decay rate of a system in the excited state inside a cavity [40]. The pioneering experimental studies dates back to about 1970, when Drexhage showed that the fluorescence lifetimes of emitters placed very close to a flat mirror depended on their separations from it [41]. During the 1980s and 1990s, several groups demonstrated the possibility of controlling radiative decay rates and emission patterns by placing emitters in confined geometries, such as the spaces between two flat substrates, between the mirrors of high-finesse cavities, and in whispering gallery mode resonators [5, 42].

A theoretical description of dipole decay in multilayer structures has been developed by Chance *et al.* [43] and expanded by many authors to cover numerous situations. In the next two subsections, recent studies are presented of such a configuration for a single-molecular dipole and a single-plasmonic antenna dipole. The main mechanism at hand can be most easily understood for a perfect metal interface. In this case, an oscillating point dipole experiences a dipole–dipole interaction with its image in the mirror. Close to the mirror, the dipole and its image are perfectly in phase, such that a $1/r^3$ distance dependence is obtained. However, as the dipole is placed farther from the mirror, retardation causes an oscillatory modulation of the lifetime that decays as $1/r$. The strength of the dipole image also depends strongly on its orientation. For a dipole parallel to the surface, its image becomes antiparallel to it, so that at very small separations these cancel and the decay rate approaches zero. A dipole normal to the mirror, on the other hand, gives rise to an image that is parallel to it, so that close to the mirror the decay rate is enhanced. For a dipole next to a lossless dielectric surface or for curved surfaces, images can also be found [3] and this formalism extended. Lossy boundaries, however, are not easily treated with image dipoles. In this case, the energy of the dipole becomes dissipated in the form of heat, leading to quenching of its radiation.

9.4.1
A Molecule Close to a Metallic Surface

Here again, scanning probe control is employed, but this time for studying single-molecule fluorescence at room temperature in the configuration shown in

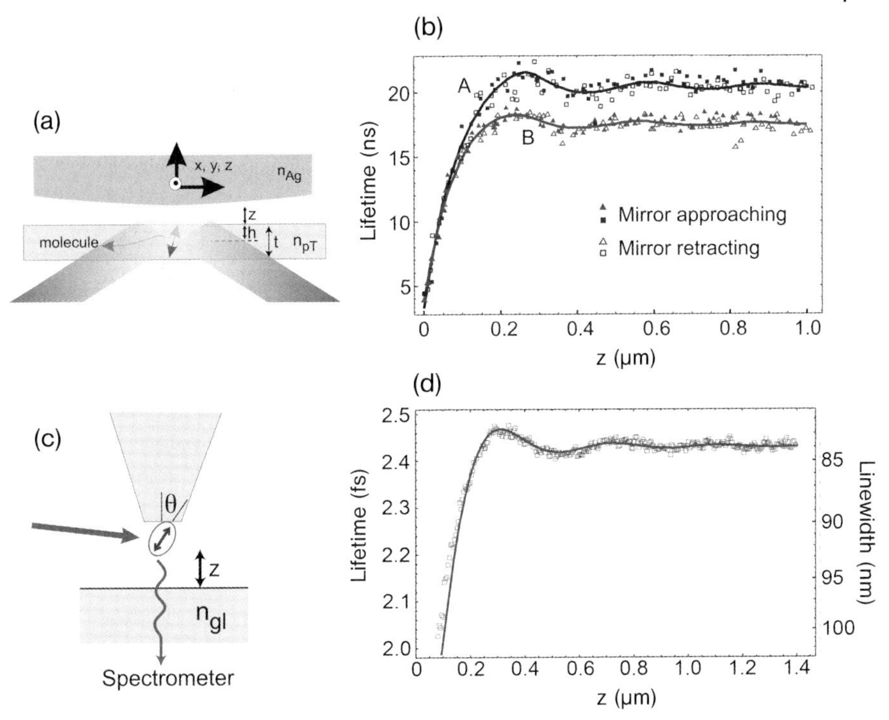

Figure 9.5 (a) A dye molecule embedded in a thin *para*-terphenyl (pT) film was approached by a silver micro-mirror with a 3-D piezo stage (not shown). (b) Lifetime as a function of mirror position (z) for two molecules A (squares) and B (triangles). (c) A gold nanoparticle is attached to a glass tip and moved in front of a dielectric glass substrate. (d) The full-width at half-maximum of the particle plasmon resonance as a function of z. The solid curves display theoretical fits. Modified from Ref. [22].

Figure 9.5a. Samples of terrylene molecules embedded in thin pT crystalline films were prepared following the method described in Ref. [44]. The crystalline character of these films has two useful consequences.

First, terrylene molecules are oriented almost perpendicular to the film plane; and second, photobleaching is essentially eliminated. The thickness (t) of the pT layer was directly measured as 35 ± 5 nm by using shear-force microscopy. The laser light was focused in the back focal plane of the objective to produce a nearly collimated beam at the pT layer, which resulted in total internal reflection at the pT–air interface. The fluorescence of the molecule (λ ~580 nm) was collected confocally through the objective and sent to an avalanche photodiode (APD) for lifetime measurements, or to a CCD camera for imaging. Above the pT layer a micromirror was mounted that had been prepared by melting the end of a tapered optical fiber to produce a ball of diameter 40 µm that was then coated with 200 nm of silver. Using the scanning stage, the mirror was lowered to a height z_0 above the pT layer. The calibrated piezo used in the z-axis of the system thus provided an accurate measurement of the mirror–pT distance (z) relative to the zero defined by z_0. The fluorescence lifetimes (τ) of the single molecules were then measured via time-correlated single-photon counting at each z.

The outcomes of the two different measurements are shown in Figure 9.5b. These data show clearly the near-field shortening of the lifetime dominated by nonradiative coupling to the metal, as well as far-field oscillations caused by the retarded interaction of the dipole with its mirror image [41, 45]. In bulk pT, the lifetime of terrylene is $\tau_0 = 4.1 \pm 0.1$ ns [46], whereas in the present thin-film system (without the micromirror) the lifetimes were increased to the range of 15 to 25 ns [47]. The solid curves in Figure 9.5 show fits to the data using a classical theory [22, 43]. The lifetime was monitored as the mirror was moved towards and away from the molecule, thereby verifying the mechanical stability of the system and the lack of any drifts.

The manipulation of a movable external mirror at large distances modifies the dipole's radiative lifetime rate, but leaves its intrinsic noradiative dissipation channels unaffected. This enables us to extract

$$\eta_0 \frac{1/\tau_r}{1/\tau_r + 1/\tau_{nr}} \tag{9.8}$$

which denotes the quantum efficiency of the emitter, with τ_r and τ_{nr} representing the radiative and nonradiative decay times of the excited state, respectively. The very good agreement between the theoretical and measured curves then allows the unknown parameters η_0, h, and z_0 to be determined. For molecule A, $0 < h < 13$ nm, $0.9 < \eta_0 < 1$ and $z_0 = 12 \pm 2$ nm, whereas for molecule B, $12 < h < 32$ nm, $0.9 < \eta_0 < 1$ and $z_0 = 11 \pm 5$ nm. As the measurements of thermal energy are less accurate than for optical fields, as assessment of η_0 is nontrivial even for ensembles.

9.4.2
A Metallic Nanoparticle Close to a Dielectric Surface

At this point, the second experimental system sketched in Figure 9.5c will be considered, where a gold nanoparticle is positioned in front of a glass substrate. Details of the set-up and the spectroscopy procedure are similar to that outlined in Section 9.3.2. By performing a tomographic measurement as described in Ref. [31], the long axis of the particle was determined to be oriented at $\theta = 10° \pm 2°$ (see Figure 9.5c). The tip was then rotated about its axis, and the illumination polarized so as to only excite the plasmon resonance associated with the long axis. The point of closest approach (z_0) was found, as in the case of the single molecules, by using a shear force signal. The plasmon decay time was found according to $\tau = 1/(2\pi\gamma)$ [48], where γ (in Hz) is the full-width at half-maximum of the resonance.

A plot of the plasmon lifetime (and linewidth on the right-hand axis) as a function of z is shown in Figure 9.5d. This result qualitatively resembles that of the molecular dipole. In particular, for a large z there were slowly dying oscillations of the lifetime that eventually relaxed to a value of 2.43 fs. As shown by the solid curve in Figure 9.5d, the data were well reproduced by a theoretical fit with parameters $z_0 = 9 \pm 6$ nm and $\eta_0 = 0.64 \pm 0.07$. The quantum efficiency η_0 for this system is a measure for the power that is scattered compared to the sum of the scattered and absorbed powers. Thus, it can be seen again that the modification of the near-field environment of a dipole can change its radiative properties and its quantum efficiency.

9.4.3
A Dipole Senses its Nanoenvironment via the Modification of its Resonance Spectrum

In the previous section, the lifetime of a dipolar transition was seen to be modified in the near field of a flat substrate. Given the short interaction range, it might also be expected that a dipole would undergo spectral modifications if the local index of refraction of the substrate were to change laterally. The results of analytical calculations of the variations in linewidth and resonance frequency of a dipole scanned across a dielectric sample are shown in Figure 9.6a,b; small regions with a higher index of refraction are illustrated in Figure 9.6c. Indeed, it is found that the spectrum of a dipolar emitter is very sensitive to even small modulations in

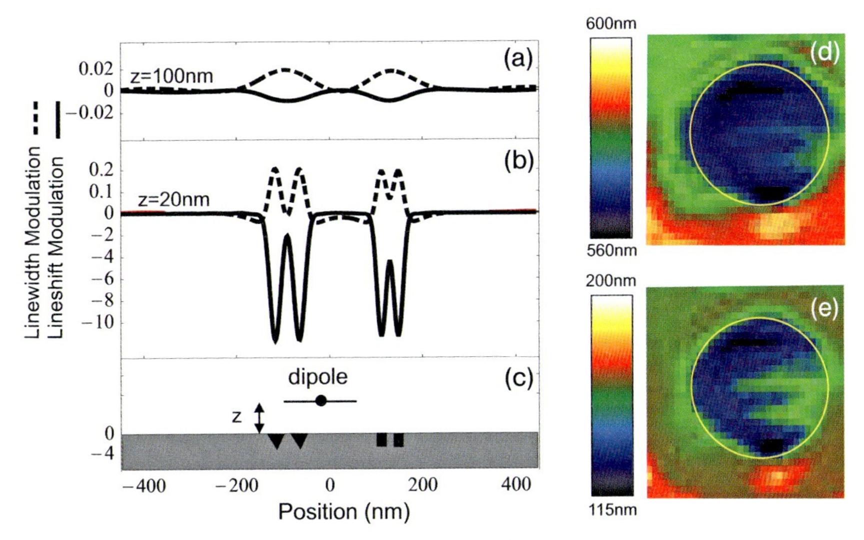

Figure 9.6 (a, b) Solid lines and dashed lines show the variations of the linewidth and lineshift, respectively, in units of the dipolar resonance linewidth. Two dipole-surface separations $z = 100$ and 20 nm are plotted in panels (a) and (b), respectively. (c) The dipole is scanned across a dielectric surface with refractive index $n = 1.5$, embedding four nano-objects with $n = 2.5$. (d, e) The central wavelength (d) and the full-width at half-maximum (FWHM) (e) of the plasmon resonance of a gold nanoparticle as it is scanned above a circular opening with a diameter of 2 μm in a thin chromium film. Color scales quantify these values. The circles are traced as a guide to the eye and mark the opening in the film. Modified from Refs [17, 49].

the refractive index [49]. The spatial resolution of this procedure is given by the separation of the dipole from the sample. As this distance approaches zero, the resolution is expected to reach the molecular scale. In other words, it should be possible to resolve individual molecules on the substrate by using a single molecule as a sensor. In the particular case that the sensor molecule and the target molecule have close resonances, they undergo a strong coupling. Depending on the details of their spectral overlap and dephasing by their surrounding, they might experience a coherent dipole–dipole coupling or an incoherent energy transfer, as was discussed in Section 9.3.1.

The experimental realization of a single-emitter near-field sensor is quite challenging, because a photostable emitter must be placed very close to the sample [50, 51]. However, the basic principles of the scheme discussed in Figure 9.6a–c have been demonstrated by monitoring the plasmon resonances of a gold particle acting as a nanoscopic dipole. Here, a sample was used that contained a circular opening in a thin semi-transparent chromium film with very sharp edges that rose within less than 10 nm. Figure 9.6d,e show the center wavelength and linewidth of the plasmon resonance as a gold nanoparticle attached to the end of a glass fiber tip was scanned in the immediate vicinity of the sample [17]. It is evident that the circular aperture is imaged by the changes in the plasmon resonance frequency and its width. In this experiment, the resolution was limited by the size of the nanoparticle, and thus the minimal distance that the dipole could assume in front of the sample. An extension of these experiments to molecular emitters [50, 51] or to very small nanoparticles [52] would provide access to more local interactions.

9.5 A Single Molecule and a Single Nanoparticle

It has been shown that the spectrum of a dipolar emitter changes close to surfaces, and can be very sensitive to tiny modulations of its optical contrast. Next, attention will be focused on the interaction of a molecule with an isolated nano-object, which is modeled as a nanosphere. The emission of the molecule is altered in the presence of the sphere [53, 54] via a change in the radiative decay rate $\gamma_r = 1/\tau_r$ [55–58], and in the nonradiative rate $\gamma_{nr} = 1/\tau_{nr}$ if the material of the sphere absorbs light [59]. In an oversimplified intuitive picture, the situation is just as in the case of a planar interface: γ_r is modified owing to the interaction of the molecular dipole with its image dipole, whereas γ_{nr} is increased because the lossy electromagnetic oscillations induced in the sphere extract and dissipate energy from the molecular dipole.

First, $\eta = \eta(\hat{\mathbf{d}},\mathbf{r})$ is defined as the apparent quantum yield for a given particle-emitter displacement $\mathbf{r}$ and dipole orientation $\hat{\mathbf{d}}$ (see Figure 9.7). The presence of a nanoparticle can change the apparent quantum yield of the joint molecule-sphere system [57, 60, 61]. It is easy to see that η may be increased for a molecule with a small initial yield $\eta_0 < 1$ [61], but can only decrease if $\eta_0 \approx 1$. The spectral depend-

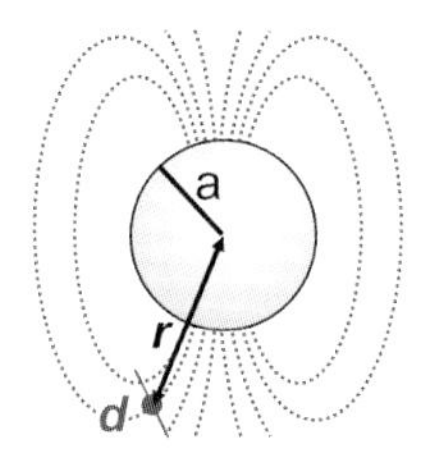

Figure 9.7 A nanosphere of radius *a* placed close to a dipolar emitter located at a distance *r* from the sphere center.

ence of $\varepsilon(\lambda)$ of the particle dictates the spectral variations of $\gamma_r(\lambda)$, $\gamma_{nr}(\lambda)$, and $\eta(\lambda)$. For certain materials and wavelengths, the polarizability of the particle can experience a resonance (see Equation 9.1) which, for metals, amounts to the plasmon resonance. As discussed earlier, such a plasmonic nanoparticle acts as a dipolar nanoantenna with a well-defined resonance frequency. Thus, in analogy with radiowave and microwave engineering [62], the particle might be expected to improve the reception and transmission of a molecule that is placed close to it; that is, it can act as a simple dipolar antenna.

If the fluorescence signal of the molecule is denoted by S_f, then:

$$S_f(\hat{\mathbf{d}},\mathbf{r}) = c\xi(\hat{\mathbf{d}},\mathbf{r})K(\hat{\mathbf{d}},\mathbf{r})\eta(\hat{\mathbf{d}},\mathbf{r}) \tag{9.9}$$

for the weak excitation regime far from saturation. Here, the coefficient $c = S_0/(\xi_0\eta_0)$ normalizes the signal to the fluorescence S_0 in the absence of a nano-object, ξ is the collection efficiency which can also depend on the system parameters, and K is the ratio of the excitation rates in presence and absence of the nano-object given by

$$K(\hat{\mathbf{d}},\mathbf{r}) = |\mathbf{d}\cdot\mathbf{E}_{loc}(\mathbf{r})|^2/|\mathbf{d}\cdot\mathbf{E}_{inc}(\mathbf{r})|^2 \tag{9.10}$$

where $\mathbf{E}_{inc}(\mathbf{r})$ and $\mathbf{E}_{loc}(\mathbf{r})$ are the excitation electric fields at position $\mathbf{r}$ in the absence and presence of the nanoparticle, respectively. The magnitude of the excitation enhancement depends on the position $\mathbf{r}$ and the relative orientations between $\mathbf{E}_{loc}$, and the molecular dipole moment $\mathbf{d}$.

9.5.1 A Molecule Close to a Dielectric Nanoparticle

A number of groups have studied the effect of various dielectric tips on single emitters [63–66]. In this section, well-controlled experiments on aligned single molecules and a glass tip are discussed. It is shown that a simple theoretical model can provide a satisfactory agreement with the measurements. Further details of these investigations can be found in Ref. [67].

A bare heat-pulled glass tip was approached and scanned over a pT thin film (doped with terrylene molecules), while the excited state lifetime $\tau = \gamma_{tot}^{-1}$ was

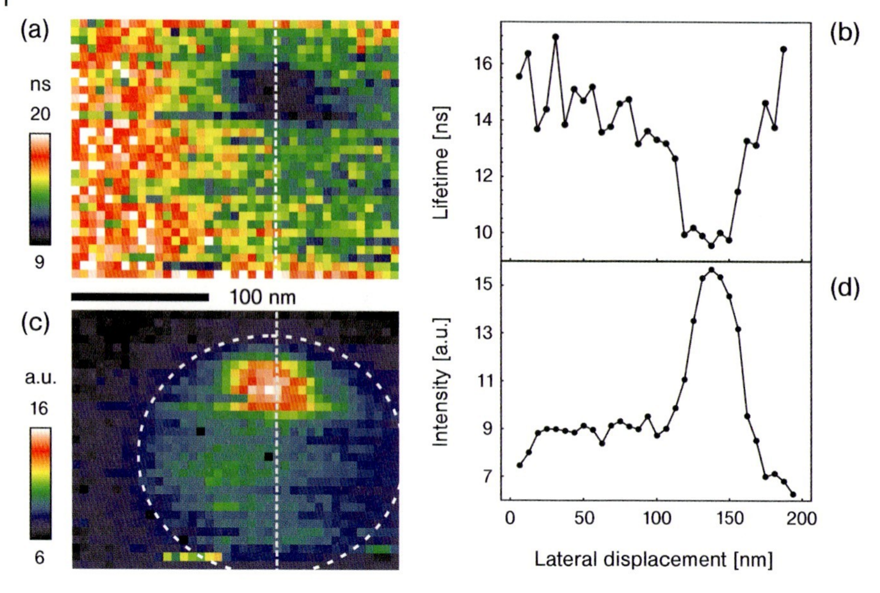

Figure 9.8 (a) Fluorescence lifetime of a single molecule as a glass tip was scanned in its vicinity. (b) A profile along the cut shown in panel (a). (c) Fluorescence near-field image recorded simultaneously with the data in panel (a). (d) Intensity profile along the cross-section indicated in panel (c). Reproduced from Ref. [67].

recorded at each scan pixel. As shown in Figure 9.8a,b, the fluorescence lifetime drops from 18 ns for a single molecule to 16 ns in the larger area under the tip, and to 9 ns in the upper part of the image. Given that glass has a negligible absorption at the molecular emission wavelength, the shorter lifetime provides direct evidence for the accelerated radiation, and can be interpreted as an enhancement of the spontaneous emission rate. Simultaneously with the fluorescence lifetime measurement, the fluorescence signal S_f was also mapped. The enhancement of total fluorescence accompanying a shortening of the lifetime is displayed in Figure 9.8c,d.

The magnitudes of the enhancement and lifetime change are well reproduced in a model where the tip was approximated by a glass sphere with a diameter of 100 nm and a refractive index of 1.5. The broken lines in Figure 9.9a show the results for a molecule oriented perpendicular to the surface. In the absence of absorption, η remains constant throughout assuming no intrinsic nonradiative process in the emitter, while γ_{rad} increases by up to a factor of 2 at a distance $z = 10$ nm. The lateral extent of the enhancement roughly corresponds to the radius of the particle. The formalism can also be applied to a tangentially oriented molecule. The results are shown in Figure 9.9b, again as broken lines.

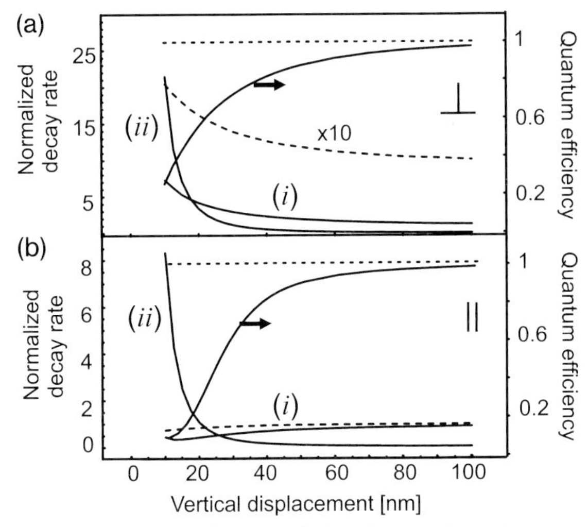

Figure 9.9 Radiative (*i*), nonradiative (*ii*) normalized decay rates, and the quantum efficiency for chromium (solid) and glass (dashed) spheres of diameter 100 nm. (a, b) Results for radially and tangentially oriented dipoles, respectively. Modified from Ref. [67].

9.5.2
A Molecule Close to a Metallic Nanoparticle

Metallic structures are particularly attractive for modifying the fluorescence properties of emitters because they have large dielectric constants $|\varepsilon(\omega)|$, leading to a strong lightning-rod effect [68]. For chromium, $\varepsilon \approx -8.5 + 29i$ [69] leads to an increase by a factor of 20 compared to glass in the limit of an infinitely thin needle [70]. However, this gain in excitation can be overshadowed by quenching caused by the imaginary part of ε [60, 71]. The shape of the structure, the orientation of the molecule, and their separations, and the wavelength of interest determine the final fluorescence modification.

The simple model of a Mie particle can be used to explain these observations. Again, a sphere of 100 nm was chosen, but now with the dielectric constant of chromium. The results are displayed by the solid curves in Figure 9.9a,b for radially and tangentially oriented dipoles, respectively. In the first case, already at a distance of 19 nm the ratio between γ_r and γ_{nr} reverses in favor of the nonradiative decay, diminishing η. At a separation of 10 nm, it drops to approximately 20%. However, the situation is different for the tangential orientation, where γ_r falls below its initial values at any distance z. The experimental data for these studies may be found in Ref. [67].

Previously, the effect has been considered of a sharp dielectric or metallic object on the fluorescence of a single molecule. In the tangential orientation, fluorescence will always drop in the near field of a tip, while its magnitude will depend on the presence of absorption in the tip. For the vertical orientation, whilst enhancement will occur for dielectric tips, quenching will extinguish the fluorescence at small separations from metallic tips. At this point, gold nanoparticles will be considered, where the plasmon resonance of the particle will lead to a strong modification of γ_r and γ_{nr}. As the details of these studies are available in Refs. [18, 72], only a brief account concerning the modification of molecular emission will be provided.

Figure 9.10a shows a map of the excited decay rate $\gamma = \tau^{-1}$ as the particle was scanned laterally across a single terrylene molecule embedded in a thin pT film. The fluorescence lifetime is shortened approximately 22-fold when the particle is on top of the molecule. Figure 9.10b shows a cross-section from Figure 9.10a, where γ is expressed in units of ns^{-1}. In Figure 9.10c, the measured variation of the lifetime is displayed as a function of z. As shown in Figure 9.10d, the calculations predict an increase in both γ_r and γ_{nr} as z is decreased, but γ_{nr} is expected to dominate at very small distances. The symbols in the experimental data plot the measured total decay rate (corresponding to the data in Figure 9.10c) as the particle approached the molecule [18]. It should be noted that simple fluorescence lifetime

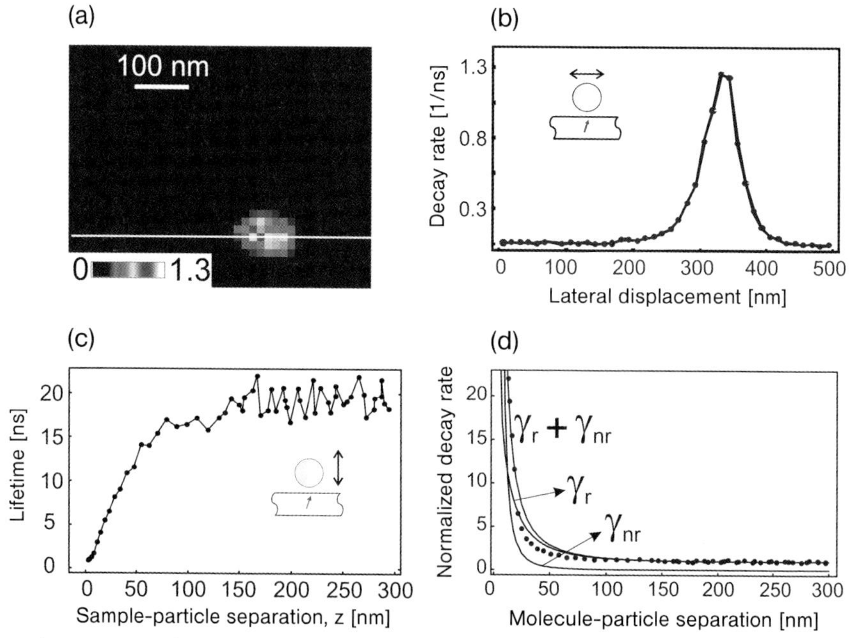

Figure 9.10 Single molecule near-field image with a resonantly excited gold-nanoparticle. (a) Map of the fluorescence decay rate γ. (b) A cross-section from panel (a). (c) Fluorescence lifetime for a $\hat{\mathbf{z}}$ approach. (d) Fluorescence decay rate for a $\hat{\mathbf{z}}$ approach. The symbols show the experimental decay rate γ, while the lines indicate the calculated values of γ_r, γ_{nr}, and the total decay rate $\gamma_r + \gamma_{nr}$. Modified from Ref. [18].

measurements do not provide any information on the radiative and nonradiative contributions separately. The agreement of the total decay rate with the theory is very good, and suggests that γ_{nr} starts to dominate at $z \leq 12\,\text{nm}$ and leads to a decline in η. The equality of γ_{rad} and γ_{nr} marks the point of $\eta = 50\%$. Notably, the nonradiative coupling of molecular energy to the metallic nanoparticle may also be seen as an incoherent energy transfer process, as discussed for FRET. However, in this case the acceptor dissipates its energy in the form of heat instead of radiation.

According to Fermi's Golden Rule, $\gamma_r(\omega)$ is proportional to the photonic density of states $\rho(\omega)$ [70, 73]. Similar to the case of an emitter placed close to a perfect mirror, the density of states, and thus γ_r, are expected to be modified for emitters close to resonant nanostructures [57, 74, 75]. In order to investigate the modification of the molecular emission spectrum under the influence of a gold nanoparticle, the fluorescence spectrum was recorded of a single molecule, using a grating spectrometer as the tip was scanned across the molecule. The integrated spectra are shown in Figure 9.11c, and constitute a total near-field image of S_f. In these measurements, the enhancement reached a factor of 25 just before the molecule underwent photobleaching. In Figure 9.11a are shown the normalized spectrum

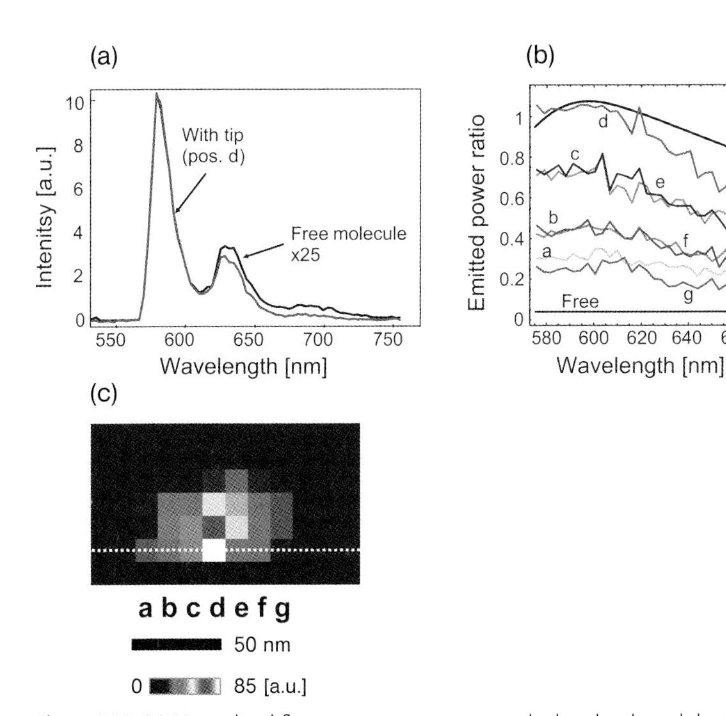

Figure 9.11 (a) Normalized fluorescence spectra of the unperturbed molecule (black) and at maximal enhancement (pixel "d" of the line indicated in panel (c)). (b) Ratio of the spectra measured at pixels a–g to the unperturbed molecule and the theoretical estimate $\gamma_r^{gold}(\lambda)/\gamma_r^{free}(\lambda)$ (black line). (c) Map of the integrated fluorescence intensity as the gold particle is scanned across a terrylene molecule. Modified from Ref. [72].

for the most intense pixel (labeled "d" in Figure 9.11c) and the spectrum of the same unperturbed molecule. It is evident that the enhanced spectrum falls below the reference spectrum at longer wavelengths (630–750 nm).

The fluorescence spectrum of terrylene in Figure 9.11a shows three peaks due to the emission from the lowest vibrational state of the electronic upper state to different vibrational levels of the electronic ground state. The spectral power density $p(\lambda)$ is proportional to $\rho_{up}\gamma_r(\lambda)$, where ρ_{up} denotes the common upper state population. Although ρ_{up} is not known in the absence and presence of the gold particle (because it is the same for all emission channels), the effect of $\gamma_r(\lambda)$ can be probed by comparing the fluorescence enhancement at different wavelengths. In order to achieve this, the ratio $\gamma_r^{gold}(\lambda)/\gamma_r^{free}(\lambda)$ was computed by dividing the spectra obtained with and without the gold particle. This ratio is shown in Figure 9.11b for a selection of pixels from Figure 9.11c. The imbalance in the emission enhancement increases gradually as the distance between the molecule and nanoparticle decreases. At the maximum of fluorescence intensity (curve d in Figure 9.11b), it is found that the highest emission enhancement is assumed at approximately 600 nm – that is, red-shifted from the scattering resonance at 550 nm [75, 76]. The overall tendency of the experimental data is in reasonable agreement with the theoretical ratio of $\gamma_r^{gold}(\lambda)/\gamma_r^{free}(\lambda)$, as plotted by the solid black curve in Figure 9.11b. Recently, Rignler *et al.* have reported similar effects on ensembles of molecules placed between two gold nanoparticles [77].

9.6 Modification of the Spontaneous Emission and Quantum Efficiency by Nanoantennae

It has been seen that the emission of dipolar radiators can be enhanced by engineering the boundary conditions in their near fields. However, it has also been shown that the absorption of light by metals causes quenching, such that the greatest excitation and spontaneous emission enhancements cannot be harvested at very small separations. The question arising then, is whether an arrangement could be devised for achieving a very high spontaneous emission rate, without sacrificing the quantum efficiency. By using full electrodynamics simulations [78], some simple design rules have been applied to plasmonic nanoantennae [75]. For example, by varying the size, aspect ratio, distance, and background medium, quenching can be avoided to a large extent. Moreover, radiative decay enhancements of up to three order of magnitudes can be achieved in the NIR spectral range [61, 75, 79, 80].

Figure 9.12a (and inset) display the behavior of the radiative and nonradiative decay rates of a molecule in front of a gold ellipse as a function of the particle–molecule separation and wavelength, respectively. In contrast to the data in Figure 9.10d, quenching does not dominate the change in the fluorescence lifetime, even at a particle–molecule separation as small as 5 nm. The reason for this paradigm shift is that an elongation of the gold antenna has moved the plasmon resonance

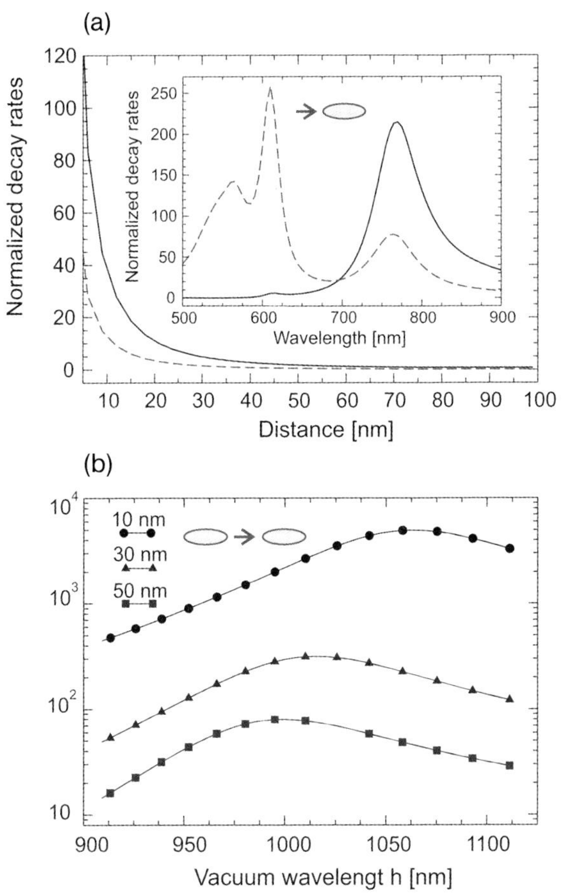

Figure 9.12 (a) Two-dimensional calculation of the normalized radiative (solid) and non-radiative (dashed) decay rates of a molecule as a function of its separation from an elliptical gold particle with long and short axes of 60 nm and 10 nm, respectively. The inset shows the wavelength dependence of the two decay rates; (b) Three-dimensional calculation of the radiative decay rate as a function of wavelength. The molecule is centered between two ellipsoidal particles (long axis 120 nm, short axis 38 nm) separated by 10, 30, and 50 nm. Modified from Ref. [75].

to the NIR regime, where gold absorbs more weakly. Figure 9.12b shows that, by sandwiching a molecule between two ellipsoidal gold nanoparticles, it is possible to achieve enhancements in spontaneous emission in excess of 1000, whilst maintaining a quantum efficiency on the order of 80%. Gold and silver are the best-known materials in the context of plasmonics in the visible range, while aluminum and copper may yield useful resonances in the ultraviolet or visible ranges [80]. Although a wide range of plasmonic geometries has already been proposed [81], the engineering of antenna structures is by no means optimized and will surely remain a topic of research in the coming years.

9.7 Conclusions

In this chapter, some of the most elementary interactions have been discussed that lead to coherent and incoherent exchange of energy at optical frequencies. In particular, attention was focused on the coupling between an emitter and its environment in the near field. Whether the emitter is coupled to a second emitter, a nanoparticle, a complex antenna, or an extended surface, it has been shown that the dipolar couplings play a central role. Alternatively, the boundary conditions imposed by the nearby surfaces modify the density of the photonic states, such that the radiative properties of an emitter are modified.

Extending the concepts proposed in this chapter, to more complex geometries, promises a myriad of interesting photonic architectures for manipulating the optical properties of matter. The realization of nanoantennae would allow the highly efficient coupling of light both into and out of molecules, while nanowires and interfaces would provide nanoscopic avenues for processing optical signals, in similar fashion to the light-harvesting complexes used by Nature in photosynthesis. Although the absorption of material, and hence its dissipation, places real limits on the applications of such devices, the optimization of geometries and correct choice of wavelength of operation should permit efficient optical communications over short distances.

Acknowledgments

The author thanks M. Agio, B. Buchler. I. Gerhardt, S. Götzinger, M. Graf, U. Hakanson, C. Henkel, C. Hettich, T. Kalkbrenner, F. Kaminski, S. Kühn, P. Olk. J. Michaelis, A. Mohammadi, R. Pfab, M. Ramstein, A. Renn, L. Rogobete, C. Schmitt, J. Zimmermann, and J. Zitzmann for their contributions over the years to the information presented here. These studies were supported by the Swiss Ministry of Education and Science (EU IP-Molecular Imaging), the ETH Zurich initiative on Composite Doped Metamaterials (CDM), and the Swiss National Foundation (SNF).

References

1 Nielsen, M.A. and Chuang, I.L. (2000) *Quantum Computation and Quantum Information*, Cambridge University Press.
2 Bouwmeester, D., Ekert, A.K., and Zeilinger, A. (2000) *The Physics of Quantum Information: Quantum Cryptography, Quantum Teleportation, Quantum Computation*, 1st edn, Springer.
3 Jackson, D. (1999) *Classical Electrodynamics*, John Wiley & Sons, Inc.
4 Zumofen, G., Mojarad, N.M., Sandoghdar, V., and Agio, M. (2008) *Phys. Rev. Lett.*, **101**, 180404.
5 Berman, P.R. (1994) *Cavity Quantum Electrodynamics*, Academic Press.
6 Götzinger, S., Menezes, L.de S., Mazzei, A., Kühn, S., Sandoglidar, V., and Benson, O. (2006) *Nano Lett.*, **6**, 1151.
7 Gerhardt, I., Wrigge, G., Bushev, P., Zumofen, G., Agio, M., Pfab, R., and

Sandoghdar, V. (2007) *Phys. Rev. Lett.*, **98**, 033601.

8 Wrigge, G., Gerhardt, I., Hwang, J., Zumofen, G., and Sandoghdar, V. (2008) *Nat. Phys.*, **4**, 60.

9 Basche, T., Moerner, W.E., Orrit, M., and Wild, U. (1999) *Single Molecule Spectroscopy*, John Wiley & Sons, Inc.

10 Moerner, W.E. and Orrit, M. (1999) *Science*, **283**, 1670.

11 Tamarat, P., Maali, A., Lounis, B., and Orrit, M. (2000) *J. Phys. Chem.*, **104**, 1.

12 Mie, G. (1908) *Ann. Phys.*, **25**, 378.

13 Bohren, C.F. and Huffman, D.R. (1983) *Absorption and Scattering of Light by Small Particles*, John Wiley & Sons, Inc.

14 Kreibig, U. and Vollmer, M. (1995) *Optical Properties of Metal Clusters*, Springer, Berlin.

15 Hakanson, U., Agio, M., Kühn, S., Rogobete, L., Kalkbrenner, T., and Sandoghdar, V. (2008) *Phys. Rev. B*, **77**, 77.

16 Meier, M. and Wokaun, A. (1983) *Opt. Lett.*, **8**, 581.

17 Kalkbrenner, T., Håkanson, U., Schädle, A., Burger, S., Henkel, C., and Sandoghdar, V. (2005) *Phys. Rev. Lett.*, **95**, 200801.

18 Kühn, S., Håkanson, U., Rogobete, L., and Sandoghdar, V. (2006) *Phys. Rev. Lett.*, **97**, 017402.

19 Loudon, R. (2000) *Quantum Theory of Light*, Oxford University Press.

20 Haroche, S. (1992) Cavity quantum electrodynamics, in *Fundamental Systems in Quantum Optics* (eds J. Dalibard, J.-M. Raimond, and J. Zinn-Justin), North-Holland, Amsterdam, pp. 767–940.

21 Mazzei, A., Götzinger, S., Menezes, L. de S., Benson, O., and Sandoghdar, V. (2007) *Phys. Rev. Lett.*, **99**, 173603.

22 Buchler, B.C., Kalkbrenner, T., Hettich, C., and Sandoghdar, V. (2005) *Phys. Rev. Lett.*, **95**, 063003.

23 Sondermann, M., Maiwald, R., Konermann, H., Lindleinl, N., Peschel, U., and Leuchs, G. (2007) *Appl. Phys. B*, **89**, 489.

24 Hettich, C., Schmitt, C., Zitzmann, J., Kühn, S., Gerhardt, I., and Sandoghdar, V. (2002) *Science*, **298**, 385.

25 Varada, G.V. and Agarwal, G.S. (1992) *Phys. Rev. A*, **45**, 6721.

26 Dicke, R.H. (1954) *Phys. Rev.*, **93**, 99.

27 Akram, U., Ficek, Z., and Swain, S. (2000) *Phys. Rev. A*, **62**, 13412.

28 DeVoe, R.G. and Brewer, R.G. (1996) *Phys. Rev. Lett.*, **76**, 2049.

29 Orrit, M. and Bernard, J. (1990) *Phys. Rev. Lett.*, **65**, 2716.

30 Kalkbrenner, T., Ramstein, M., Mlynek, J., and Sandoghdar, V. (2001) *J. Microsc.*, **202**, 72.

31 Kalkbrenner, T., Håkanson, U., and Sandoghdar, V. (2004) *Nano Lett.*, **4**, 2309.

32 Quinten, M. (1998) *Appl. Phys. B*, **67**, 101.

33 Rechberger, W., Hohenau, A., Leitner, A., Krenn, J.R., Lamprecht, B., and Aussenegg, F.R. (2003) *Opt. Commun.*, **220**, 137.

34 Dahmen, C., Schmidt, B., and von Plessen, G. (2007) *Nano Lett.*, **7**, 318.

35 Sönnichsen, C., Reinhard, B., Liphardt, J., and Alivisatos, A.P. (2005) *Nat. Biotechnol.*, **23**, 741.

36 Reinhard, B.M., Siu, M., Agarwal, H., Alivisatos, A.P., and Liphardt, J. (2005) *Nano Lett.*, **5**, 2246.

37 Nordlander, P., Oubre, C., Prodan, E., Li, K., and Stockman, M.I. (2004) *Nano Lett.*, **4**, 899.

38 Ekinci, Y., Christ, A., Agio, M., Martin, O.J.F., Solak, H.H., and Löffler, J.F. (2008) *Opt. Express*, **16**, 13287.

39 Sommerfeld, A. (1909) *Ann. Phys.*, **28**, 665.

40 Purcell, E.M. (1946) *Phys. Rev.*, **69**, 681.

41 Drexhage, K.H. (1974) *Prog. Optics*, **12**, 165.

42 Chang, R.K. and Campillo, A. (eds) (1996) *Optical Processes in Microcavities*, Advanced Series in Applied Physics, vol. 3, World Scientific, Singapore.

43 Chance, R.R., Prock, A., and Silbey, R. (1978) *Adv. Chem. Phys.*, **37**, 1.

44 Pfab, R.J., Zimmermann, J., Hettich, C., Gerhardt, I., Renn, A., and Sandoghdar, V. (2004) *Chem. Phys. Lett.*, **387**, 490.

45 Barnes, W.L. (1998) *J. Mod. Opt.*, **45**, 661.

46 Harms, G.S., Irngartinger, T., Reiss, D., Renn, A., and Wild, U.P. (1999) *Chem. Phys. Lett.*, **313**, 533.

47 Kreiter, M., Prummer, M., Hecht, B., and Wild, U.P. (2002) *J. Chem. Phys.*, **117**, 9430.

48 Sönnichsen, C., Franzl, T., Wilk, T., von Plessen, G., Feldmann, J., Wilson, O.,

and Mulvaney, P. (2002) *Phys. Rev. Lett.*, **88**, 077402.
49 Henkel, C. and Sandoghdar, V. (1998) *Opt. Commun.*, **158**, 250.
50 Michaelis, J., Hettich, C., Mlynek, J., and Sandoghdar, V. (2000) *Nature*, **405**, 325.
51 Kühn, S., Hettich, C., Schmitt, C., Poizat, J.-P., and Sandoghdar, V. (2001) *J. Microsc.*, **202**, 2.
52 Lindfors, K., Kalkbrenner, T., Stoller, P., and Sandoghdar, V. (2004) *Phys. Rev. Lett.*, **93**, 037401.
53 Chew, H., Wang, D.-S., and Kerker, M. (1979) *Appl. Opt.*, **18**, 2679.
54 Lakowicz, J.R. (2005) *Anal. Biochem.*, **337**, 171.
55 Bian, R.X., Dunn, R.C., Xie, X.S., and Leung, P.T. (1995) *Phys. Rev. Lett.*, **75**, 4772.
56 Sullivan, K. and Hall, D. (1997) *J. Opt. Soc. Am. B*, **14**, 1149.
57 Metiu, H. (1984) *Prog. Surf. Sci.*, **17**, 153.
58 Klimov, V., Ducloy, M., and Letokhov, V. (1996) *J. Mod. Phys.*, **43**, 2251.
59 Chew, H. (1987) *J. Chem. Phys.*, **87**, 1355.
60 Das, P. and Puri, A. (2002) *Phys. Rev. B*, **65**, 155416.
61 Agio, M., Mori, G., Kaminski, F., Rogobete, L., Kühn, S., Callegari, V., Nellen, P.M., Robin, F., Ekinci, Y., Sennhauser, U., *et al.* (2007). *Proc. SPIE*, **6717**, 67170R.
62 Pohl, D.W. (2004) *Philos. Trans. R. Soc. Lond. A*, **362**, 701.
63 Trabesinger, W., Kramer, A., Kreiter, M., Hecht, B., and Wild, U. (2003) *J. Microsc.*, **209**, 249.
64 Trabesinger, W., Kramer, A., Kreiter, M., Hecht, B., and Wild, U. (2002) *Appl. Phys. Lett.*, **81**, 2118.
65 Protasenko, V. and Gallagher, A. (2004) *Nano Lett.*, **4**, 1329.
66 H'dhili, F., Bachelot, R., Rumyantseva, A., and Lerondel, G. (2002) *J. Microsc.*, **209**, 214.
67 Kühn, S. and Sandoghdar, V. (2006) *Appl. Phys. B*, **84**, 211.
68 Ermushev, A., Mchedlishvili, B., Olcinikov, V., and Petukhov, A. (1993) *Quant. Electron.*, **23**, 435.
69 Lide, S.R. (1995) *Handbook of Chemistry and Physics*, 75th edn, CRC Press.
70 Klimov, V., Ducloy, M., and Letokov, V. (2001) *Quant. Electron.*, **31**, 569.
71 Azoulay, J., Debarre, A., Richard, A., and Tchenio, P. (2000) *Europhys. Lett.*, **51**, 374.
72 Kühn, S., Mori, G., Agio, M., and Sandoghdar, V. (2008) *Mol. Phys.*, **106**, 893.
73 Joulain, K., Carnlinati, B., Millet, J.-P., and Greffet, J.-J. (2003) *Phys. Rev. B*, **68**, 245405.
74 Käll, M., Xu, H., Johunsson, P., and Raman, J. (2005) *Spectroscopy*, **36**, 510.
75 Rogobete, L., Kaminski, F., Agio, M., and Sandoghdar, V. (2007) *Opt. Lett.*, **32**, 1623.
76 Anger, P., Bharadwaj, P., and Novotny, L. (2006) *Phys. Rev. Lett.*, **96**, 113002.
77 Ringler, M., Schwemer, A., Wunderlich, M., Nichtl, A., Kürzinger, K., Klan, T.A., and Feldmann, J. (2008) *Phys. Rev. Lett.*, **100**, 203002.
78 Kaminski, F., Sandoghdar, V., and Agio, M. (2007) *J. Comput. Theor. Nanosci.*, **4**, 635.
79 Mohammadi, A., Sandoghdar, V., and Agio, M. (2008) *New J. Phys.*, **10**, 105015.
80 Mohammaldi, A., Sandoghdar, V., and Agio, M. (2009) *J. Comput. Theor. Nanosci.*, **6**, 1.
81 Blanco, L. and Garcia de Abajo, F. (2004) *Phys. Rev. B*, **69**, 205414.

10
Energy Transfer in Single Conjugated Polymer Chains

Manfred J. Walter and John M. Lupton

10.1
Introduction

Conjugated polymers constitute a fascinating class of materials with many unique optical and electronic properties, along with diverse structure–property relationships. Alternating single and double electron bonds between carbon atoms enable electrons to delocalize between carbon atoms in a molecule, resulting in the formation of large molecular orbitals with a high degree of spatial anisotropy. These delocalized orbitals give rise to remarkable electronic characteristics of the material, both on the level of the individual chain and in a bulk film, leading to semiconducting or even metallic behavior, depending on the level of doping. These key discoveries which opened the field of conducting conjugated polymers, eventually led to the award of the 2000 Nobel prize in Chemistry. Conjugated polymers are often considered a subset of organic semiconductors [1], with a wide range of applications in light-emitting diodes, displays, solar cells, lasers, and field-effect transistors [2–6]. This simple categorizing, however, risks understating the inherent fascination emanating from intrinsically disordered and poorly defined systems such as polymers. How does a chemical formula relate to the actual microscopic structure of the molecular chain? How do chains interact with each other in the solid? What is the underlying electronic structure of a chain? Are transitions best described using a band structure picture or considering discrete excitations [7, 8]?

A simple calculation hints at the complexity facing the spectroscopist when dealing with conjugated polymers. Most carbon-based dielectrics have low refractive indices ($n \sim 2$), and consequently dielectric constants of order 3. If a photon with suitable energy impinges on the material, it can raise an electron from an occupied to an unoccupied molecular orbital, thereby changing the charge configuration and thus the nature of the interatomic bonding within the molecule. This transition effectively leads to a vacancy in the original occupied orbital, often referred to as a "hole" in the picture of conventional semiconductor physics. This vacancy, a net positive charge, can form an electrostatic charge pair with the excited electron. Due to the low dielectric screening, coulombic interactions between the opposite charges are strong, so that the resulting charge pair is tightly bound

Single Particle Tracking and Single Molecule Energy Transfer
Edited by Christoph Bräuchle, Don C. Lamb, and Jens Michaelis

ISBN: 978-3-527-32296-1

[9, 10]. The typical dimensions of such a pair are a few nanometers, whereas the polymer chain itself may be hundreds of nanometers in length. How then does the remainder of the polymer chain influence this elementary photoexcitation? Can the excitation move along the chain, or is it localized?

It turns out that conjugated polymers are best thought of as linear arrangements of strongly bound, optically active units of random length, termed *chromophores* [11–20]. The interaction of these chromophores with each other, through dipole–dipole coupling or electron transfer [21, 22], bears many similarities with processes studied in depth in natural light-harvesting complexes [23–28]. The key difference, however, lies in the restricted dimensionality and the intrinsically strong covalent bonding of optically active units to each other. Such interchromophoric energy transfer is crucial to the operation of devices. For example, a polymer chain may contain a simple defect which can quench luminescence. In an organic light-emitting diode (OLED), such a defect may be extremely detrimental to the operation of the device if excitations tend to migrate along the chain and are then swallowed up by the defect [29–32]. On the other hand, there may be a desire to employ one and the same polymer architecture to generate the emission of different colors. Attaching different dye molecules to the ends of the polymer chain can then enable such color tuning, without changing the physical structure of the material [33, 34], provided that the excitations migrate rapidly along the backbone to find the endcap acceptor from which emission occurs [35–38]. Energy transfer may depend very sensitively on the shape of the polymer chain, which can be controlled by the polarity of the solvent employed [29, 39], or even by reversible (bio)chemical reactions performed on the periphery of the chain [40]. Finally, interchromophoric energy transfer is crucial in photovoltaics, where – in analogy to natural photosynthetic complexes – light absorption and charge generation must be spatially separated. Most recently, the usefulness of nonradiative energy transfer has been demonstrated in the context of solar light concentrators; these are passive devices which promise to lower the cost of photovoltaics by guiding the incident light to the solar cell [41]. Such an approach would be particularly appealing by tailoring both intermolecular – and especially long-range intramolecular – energy transfer in complexes based on conjugated polymers.

Unfortunately, conjugated polymers pose severe limitations to in-depth spectroscopy. The nature of the chromophore is ill defined, every polymer chain has a slightly different length, and the actual physical shape of both the chromophore and the overall polymer chain, which will vary from molecule to molecule, will influence interactions between chromophores. Time domain spectroscopy can provide some insight into elementary interchromophoric interaction processes [42, 43], but often fails to reveal the true complexity of the individual chains which is masked by intrinsic disorder. Single-molecule spectroscopy can help to isolate individual molecules making up an ensemble, thereby bypassing energetic and structural disorder which tends to limit spectroscopy [14, 15, 28, 31, 37–39, 44–72]. An individual chromophore can be identified, for example, by a distinct polarization anisotropy, provided that the chromophore is linearly extended in space [37, 38, 56, 59, 60, 71, 73–76]. Resorting to cryogenic temperatures limits dynamic disorder such as spectral diffusion, conformational fluctuations and electron phonon coupling, and can even help to identify individual chromophoric units in the frequency domain by their distinct spectroscopic signature [15, 77]. Following this route, the intrinsic couplings between chromophores on a single polymer chain can be isolated, illuminating the truly microscopic nature of intramolecular interchromophoric energy transfer.

In this chapter we review recent progress in applying single-molecule spectroscopy to conjugated polymer systems to improve our understanding of intramolecular energy transfer. The results are of significance to materials science, where microscopic experimental data are crucial to parameterizing computational models used in heuristic novel material design. Nanoscale energy-transfer phenomena are equally important in biophysical processes, and conjugated polymers can even serve as model systems to illuminate and simulate conformational fluctuations, and folding and packing phenomena in large biological molecules.

First, a brief motivation is provided for applying single-molecule spectroscopy to the field of conjugated polymers, outlining the basic experimental techniques. Before addressing experimental studies of energy transfer phenomena, the elementary aspects of the photophysics of single conjugated polymer chains are reviewed, such as the formation of spectrally distinct chromophores, random spectral and intensity fluctuations of the single emitting state, and the influence of nanoscale shape of the molecule on the emission properties. Energy transfer in single chains manifests itself in a number of observables, including: spontaneous intensity fluctuations due to funneling to quenching sites; a loss of polarization memory; a change in emission color; and transient spectral shifts over the lifetime of the excited state. In addition, a new technique is discussed to simultaneously probe the absorbing and emitting units of the macromolecule by combining single-molecule resonance Raman spectroscopy with fluorescence, and studying the distinct vibronic fingerprints in both cases. Energy transfer is generally considered to be a "cold" process, occurring after internal conversion and nonradiative energy dissipation following excitation. This assumption is tested in the context of conjugated polymers – strongly-bound donor–acceptor systems – by investigating the dependence of energy-transfer characteristics on the initial excitation energy. It has been found that, even at cryogenic temperatures, the intrinsic molecular excitation spectrum is surprisingly broad in the presence of energy transfer. The initial donor excitation energy influences the ultimate acceptor emission spectrum, demonstrating that nonradiative energy dissipation – cooling – is not a local process in these large macromolecules.

10.2 Why Single Chain Spectroscopy?

Conjugated polymers are inherently disordered materials. The chain length distribution is described by the polydispersity in molecular weight, yet many different conformations of both the individual chain and of the chromophore on the chain

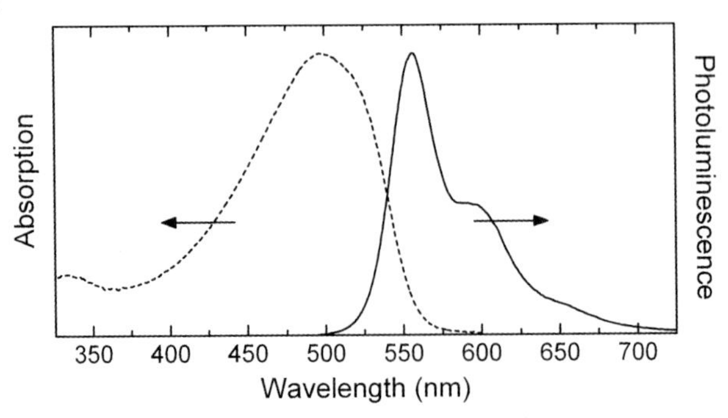

Figure 10.1 Absorption (dashed line) and photoluminescence (solid line) spectra of a toluene solution of the conjugated polymer MEH-PPV (molecular weight $M_N = 200$ kDa).

exist. These structural variations lead to an energetic spread of excited state energies so that the elementary interaction with light appears over a very broad spectral region. Figure 10.1 shows an example of absorption and emission spectra of the commonly used conjugated polymer poly[2-methoxy-5(2′-ethylhexyloxy)-phenylene-vinylene] (MEH-PPV). The absorption spectrum is broad and featureless, and spans a wide range of the higher energy visible spectrum. The emission spectrum is somewhat narrower and exhibits some vibronic progression, but nevertheless still spans a large region of the visible spectrum. Time-resolved luminescence of solutions and films of such materials display two interesting features. First, the emission energy tends to shift to the red within the first few picoseconds after excitation [42, 43, 78]. This spectral transient is a consequence of efficient interchromophoric dipole–dipole coupling, allowing the primary photoexcitations to thermalize by migrating to ever lower-energy-emissive sites in the disordered material. This interchromophoric energy transfer process leads to a second–related–effect, namely an ultrafast depolarization of the fluorescence of the material–that is, a loss in polarization memory [8, 74, 79]. These spectral dynamics yield only limited information on the elementary excited state characteristics. What, for example, is the elementary transition linewidth of a single optically active unit? Measurements of the electronic coherence, using speckle patterns [43], coherent control [43], or photon echo [80], have tended to indicate very substantial homogeneous broadening. These observations would suggest that the absorption spectrum of the material is primarily homogeneously, and not disorder, broadened.

Single-molecule spectroscopy, when performed at cryogenic temperatures, can allow the individual chromophores of the polymer chain to be addressed one at a time, and reveal the elementary transition linewidth, as well as potential correlations of the transition with molecular shape, which can be accessed by the polarization anisotropy. Figure 10.2 shows an example of three low-temperature (5 K) single-molecule (single-chromophore) emission spectra, overlaid on the ensemble photoluminescence spectrum of MEH-PPV. The single-molecule spectra are

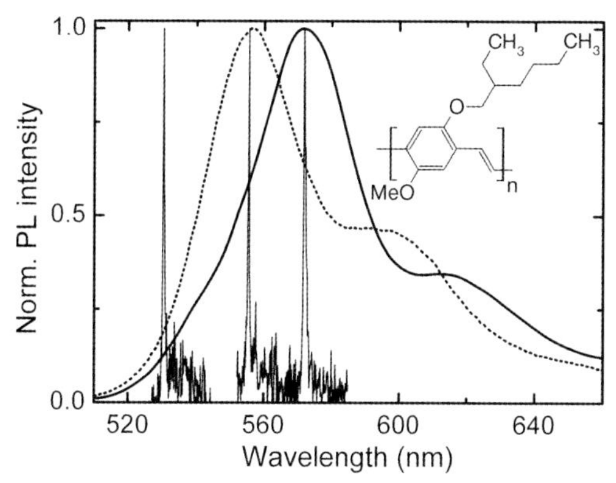

Figure 10.2 Ensemble photoluminescence spectra of MEH-PPV dispersed in a polystyrene matrix (10^{-4} mg ml^{-1}) at 5 K (solid line) and in toluene solution (100 μg l^{-1}) at room temperature (dashed line). Three single-molecule spectra recorded for three different molecules at 5 K are overlaid. The chemical structure of MEH-PPV is also shown. Adapted from Ref. [81].

much narrower than the ensemble spectrum, and tend to scatter within the spectral range covered by the ensemble. The spectra display a narrow peak, the zero phonon line, followed by multiple vibronic side peaks. These can be clearly resolved above the background level through careful integration and summation of spectra, yielding a signal-to-noise ratio (SNR) in excess of 100 : 1 [81]. Electron phonon coupling in the emission spectra of single polymer chains is discussed in more detail below. Clearly, the ensemble spectrum itself is made up of a superposition of discrete emission colors, arising from different chromophores. In order to investigate the intrinsic interchromophoric couplings on the chain, it is crucial to be able to isolate the spectral signatures of individual chromophores. This can be achieved by using single-molecule luminescence spectroscopy.

Site-selective fluorescence spectroscopy (fluorescence line narrowing) [1, 8, 82, 83] and spectral hole burning [84] have, in the past, yielded some insight into the electronic structure of conjugated polymers, but have not proven applicable to highly disordered materials such as MEH-PPV. Most importantly, these steady-state techniques fail to differentiate between different optically active centers being located within one chain or on different chains. Chain selection is achieved in single-molecule spectroscopy simply through spatial isolation. Spectral isolation then reveals the individual chromophores.

10.3
Experimental Approach and Material Systems

The original single-molecule fluorescence experiments were carried out in a photoluminescence excitation configuration; that is, by detecting the spectrally integrated emission and scanning a narrow excitation laser over the molecular absorption [72]. In the following sections, attention will be focused primarily on spectrally dispersed luminescence detection following excitation by a single laser wavelength. This approach is only of limited applicability to single molecules in crystalline matrices, as these materials tend to exhibit very narrow absorption lines and may randomly drift in and out of resonance with a narrow exciting laser line. The elementary transitions in conjugated polymers appear to be broader, not least due to the presence of multiple coupled chromophores, so that reasonably constant molecular excitation densities and emission intensities can be reached as a function of time. This approach of single-wavelength excitation and spectrally dispersed detection was initially pursued in the spectroscopy of single- semiconductor nanocrystals, which exhibit a broad continuum-like absorption [85, 86].

Figure 10.3 illustrates the basic experimental set-up together with typical single-molecule data. A laser impinges at an angle of approximately 30° to the normal on an aluminum-backed quartz substrate covered with the single-molecule sample. Typical single-molecule samples consist of a low concentration (micromolar) dispersion of the conjugated polymer in a highly pure toluene : polystyrene (or zeonex) solution, spin-coated on top of the quartz substrate. The substrate is then mounted on the cold finger of a microscope cryostat under vacuum. The luminescence is detected in the forward direction, collected by a long working distance microscope objective lens, filtered to remove the excitation light and focused into a monochromator with optional mirror to image either the entire spatial information or the spectrally resolved luminescence onto a cooled charge-coupled device (CCD) camera. The overall set-up is summarized in Figure 10.3a.

Figure 10.3b,c displays typical data recorded with the experimental set-up. The fluorescence microscope image is shown in Figure 10.3b, the restriction in the x-direction is by a slit which selects a column of single molecule spots. This image is recorded by reflecting the light incident on the spectrometer straight onto the CCD camera, using a mirror. Exchanging the mirror for a grating leads to a loss of the spatial x-coordinate information, but provides the emission spectra as a function of y-coordinate (Figure 10.3c). Different single-molecule luminescence spectra are seen in this example, which peak at slightly different wavelengths.

Information on the orientation of the absorbing and emitting dipole is available by placing a half-wave plate in the pathway of the laser, or a polarization filter in the emission pathway, respectively. Most experiments discussed in the following sections were performed under steady-state detection. An ultrafast image intensifier, placed in front of the CCD camera, can enable limited time resolution which is sufficient to reveal slow intramolecular energy transfer processes. In this case, the intensifier is activated for a brief period of time (typically 200 ps), and the delay of this time window following the laser pulse is varied. Alternatively, time-resolved detection can be carried out using time correlated single photon counting, in which only one spatial position at a time is probed rather than recording the temporal evolution of the entire image.

Many conjugated polymer systems exhibit universal spectroscopic features on the single-molecule level, but subtle differences exist in terms of chain morphol-

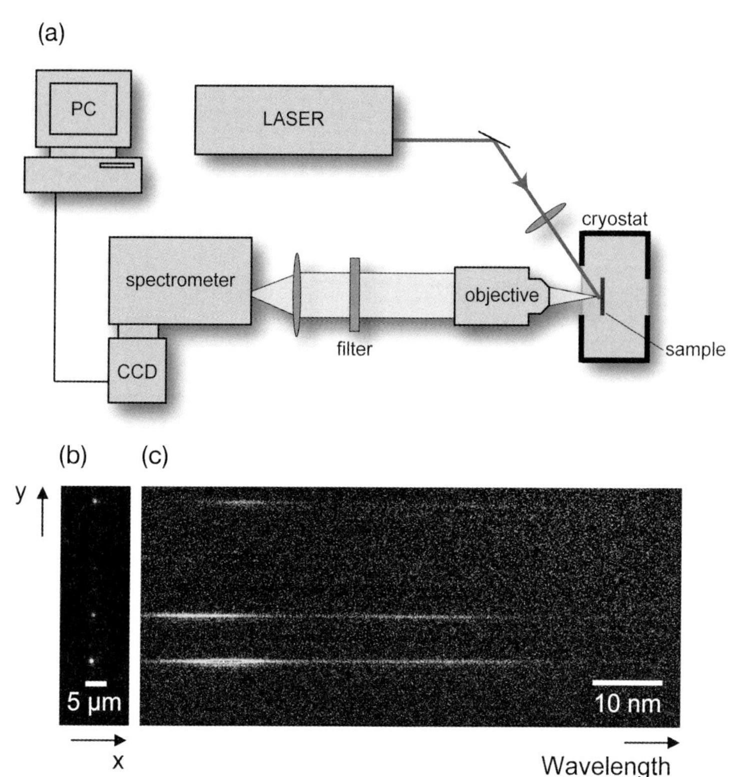

Figure 10.3 (a) Single-molecule spectroscopy set-up consisting of a laser source (which can either be a continuous wave Ar^+ laser or a tunable frequency-doubled 80 MHz Ti : sapphire laser with wavelength range 340–540 nm), the sample being mounted in a cold-finger helium cryostat, a long-working distance microscope objective (0.55 numerical aperture) for fluorescence collection, emission filters, imaging spectrograph and cooled back-illuminated CCD camera. (b) Fluorescence image of a narrow sample region showing three molecules as diffraction-limited spots. (c) The emission of the three molecules can be spectrally resolved by inserting a diffraction grating. The spatial information along the *y*-axis is preserved. Panels (b) and (c) adapted from Ref. [87].

ogy and conformation. The main materials discussed in the context of this review are summarized in Figure 10.4. Figure 10.4a shows the structure of MEH-PPV, which consists of alternating phenylene and vinylene units. This material is rather flexible, as the phenylene rings can twist about the vinylene linkers. Consequently, the ensemble spectra are dominated by energetic disorder broadening. In contrast, methyl-substituted ladder-type poly(*para*-phenylene) (MeLPPP; Figure 10.4b) has a very rigid backbone as the phenylene units are bridged in a full ladder-type configuration. The ensemble disorder broadening of this material is dramatically reduced, resulting in narrower spectral characteristics. Polyfluorene (Figure 10.4c) is viewed as being between these two materials in terms of structural rigidity. Remarkably, this material exists in two spectrally distinct conformations (Figure

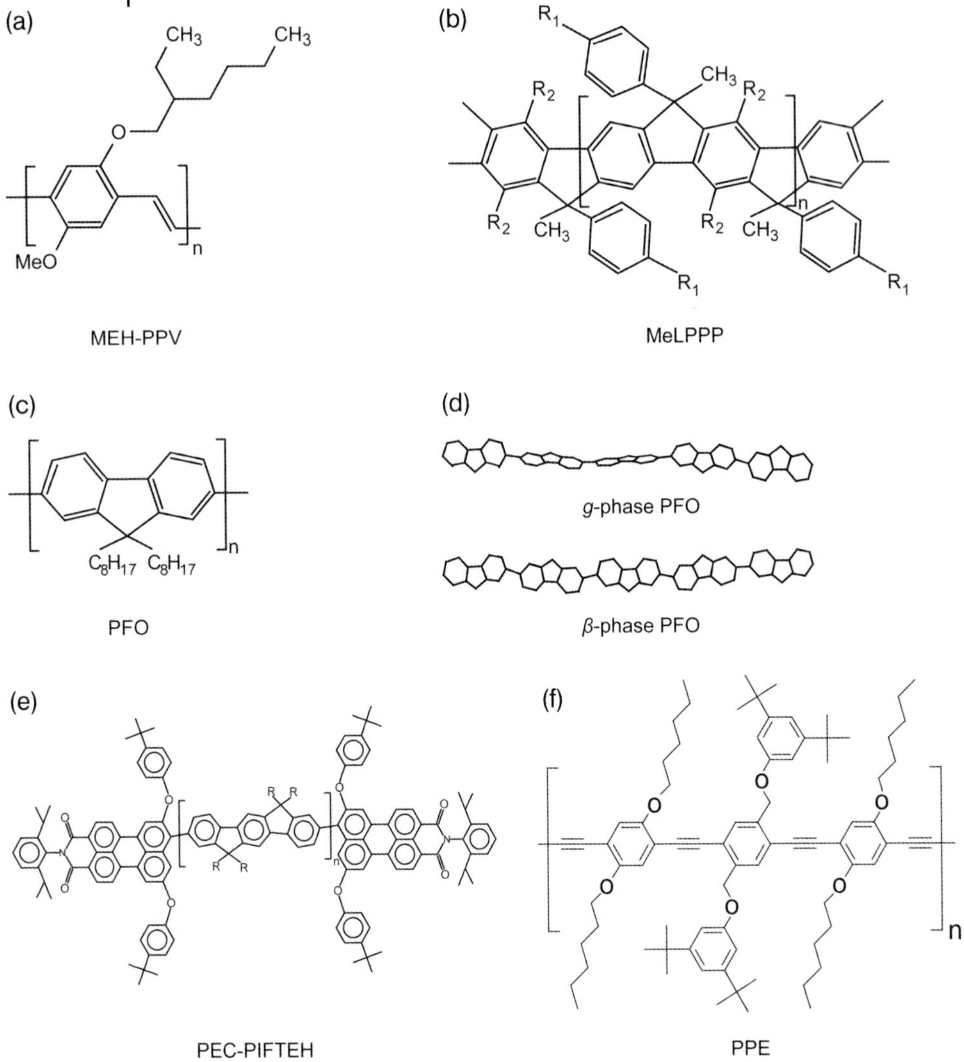

(a) MEH-PPV (b) MeLPPP (c) PFO (d) *g*-phase PFO, *β*-phase PFO (e) PEC-PIFTEH (f) PPE

Figure 10.4 Conjugated polymers used in the studies described in this chapter. (a) MEH-PPV [14]: poly(2-methoxy-5-(2′-ethyl-hexyloxy)-1,4-phenylene-vinylene). (b) MeLPPP [14]: methyl-substituted ladder-type poly(*para*-phenylene); $R_1 = C_{10}H_{21}$, $R_2 = C_6H_{13}$. (c) PFO [62]: poly(9,9-dioctyl-fluorene) which exists in two structural phases. (d) A disordered glassy phase (g-phase PFO) and the more ordered planarized *β*-phase (*β*-phase PFO). (e) PEC-PIFTEH [35, 37]: perylene-endcapped polyindenofluorene; R = ethylhexyl. (f) PPE: poly(phenylene-ethynylene-butadiynylene).

10.4d), which can be identified using X-ray scattering [88]. The intramolecular disordered conformation, in which the repeat units are twisted with respect to each other, is referred to as the "glassy phase," whereas the fully planarized conformation, in which all phenylene units lie in a plane, is termed the "*β*-phase." Polyindenofluorene (PIF) is halfway between polyfluorene and LPPP, in that the phenylene units are bridged in a step-ladder configuration. This material has served as a model system of intramolecular energy transfer as it proved possible to terminate the polymer chains with perylene dye endcaps (PEC-PIFTEH; Figure 10.4e) [35]. As the backbone and endcap have distinct absorption spectra, energy transfer along the polymer chain can be monitored directly by recording the final luminescence of the perylene dye at the end of the chain [35–38, 71]. Figure 10.4f shows the structure of poly(phenylene-ethynylene-butadiynylene), a conjugated polymer which has turned out to be exceptionally useful in applying the techniques of single-molecule Raman scattering for investigating intramolecular energy transfer, as the carbon–carbon triple bonds result in a unique spectral signature at 2200 cm^{-1} [69]. This mode is usually not overlaid by the contaminant bands which tend to plague single-molecule Raman experiments.

10.4 Photophysics of Single Conjugated Polymer Chains

10.4.1 Single Chromophores

When considering the chemical structure of a conjugated polymer on paper, it is tempting to expect the chain to behave as a perfect quantum wire. Although such systems do exist – most notably in polydiacetylene polymerized in its monomeric matrix [50, 89], – most polymer chains do not form quantum wires due to a disruption of π-electron conjugation through chemical or structural defects. The first experiments to establish this important intramolecular scission, which led to the formulation of the polymeric chromophore model, were based on resonance Raman spectroscopy and were carried out some 30 years ago [90, 91]. Scanning the laser wavelength and monitoring the intensity and frequency of the Raman modes demonstrated that the ensemble of polyacetylene is, in general, inhomogeneously broadened. Similar conclusions may be arrived at by comparing the absorption, luminescence and site-selective fluorescence of well-defined model oligomer systems with the polymeric ensemble [92]. These initial studies led to the conclusion that the excited states of conjugated polymers are less band-like, in contrast to many inorganic semiconductors, but are more localized, based on strongly correlated, spatially confined electron-hole pairs – the excitons [7]. Single-molecule spectroscopy provides a vivid illustration of the excitonic and chromophoric nature of conjugated polymers through the comparison of polymers of different chain lengths with well-defined oligomers [15].

Figure 10.5 illustrates three typical single chain spectra of MeLPPP, recorded at 5 K, and shown for different chain lengths. The undecamer oligomer was chosen as a reference system because the ensemble absorption and emission spectra are nearly identical to that of the polymer. The single chain spectrum of the undecamer (Figure 10.5a) displays a peak at 448 nm of approximately 2 nm width. The low-molecular-weight polymer (on average 62 phenylene rings length) displays a

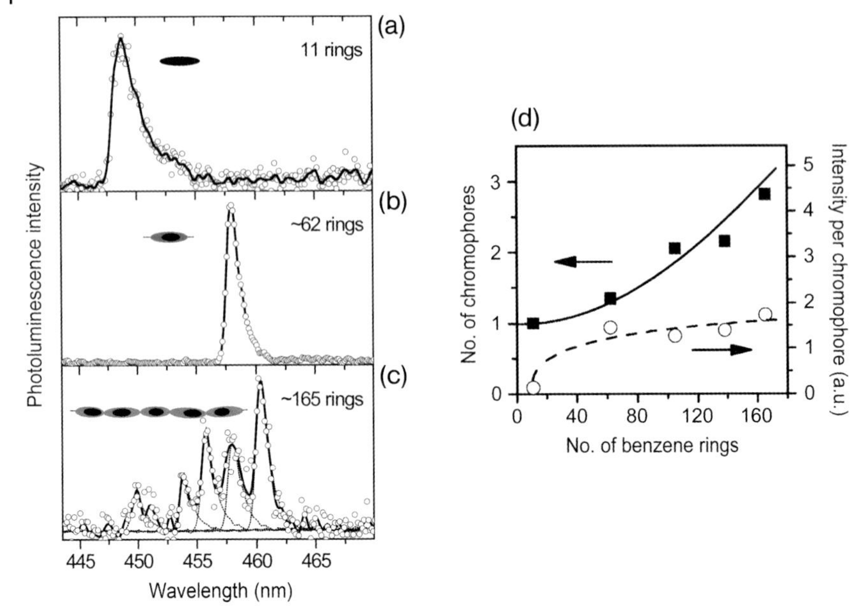

Figure 10.5 Counting chromophores in single conjugated polymer chains. Single-molecule photoluminescence emission spectra at 5 K of the model oligomer and the conjugated polymer MeLPPP. (a) The undecamer; (b) a short-chain conjugated polymer ($M_r \approx 25$ kDa); (c) a long-chain conjugated polymer ($M_r \approx 67$ kDa). Each of the individual peaks in the emission from the long-chain conjugated polymer is identical in shape to the single peak in the emission from the short-chain conjugated polymer (this single peak is superimposed for each peak in panel (c); dotted line). The sketches indicate the difference between exciton size (black) and conjugation length (gray). (d) ■ Variation of the average number of peaks (chromophores) in the photoluminescence spectra with chain length (i.e., number of benzene rings). ○ Average intensity per peak for a total of 108 molecules. Adapted from Ref. [15].

similar peak around 458 nm which is somewhat narrower (Figure 10.5b). As the chain length increases, the number of peaks observed increases dramatically. Figure 10.5c displays a typical spectrum for the high-molecular-weight sample with, on average, 165 rings per chain. In this case, five distinct peaks are observed, which scatter in wavelength between the maxima of the undecamer and the short polymer luminescence, but have a similar spectral width to the short polymer peak [15]. The observation of an increased probability of interrupting the π-electron system and the resulting formation of distinct chromophores can be substantiated by plotting the average number of peaks (chromophores) observed as a function of molecular weight (chain length). This relationship is shown in Figure 10.5d for 108 single chains. The number of peaks increases monotonously with chain length, whilst at the same time the average luminescence intensity, the brightness of a single chromophore on a chain, remains constant. The brightness of a chromophore is determined primarily by the oscillator strength of the π-electron system, which in turn depends on the number of electrons involved in the formation of the molecular orbital. The average length of a chromophore is therefore independent of the overall chain length of the polymer [15]. A further important observation can be made from Figure 10.5, namely that the undecamer is consistently less bright than the polymer chromophore. Following a simple "particle-in-a-box" picture, it might be expected that the chromophore color would be determined by the conjugation length of the system, the spatial extent over which π-electron delocalization occurs. As the undecamer and polymer emit and absorb at approximately the same wavelength, should then not the extent of the π-conjugation be the same? It turns out that the oligomer length is actually equivalent to the spatial extension of the exciton, but not to the spatial range which the exciton can effectively probe. The size of the exciton is primarily given by the extent of coulombic correlation, which is substantial in these low-dielectric materials. Although the polymer chromophore supports an exciton which is approximately 11 rings in size, this exciton can move along the π-system, thereby gaining more oscillator strength [15, 93].

Further proof for the increase in oscillator strength upon going from the oligomer to the polymer derives from the fact that the fluorescence lifetime is almost halved in the polymer with respect to the oligomer [15]. Independent evidence for the fact that the exciton can actually move along the π-system of the chromophore comes from ultrafast (sub-50 fs) fluorescence depolarization measurements [74]. These studies reveal that the excitation can lose its polarization memory on timescales much shorter than those characteristic of interchromophoric energy transfer. The loss of polarization memory is decelerated by including topological defects in the polymer chain – that is, by forcing the formation of chromophores of shorter conjugation length [74].

10.4.2
Interchromophoric Coupling: Energy Transfer

Photoexcitations in conjugated polymers are localized on chromophores, unless the polymer forms a perfect linear π-electron system. The excitations are mobile within the conjugated chromophore, and can transfer from one chromophore to the other by energy transfer [31, 32, 36–38, 56, 60, 63, 69, 71, 82, 94–97]. The most common process for this energy transfer is a Förster-type, dipole–dipole coupling in which one chromophore acts as a donor and the adjacent chromophore constitutes the acceptor. Förster energy transfer is a resonant process, requiring that the acceptor has an energy level which matches the donor energy level from which emission would otherwise occur. In other words, the donor emission spectrum must overlap the acceptor absorption spectrum. In addition, the transition dipoles must be close to parallel in order for the oscillating dipole of the donor to induce a charge density oscillation in the acceptor. These general requirements are illustrated in Figure 10.6. Such energy transfer is often referred to as "through-space energy transfer," as it is strictly the electromagnetic field which mediates the interaction between the two chromophores [21, 22, 27, 30, 36, 54, 77, 79, 94, 98–107].

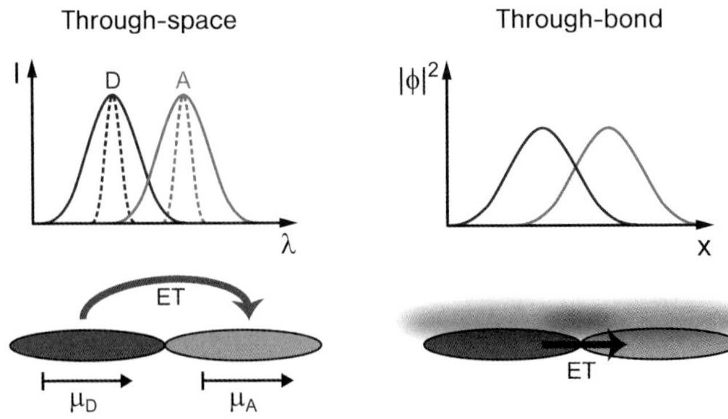

Figure 10.6 The two contributions to interchromophoric energy migration in conjugated polymers. Left: Through-space energy transfer (ET) requires coupling of donor (D) and acceptor (A) dipoles (μ) of adjacent chromophores. According to Förster theory, this necessitates sufficient spectral overlap of donor emission and acceptor absorption to ensure an electrodynamic resonance. Spectral narrowing, which occurs for decreasing temperatures, reduces this spectral overlap and lessens interchromophoric coupling (dashed lines). Right: Through-bond energy transfer is understood in terms of a tunneling-type interaction determined by the wavefunction overlap of the π-electrons in the two chromophores.

The overlap between acceptor absorption and donor emission depends on the relative energetic separation between the two transitions and the individual linewidths, which typically depend on the temperature. As the temperature is increased, more molecular vibrational modes as well as phonon modes associated with the environment are populated, so that the apparent transition linewidth increases. In addition, the magnitude of random spectral fluctuations increases, leading to further line broadening. As illustrated in Figure 10.6 it is therefore expected that, as the temperature is reduced, the strength of interchromophoric coupling decreases, making the excitations on the polymer backbone less mobile [56].

Alternatively, energy transfer can also be mediated by an exchange-type interaction, in which an electron in the excited state of the donor effectively passes its energy to the acceptor by switching with a ground-state electron of the acceptor. This process, which can also be thought of in terms of quantum mechanical tunneling, depends sensitively on distance and will only occur for the closest of proximities of two two-level systems [21, 22, 104]. As the nature of a polymer chromophore is poorly defined, it is often impossible to assess how efficient electron delocalization between chromophores is across the nominal disruptor of the π-system. It is, therefore, usually not trivial to distinguish between pure through-bond and through-space coupling [21, 22, 38, 68, 104]. Single-molecule spectroscopy can help to provide a lower estimate for the strength of dipole–dipole coupling by revealing the intrinsic linewidths of the optical transitions. Here, it will be shown that, in the case of perylene endcaps covalently bound to the polymer backbone, donor dipole–dipole coupling is expected to be rather weak, although an efficient population of the endcap following excitation of the backbone is observed. In this case, the through-bond tunneling-type activation of the acceptor from the final polymer chromophore linking to the fluorescent endcap appears to be dominant.

10.4.3 Blinking and Spectral Diffusion

Single molecules typically exhibit two important characteristics: random fluctuations in emission intensity down to the level of complete modulation of the light signal (blinking) [14, 31, 44–46, 49, 51, 53, 55, 56, 58, 62, 98, 108, 109]; and a random jitter of the emission energy (spectral diffusion) [14, 46, 55, 58, 62, 67, 98].

10.4.3.1 Blinking

"Blinking," or intermittency, is a universal phenomenon and occurs in any saturable system in which the initial conditions depend on the system's history [110–112]. Blinking is observed in many two-level systems, such as quantum dots (QDs) and single atoms [86, 113–116]. In a molecule, an excited state singlet may convert to an excited state triplet by intersystem crossing. The transition from the triplet excited state to the singlet ground state is dipole-forbidden, so that the molecule is effectively blocked in the excited triplet state. As the electron is removed from the ground state, renewed photon absorption cannot occur and the single molecule will appear dark [51, 58, 117]. Alternatively – and more commonly in larger molecules embedded in matrices – the excited electron may be emitted from the molecule to the environment, such as the matrix. The molecule is then charged and unable to absorb or emit subsequently. This process leads to the formation of unpaired electrons, the presence of which can be confirmed by using optically detected magnetic (electron spin) resonance [117]. Electron emission to the environment is a common process in semiconductor nanocrystals [86, 113, 114, 116, 118].

What happens then in a multichromophoric system if one chromophore turns dark because of the formation of a triplet excited state or a metastable charge transfer state? Although a triplet cannot readily relax down to the ground state, it can still absorb a photon, thereby populating a higher-lying triplet excited state. If one chromophore on the polymer chain is in the triplet excited state and an adjacent chromophore absorbs a photon to populate a singlet excited state, the excited singlet may dissipate its energy by nonradiative energy transfer to the adjacent excited triplet [1, 51, 63, 95, 101, 102]. A similar phenomenon can occur if the adjacent chromophore is temporarily dark due to the formation of a radical (pair) state. A radical has a characteristic absorption spectrum, and may simply act as an acceptor of energy, subsequently relaxing following nonradiative pathways [101, 119, 120]. Mechanistically, these processes are equivalent to Auger (secondary) electron emission in atoms, or the related process of nonradiative Auger recombination in electron bands in metals and semiconductors [113, 118]. In a semiconductor, for example, with two electrons in the conduction band and one hole in the valence band, electron-hole recombination may occur nonradiatively by passing

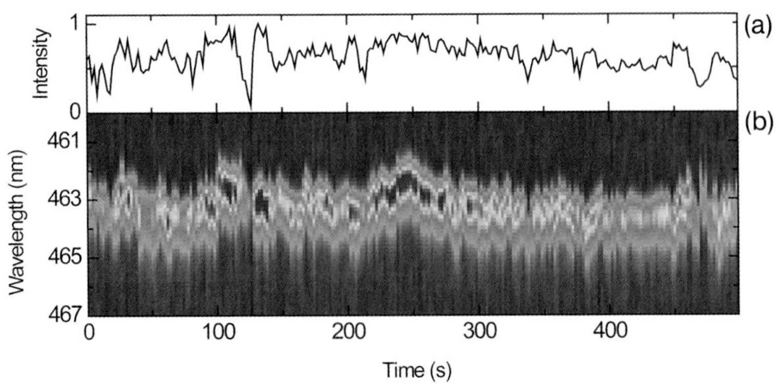

Figure 10.7 Spectral diffusion in single-chain emission. (a) The spectrally integrated time-trace shows intensity fluctuations – "blinking" – which are typical for single-molecule emission. (b) The time-dependent photoluminescence spectrum of a single MeLPPP molecule at 5 K shows strong "spectral diffusion" (temporal resolution 2 s; only the 0-0 peak is shown). Adapted from Ref. [14].

the entire energy of the electron-hole pair onto the second electron, raising its energy in the conduction band.

Blinking is a common phenomenon in conjugated polymers, and is usually related to the formation of a metastable charge-separated state which subsequently quenches excitations on adjacent chromophores [95, 116]. Figure 10.7a shows a typical evolution of single polymer chain fluorescence with time. Interestingly, conjugated polymers often display very weak intersystem crossing, making the formation of triplet excitons rather improbable following the absorption of a photon. This is readily seen in the comparison of the triplet-mediated phosphorescence between electrically (i.e., injection of uncorrelated spin pairs, which lead to 75% triplet formation) and optically generated excitations [121, 122]. In addition, the triplet lifetime in large conjugated polymers such as polyfluorenes can be as great as several seconds, as determined from the single exponential phosphorescence decay in dilute solutions [123]. Polyfluorenes are an interesting case, as they exhibit very little blinking (if any at all), provided that the chain exists in the fully planarized β-phase conformation, which is spectroscopically stable at low temperatures for hours [62]. Given the long triplet lifetime and the comparatively low detection efficiency of the microscope, even a small intersystem crossing yield would lead to the molecule spending most of the time in the triplet excited state, thus quenching luminescence. This is clearly not observed experimentally and suggests that, at least in rigid molecules such as LPPPs and planarized polyfluorenes, intersystem crossing is rather inefficient.

10.4.3.2 **Spectral Diffusion**

Nonresonant excitation of the molecule inevitably leads to nonradiative energy dissipation which can trigger subtle changes to the molecular configuration or its environment. Typically, these effects are stronger, the less-ordered the molecule and the embedding matrix [55]. The result is a random spectral jitter – termed *spectral diffusion* – as illustrated in the example in Figure 10.7b for the polymer MeLPPP. Spectral diffusion is an interesting phenomenon in itself, as it can be quantified statistically in terms of average energetic jumps per spectrum detected. The statistics of spectral diffusion can then be used to learn about the origin of the emitting state. Whereas, spectral diffusion due to random fluctuations of one single emitter typically follows Gaussian statistics, spectral jumps due to switching between different emissive sites, such as different chromophores, generally appears as a Lorenzian process [14]. Spectral diffusion is often driven by subtle changes in the intramolecular or local charge configuration and as such can be used to learn about the actual physical shape of the emitting nanosystem [14, 115, 124]. In the context of energy transfer, random spectral fluctuations – in effect, *dynamic disorder* – can help to overcome the limitations imposed on resonant dipole–dipole coupling by static disorder [56]. This phenomenon, which is crucial to the functioning of biological light-harvesting complexes, carefully arranged pigments in a flexible protein matrix, was recently reproduced under more controlled conditions by simulating spectral fluctuations by exploiting electric field-induced spectral shifts through the Stark effect [64, 105].

10.4.4 The Role of Chromophore Shape

Conjugated polymers are often thought of as consisting of a random spatial arrangement of effectively linear segments. An apparent verification of this picture is provided by the comparison of polarization anisotropy in excitation and in emission. Whereas, the luminescence of a polymer can be excited in many different polarization planes due to the presence of differently oriented absorbing dipoles, energy transfer to an ultimate emitting site induces linear polarization and consequently high anisotropy in emission [37]. However, can flexible systems such as PPV really be interpreted as flexible arrangements of straight segments? Well-defined large oligomers of phenylene–vinylene can serve as model systems of chromophores on PPV, as they exhibit identical spectral signatures, such as peak position, linewidth and vibronic progression, at low temperatures [70]. However, in the case of oligomers, the shape of the individual π-electron system can be accurately determined by the polarization anisotropy in excitation. Remarkably, it was found that the individual chromophore could be bent by up to 80°, leading to virtually unpolarized absorption. In contrast, exciton self-trapping due to conformational relaxation in the excited state [13, 21, 74, 125] limits the effective spatial extent of the exciton wavefunction in this flexible molecular system, thereby increasing the polarization anisotropy in emission with respect to absorption [70]. Spectroscopically, the interesting consequence of bending of the π-system lies in the fact that the transition linewidth increases [17, 70].

Figure 10.8 displays an example of two single oligomer phenylene–vinylene spectra, recorded at 5 K. The narrow spectrum originates from a molecule with a

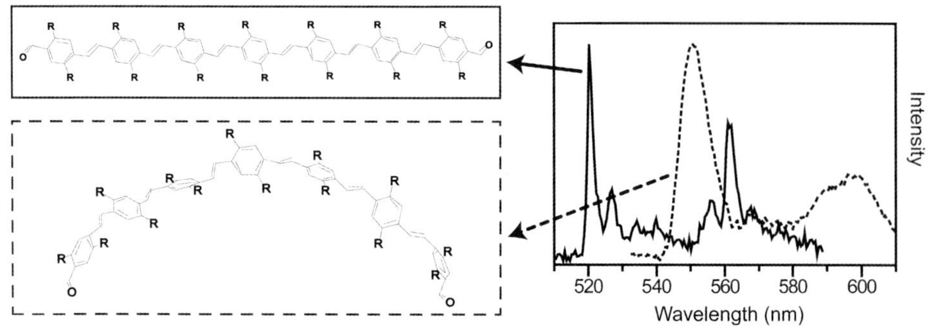

Figure 10.8 Influence of single chromophore shape on PPV emission. Photoluminescence spectrum of a single oligo(phenylene-vinylene) heptamer, which acts as a model system of a MEH-PPV chromophore. The blue-shifted, narrow spectra (solid line) can be attributed to straight chromophores; the red-shifted, broad spectra (dashed line) result from bent subunits. Adapted from Ref. [70].

well-defined transition dipole moment which exhibits linearly polarized absorption, whereas the broad spectrum corresponds to emission from a bent molecule with low polarization anisotropy [70]. As the degree of bending increases, so too does the single molecule transition linewidth. The origin of this effect again lies in spectral diffusion, which is enhanced in disordered (i.e., bent) systems. The consequence for energy transfer is, however, profound. In a multichromophoric chain of MEH-PPV, emission will always occur from bent chromophores (which are also, typically, of lower energy), provided that the excitation cannot be frozen-in on one of the higher-energy straight chromophores due to lack of a suitable acceptor. Spectral broadening in conjugated polymers has frequently been interpreted as a signature of electronic aggregation [14, 29], but in fact it originates primarily from spectral diffusion. In some cases, spectral broadening can also be induced by the emission from on-chain chemical defects [126]. In both cases, spectral broadening is a consequence of interchromophoric energy transfer. Most recently, it has become apparent that shape-controlled spectral diffusion may actually be the dominant broadening mechanism for ensembles of flexible materials such as MEH-PPV [17, 70].

10.5 Energy Transfer in Single Chains

10.5.1 Blinking

The simplest way to arrive at an experimental demonstration of energy transfer in a single conjugated polymer chain is to consider the fluorescence dynamics of the single chain as a function of time. A single MEH-PPV chain, for example, with a molecular weight of hundreds of thousands, can easily contain hundreds of individual chromophore units. As one chromophore blinks off, following the formation of a triplet excited state or a charge-separated state, statistically, another chromophore should switch on, thereby canceling the effect. After all, a bulk polymer film as used, for example, in an OLED, shows a constant fluorescence intensity as a function of time; an ensemble of emitters does not blink (although it may, under certain conditions, reveal the related process of statistical aging or reversible photobleaching). As first demonstrated by Barbara *et al.*, however, even a multichromophoric conjugated polymer chain will typically show discrete blinking, suggesting that only very few chromophores within the chain are actually optically active and contribute to emission [31]. It is straightforward to show that the emitting chromophore is not necessarily the absorbing chromophore; that is, that a chromophore may be dark on the polymer chain but can still absorb light. The conclusion is, consequently, that energy must be transferred from the absorbing chromophores to a small subset of emitting chromophores. In the extreme case, the emitting site constitutes precisely one chromophore, as is readily verified by considering the photon statistics in emission which can show signatures of photon antibunching [102, 127–129].

Huser *et al.* demonstrated elegantly how the blinking dynamics depend on the chain conformation, which in turn can be controlled by variation of the solvent polarity [39]. Extended chains in a good solvent (e.g., chloroform) typically show less blinking. In this case, the chromophores behave as independent entities, with only weak coupling present; the polymer chain can be thought of as a weakly correlated ensemble. This conclusion is also confirmed by considering the photon emission statistics, which are Poissonian. In contrast, the use of a poor solvent, such as toluene, leads to a collapsed chain architecture, forming an excitonic funnel which results in emission occurring from one or very few chromophores [32]. The photon statistics become sub-Poissonian, and the blinking dynamics increase dramatically. Folding of the chain enhances the probability of interchromophoric energy transfer in MEH-PPV. The formation of a fluorescence quencher, such as a charge transfer state, on one chromophore within the chain can then provide an efficient non-radiative relaxation pathway to the excitation on an adjacent chromophore [32, 48, 130–132].

The strength of through-space interchromphore coupling depends on the elementary transition linewidth of the chromophore, which can be much narrower than the ensemble spectral width. As the linewidth decreases, the strength of interchromophoric coupling is expected to decrease [104]. This effect is best investigated in rigid, elongated molecules such as LPPP, where a well-defined number of chromophores exist. In addition, a conformation of the rigid backbone is virtually independent of solvent polarity and temperature, providing a firm scaffold to study purely electronic effects without concern for structural variations. Figure 10.9 shows typical blinking traces of a single LPPP chain recorded at 5 K and at 300 K. The background of the detector is shown in the images for comparison. Whereas, the 300 K trace displays a complete modulation of the emission (remi-

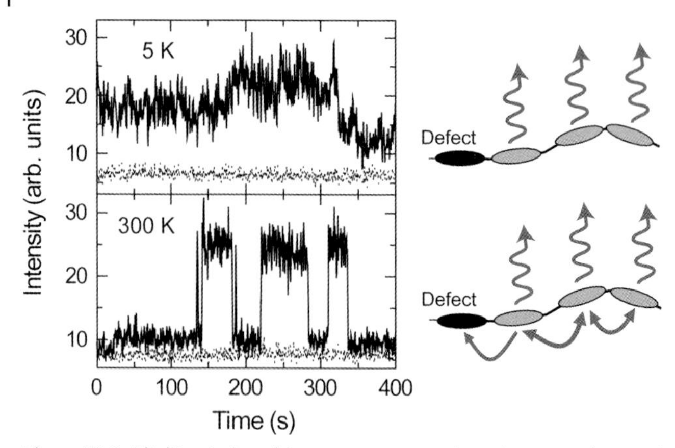

Figure 10.9 Blinking induced by energy transfer. MeLPPP single-chain fluorescence transients recorded at 5 and 300 K for 500 ms integration windows, showing fluorescence intermittency at room temperature. The dotted curve indicates the background noise. The chromophores couple via energy transfer at room temperature, so that a single defect (dark state) can quench the emission of the entire chain. Adapted from Ref. [56].

niscent of random telegraph noise), the 5 K trace reveals only random fluctuations in intensity, with no clear modulation. The noise on the 5 K curve is greater than on the 300 K curve, which suggests that some fluctuations occur on timescales shorter than the measurement window of 0.5 s. The cartoon in Figure 10.9 illustrates the main effect. At room temperature, the single chromophore linewidth exceeds the energetic disorder of the system – that is, the random variation in energy from one chromophore to the next. Resonant dipole–dipole coupling is consequently always possible between chromophores, and the excitation is free to move along the polymer chain. The formation of a quencher, such as an absorbing charge transfer state, on the backbone will then always lead to the fluorescence emission switching off, until the quencher is deactivated (e.g., by slow internal conversion). At low temperatures, the transition linewidths are narrower than the energetic disorder, so that chromophores generally tend not to couple. The formation of a quencher on one chromophore will clearly extinguish that particular chromophore for a given amount of time, but will only have a limited effect on adjacent chromophores [56].

Further evidence for the occurrence of energy transfer to charged quenching sites derives from electric field-dependent measurements of the single-molecule luminescence. The application of a small bias in the absence of charge injection can reversibly switch the single-molecule luminescence on and off, even in the absence of charge injection. In this case, it is proposed that the distance between the charged quencher and the exciton varies slightly, so that nonradiative energy transfer to the quencher is modulated [64, 119].

Interchromophoric coupling itself can enhance blinking by a phenomenon previously termed the "exciton blockade effect" [54, 106], in analogy to the well-known "Coulomb blockade." If a chromophore in a cascade is already populated, for example by a long-lived, charge-separated state, then the intramolecular energy transfer will be disrupted. The final acceptor emission will fluctuate in intensity due not only to its own intrinsic variations in emissive properties but also because of temporal changes in the coupling strength. This effect can be investigated in the dye endcapped polymers, by alternately exciting the backbone or the endcap directly, and detecting only the endcap emission [37, 71]. It is well known that increasing the excitation density of a single molecule will also increase the blinking frequency [31, 32]. Comparing endcap emission under different excitation conditions (direct excitation or energy transfer excitation) is not trivial, as the molecular absorption strength for both cases will differ. Therefore, the power dependence of intensity fluctuations must be considered for both cases of endcap excitation [71]. Under direct excitation, it is found that the perylene endcap remains virtually constant in emission intensity, independent of the excitation power, whereas the energy transfer efficiency fluctuates strongly with time. The fluctuations increase dramatically as the excitation power is raised. Blinking as a consequence of energy transfer is heavily dependent on the overall excitation density, which in turn controls the probability of forming a transient defect on the chain which can quench the fluorescence, interrupting the energy transfer pathway [71].

10.5.2 Polarization Anisotropy

Assuming that the polymer chain is not perfectly linearly extended, but rather contains kinks [76] (as is usually the case), the primary excitation will lose memory of the polarization by which it was excited as it migrates along the polymer chain. The polarization anisotropy can therefore provide a facile measurement to determine the occurrence of interchromophoric energy transfer. Even in nominally extended chains, such as LPPP, structural defects may arise during polymerization, leading to a kinking or even a branching of the chain [60]. Figure 10.10 illustrates how a measurement of the polarization anisotropy in excitation and emission of a single LPPP chain can provide insight into exciton migration. In the simplest case, the same chromophore on the polymer is responsible for emission and absorption. As shown in Figure 10.10a, the polarization anisotropy in this situation remains unchanged in excitation and emission. If energy transfer occurs to an adjacent chromophore of a different orientation, which – possibly because of its orientation in space – cannot absorb the incident radiation directly, then both absorption and emission will show a complete modulation as the halfwave plate and the polarization analyzer are rotated, respectively. However, due to the different orientations of the two chromophores, a phase difference between absorption and emission exists, as illustrated in Figure 10.10b. The situation in Figure 10.10c is more complex; here, both chromophores can absorb, but only one chromophore contributes to emission. Such a case may arise due to the low temperatures employed in this example (5 K), which only allow energetic downhill, but no uphill (phonon-assisted) energy transfer to occur. Here, the excitation is virtually unpolarized, whereas the emission is fully linearly polarized [60].

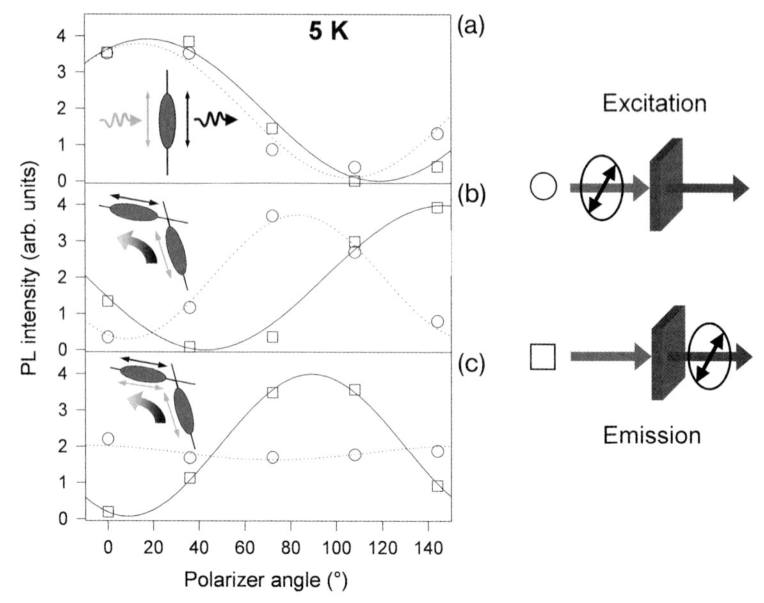

Figure 10.10 Depolarization of fluorescence by intrachain energy transfer. Polarization angle-dependent intensities in emission (□) and excitation (○) with $\cos^2(\theta)$ fits as a guide to the eye (solid line = emission, dashed line = excitation). (a) Linear polarization in excitation and emission from a single dipole. (b) Rotation of the plane of polarization between excitation and emission due to energy transfer between at least two chromophores. (c) Excitation of multiple chromophores at different angles with rapid energy transfer to a single chromophore exhibiting linearly polarized photoluminescence. The sketches show the interaction between different chromophore units on the polymer chain. Gray arrows indicate the polarization of excitation, black arrows that of emission. Adapted from Ref. [60].

It should be noted that care must be taken when interpreting experiments relating to polarization anisotropy. Most importantly, it must be established that the elementary unit itself cannot be bent, and can indeed be thought of as a straight segment supporting one single linear transition dipole [65, 70]. As discussed above, this is not strictly the case in MEH-PPV, where the flexible chromophores themselves can be substantially bent [70]. In LPPP, it was not possible to identify any signature of chromophore bending when investigating well-defined oligomers, which typically resulted in anisotropy traces equivalent to the case of Figure 10.10a [60].

As in the case of the blinking in LPPP, the polarization anisotropy should display signatures of thermally activated exciton migration in the polymer chain, provided that the temperature dependence of the single chromophore emission linewidth is comparable to the energetic disorder, the scatter in peak energies between adjacent units. This turns out to be the situation for PIF. A fully extended chain is denoted by a polarization anisotropy $M = 1$, whereas $M = 0$ corresponds to a molecule that can absorb light of all polarizations and is therefore strongly deformed.

In other words, the anisotropy M can be defined by considering the maximum and minimum emission intensity detected, either under variation of the plane of laser polarization or the plane of polarization of detection, $M = (I_{max} - I_{min})/(I_{max} + I_{min})$. The histograms of the polarization anisotropy for single PEC-PIFTEH molecules, recorded both in emission and absorption, at different temperatures, were studied [37]. As a consequence of intrachain energy transfer, the polarization anisotropy in emission is generally larger than in excitation; fewer chromophores contribute to emission than do to excitation due to energy transfer and exciton funneling. As the temperature is lowered from 300 K to 5 K the anisotropy in excitation remains virtually unchanged. This observation underlines the fact that PEC-PIFTEH has a comparatively rigid backbone structure, which does not undergo any significant structural modifications as the temperature is changed. However, as the temperature is lowered, the single chromophore transition narrows while the overall energetic scatter between chromophores remains comparable. The coupling efficiency between chromophores decreases as the average spectral overlap between adjacent chromophores is reduced, and more chromophores contribute to emission. Consequently, the average polarization anisotropy in emission is reduced as the temperature is lowered [37].

10.5.3
Steady-State Spectroscopy

As the temperature is lowered, the intrinsic chromophoric transition width becomes narrower, making it possible to identify individual chromophores spectroscopically on a single polymer chain. Energy transfer from one chromophore to another can then be identified by considering the disappearance of one spectral signature as a second feature appears in the spectrum. Although this effect is only rarely observed, and is most likely to occur in well-defined short multichromophoric chains of rigid materials (such as LPPP), Figure 10.11 illustrates a concrete example of spectral fluctuations due to energy transfer between two chromophores [14]. It is proposed that an energy transfer cascade exists within the polymer chain, and allows excitations to be passed from the higher-energy chromophore to the lower-energy chromophore [14]. This transfer process depends sensitively on the precise location of the exciton within the conjugated segment of the chromophore, which is typically much larger than the exciton dimension [15, 50, 89, 125, 133]. Small fluctuations in the environment may therefore dramatically alter the strength of interchromophoric dipole–dipole coupling. Alternatively, spectral fluctuations due to spontaneous spectral diffusion may also influence the strength of dipole–dipole coupling. It should be noted that the two lines at 460 nm and 463.5 nm are separated by significantly more than the range covered by random spectral fluctuations [14], leading to the conclusion that the two emission features do indeed arise from two distinct chromophores and not from different (two-level system) configurations of one chromophore.

The occurrence of energy transfer can also be inferred from subtle spectral signatures at room temperature, provided that the energetic separation between

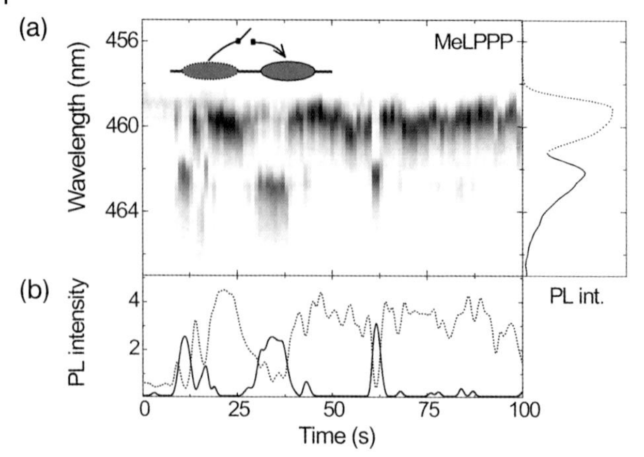

Figure 10.11 Switchable interchromophoric energy transfer in a single molecule. (a) Photoluminescence trace of a single MeLPPP molecule (1 s resolution) showing correlated emission of two chromophore units. The envelopes of the two peaks are also shown. (b) Temporal evolution of the peak intensities of the two discrete emitting units, displaying a clear anti-correlation. Adapted from Ref. [14].

the different spectral entities is sufficient (i.e., it is greater than the linewidth). An example of spectrally resolved energy transfer at room temperature lies in the PIF polymer chains combined with covalently bound perylene endcaps. As the absorption and emission spectra of the PIF backbone and the endcap are distinct, excitation of the PIF backbone and detection of the endcap emission offers direct evidence for intramolecular energy transfer [35–38, 71, 134]. Figure 10.12 shows a fluorescence micrograph of single molecules of the endcap polymer, excited in the backbone absorption at 395 nm, and detected either in a blue spectral band of the backbone emission (peak ca. 430 nm) or in a red spectral region of the endcap emission (peak ca. 570 nm). Most polymer chains (~80%) show only blue emission, indicating an absence of energy transfer from the backbone to the endcap. However, in some cases, energy transfer is almost complete, as only red emission is observed. Occasionally, both red and blue emission is observed from the same single polymer chain; this is labeled as a purple spot in the false color image [38].

The occurrence of highly efficient energy transfer in these model one-dimensional (1-D) systems is rather surprising, as ensemble measurements in solution and diluted films reveal that intramolecular energy transfer is, on average, inefficient [29, 35, 57, 94]. The donor transition is far from resonance from the acceptor transition, particularly at low temperature where the transition lines are very narrow, so that the overall spectral overlap–and hence the coupling efficiency–is very low [38]. The fact that, in approximately 20% of all cases, an efficient intrachain energy transfer does occur most likely originates from ultrafast through-bond energy transfer process between the end of the backbone and the endcap. Such ultrafast coupling in the absence of appreciable spectral overlap is not expected for more weakly bound or even spatially isolated donor–acceptor systems.

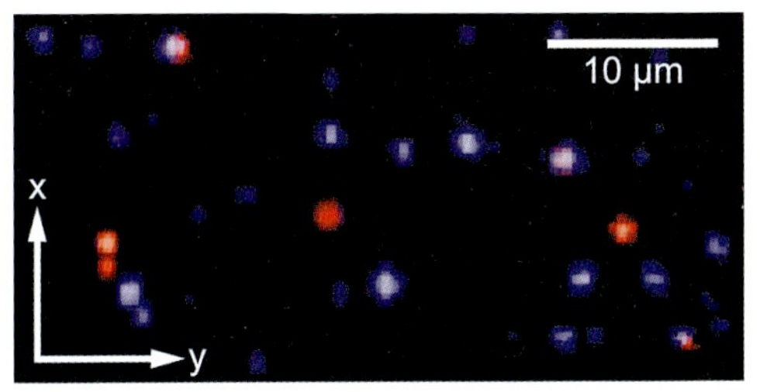

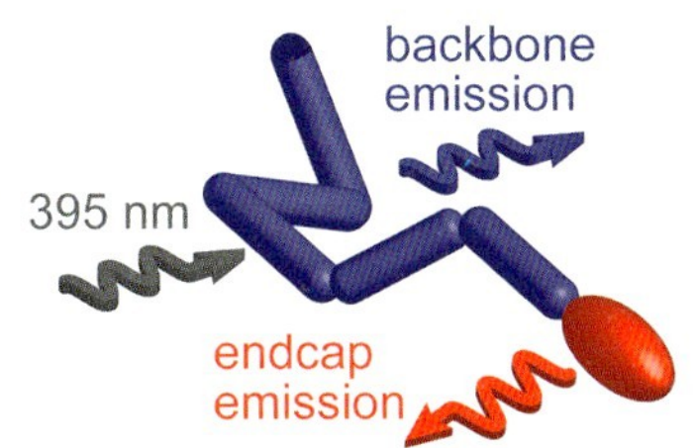

Figure 10.12 False color fluorescence microscope image of perylene-endcapped PIF at 5 K, demonstrating efficient intramolecular energy transfer from the polymer backbone to the endcap in approximately 20% of all chains. Backbone excitation at 395 nm; emission from the polymer (blue), endcap (red), or both (purple). Similar images were also recorded at 300 K. Adapted from Ref. [38].

10.5.4 Time Domain Spectroscopy

Energy transfer between single molecules is often studied using time-resolved spectroscopy techniques such as time-correlated single photon counting [36, 49, 54, 101, 108]. A transfer of energy is then manifested by a reduction in the donor fluorescence lifetime, and often also by a characteristic rise in the acceptor luminescence. Often, it is not trivial to excite donor and acceptor independently at room temperature, where the spectral absorption features may overlap. The endcapped PIF system provides a suitable model material where the spectral features of donor and acceptor emission are separated by over 200 nm. Figure 10.13 displays the results of time-resolved intrachain energy transfer studies. Immediately following excitation of the molecules by the pulsed laser in the absorption of the backbone at 395 nm, the backbone emission appears, visible as blue spots in the upper row. Significant endcap emission (red) is not detected until 300 ps after excitation (shown in the lower row). The backbone emission effectively follows the instrument response function, falling as quickly as it rises at the beginning of the gate. The endcap is much longer-lived [37].

10.5.5 Combined Fluorescence and Raman Scattering

Recording the polarization anisotropy in excitation and emission in principle provides access to the absorbing and emitting unit of the polymer chain, thereby offering evidence – at least in rigid backbone systems – of intramolecular energy transfer. There is, however, a further powerful technique to access the absorbing and emitting chromophores independently, namely *resonance Raman scattering*. Single-molecule Raman scattering is made possible by a plasmonic surface enhancement which occurs close to clusters of noble metal nanoparticles, such as silver and gold. Surface-enhanced Raman scattering (SERS) has attracted a vast

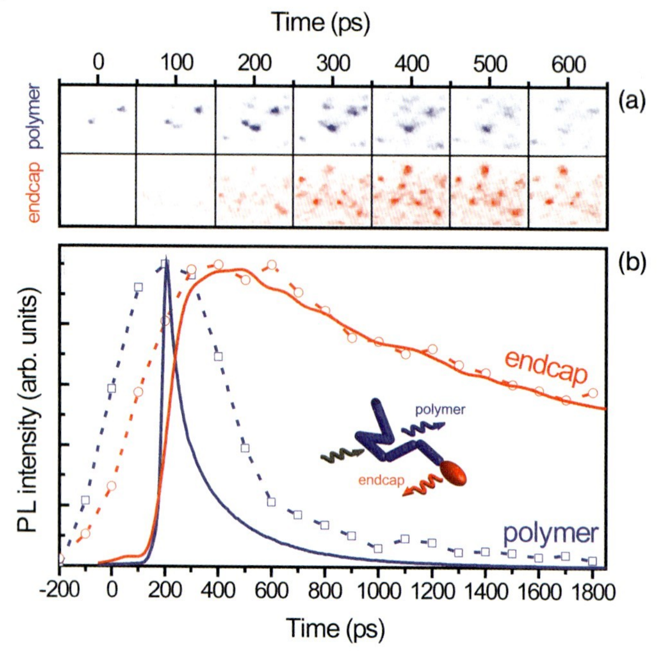

Figure 10.13 Time-resolved fluorescence measurements of PEC-PIFTEH dispersed in a Zeonex matrix at 300 K, excited at 395 nm. (a) Time-gated fluorescence microscopy images of the polymer backbone emission of single molecules (top, blue) and of the endcap emission (bottom, red). The images were obtained in gates of 200 ps width, and illustrate an area of $10 \times 10\,\mu m^2$ on the sample. Note that the two panels correspond to different positions on the sample. (b) Temporal evolution of the fluorescence of the polymer backbone (thick blue line) and the endcaps (thick red line), obtained using a streak camera on an ensemble of PEC-PIFTEH molecules at a concentration of 10^{-4} (w/w) dispersed in Zeonex. The fluorescence intensity transients extracted from fluorescence microscopy images as shown in panel (a) for 65 individual molecules for the polymer backbone emission (□) and 95 individual molecules for the endcap emission (○) are overlaid for comparison. The rise in the endcap acceptor emission is clearly discernible, along with the rapid decay of the backbone donor fluorescence. Adapted from Ref. [37].

amount of attention since its discovery over 30 years ago, as it constitutes one of the most sensitive spectroscopic tools known, based on one of the largest amplification phenomena occurring in Nature. Single-molecule SERS was first demonstrated in 1997 [135, 136], but has encountered repeated skepticism [137], the main problem being that the Raman cross-section is typically more than 14 orders of magnitude smaller than the fluorescence cross-section. The observation of SERS from a single molecule at intensities comparable to single-molecule fluorescence therefore requires an amplification of the scattering process by the same order of magnitude. While this is theoretically possible, such a vast amplification brings problems of its own [137]. Most significantly, the Raman amplitude becomes very sensitive to miniscule changes in molecular conformation, charging and variations in the metal–molecule interaction. Contaminants, such as dust, which invariably appear on the surface of the SERS probe, will often yield strong Raman signatures, as do the nanoparticles, which are usually grown with hydrocarbon ligands. The net result is that it is very easy to observe a signal which looks reminiscent of single-molecule SERS. It is not so easy to make a definite assignment that the signal observed really does originate from the intended analyte molecule [137].

Such an assignment can be made by carrying out simultaneous fluorescence and resonance SERS (SERRS) [69]. Conjugated polymers are particularly suited for this experiment. As with the electronic transitions, the vibronic mode frequencies will scatter slightly from chromophore to chromophore. Consequently, each chromophore will have its own distinct vibronic fingerprint. If SERRS and fluorescence occur from the same chromophore, then the vibronic modes recorded in the two measurements will agree; however, if energy transfer occurs between the chromophores, then a statistical deviation in mode frequencies will be observed. SERRS, which is a virtually instantaneous process, can therefore be used to probe the absorbing chromophore, whereas fluorescence reports on the emitting unit [69].

Single-chain SERRS experiments were carried out at 5 K using silver nanoparticle films grown following the Tollens silver mirror reaction [138]. In these studies, a poly(*para*-phenylene-ethynylene) was employed as an analyte (see Figure 10.4f); this material has carbon–carbon triple bonds on the backbone that provide a characteristic vibrational signature at around 2200 cm^{-1}. In contrast, most contaminant bands observed on a bare SERS substrate typically appear in the region between 1200 and 1800 cm^{-1} [137]. Figure 10.14a shows a comparison of single-molecule fluorescence and Raman spectra recorded for the conjugated polymer. The fluorescence spectrum (red) is shifted on an absolute energy scale, so that its 0-0 transition peak coincides with the origin of the plot. The zero-phonon line extends beyond the intensity scale plotted, so that the vibronic progression in fluorescence is clearly visible. Remarkably, the vibrational peaks observed in fluorescence correspond directly with the peaks seen in Raman scattering (blue). Both the Raman and fluorescence processes probe the ground-state vibrational manifold, but typically have different selection rules. The fact that Raman and fluorescence agree so well in this experiment underlines the benefit of using conjugated polymers to study SERS itself [69].

There are subtle differences between SERRS and fluorescence, illustrated in the temporal evolution of light detected shown in Figure 10.14b. The vibrational modes are much less sensitive to slight conformational changes of the molecule or the environment, which makes them more stable to spectral diffusion. The Raman peaks therefore appear as narrow, straight lines in the 2-D plot. In contrast, fluorescence exhibits spectral diffusion along with significant line broadening; the broad features which drift up and down in photon energy therefore correspond to fluorescence. As the fluorescence line shifts randomly in energy, it may be anticipated that the molecular absorption line will also vary with time. If this is the case, then the degree of resonance with the Raman laser will change with time; consequently, the spectral position of the fluorescence would be expected to correlate

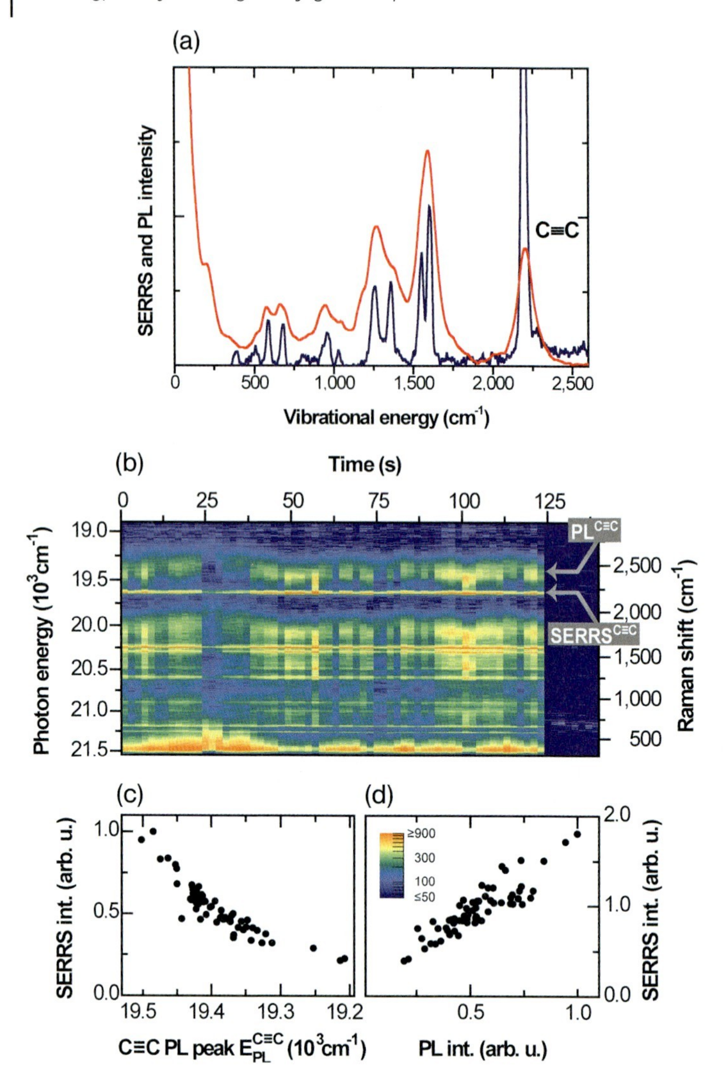

Figure 10.14 Single-chain Raman scattering. (a) Photoluminescence (red) and SERRS (blue) spectra of single poly(phenylene-ethynylene-butadiynylene) chains at 5 K under excitation at $21.84 \times 10^3\,cm^{-1}$ (457.9 nm). The energy of the photoluminescence maximum is shifted to $0\,cm^{-1}$. Note the strong peak around $2200\,cm^{-1}$ due to the C≡C mode. (b–d) Influence of spectral diffusion on simultaneous SERRS and photoluminescence from a single chromophore on a single polymer chain. (b) Intensity–time trace showing photoluminescence and resonant Raman scattering (logarithmic color coding of intensity given in panel (d)). The C≡C bands in luminescence and Raman scattering are labeled. (c) Correlation between SERRS intensity and photoluminescence photon energy of the C≡C mode. (d) Correlation between SERRS and photoluminescence intensity of the C≡C mode. Adapted from Ref. [69].

with the Raman intensity. Precisely this correlation is observed in Figure 10.14c, where the SERRS intensity is plotted as a function of the carbon–carbon triple bond fluorescence peak energy. At the same time, a drift of the molecular absorption out of resonance with the laser will also reduce the excitation density. The detected fluorescence intensity therefore also decreases as the SERRS intensity decreases, as shown in Figure 10.14d. This strong correlation can only arise if SERRS and fluorescence occur from one and the same chromophore – that is, if intramolecular energy transfer is absent. Detecting such a correlation therefore equips the spectroscopist with a means of classifying chromophoric fluorescence in terms of whether or not absorption occurred in the same unit. It should be noted that a comparison of polarization anisotropy in excitation and emission cannot provide conclusive confirmation that intramolecular energy transfer did not occur, as two chromophores may lie in parallel and still couple, without changing the polarization anisotropy [69].

Figure 10.15 illustrates the case of simultaneous Raman scattering and fluorescence under the condition of interchromophoric energy transfer. The trace in

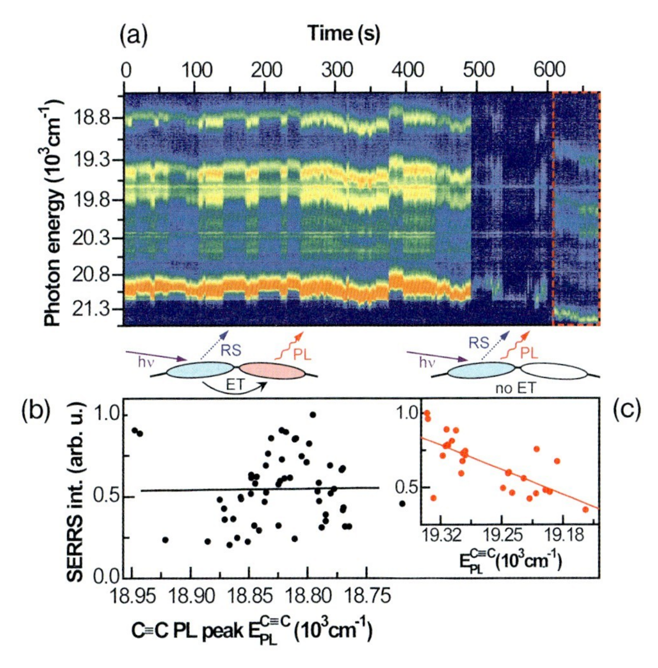

Figure 10.15 Single-chain Raman scattering as a probe of intrachain energy transfer. (a–c) Photoluminescence and SERRS occurring from different chromophores of a single polymer chain. (a) Emission and scattering spectrum (logarithmic scale) as a function of time. (b) Absence of a correlation between SERRS intensity and photoluminescence photon energy of the C≡C mode due to the occurrence of energy transfer. (c) A correlation is observed following the spectral jump to the blue at $t = 610$ s (red box), suggesting SERRS and photoluminescence now arise from the same chromophore, as indicated in the sketch. Adapted from Ref. [69].

Figure 10.15a is similar to that in Figure 10.14, although in this case no correlation between the fluorescence peak position and the SERRS intensity is observed (Figure 10.15b). At approximately 500 s, the molecule suddenly bleaches and both the fluorescence and resonance SERS vanish; the signal returns again at approximately 600 s. In this limited subsequent dataset (600–700 s), the fluorescence peak position and Raman intensity exhibit a correlation, as shown in Figure 10.15c. The absence of a correlation at the onset of the combined Raman and fluorescence trace suggests that, in contrast to the case of Figure 10.14, Raman scattering and fluorescence occur from different chromophores. This situation is illustrated in the cartoon in Figure 10.15. Suddenly, at approximately 500 s into the measurement, the acceptor chromophore photobleaches. When emission from the polymer chain returns, it occurs from the absorbing chromophore, so that a correlation is observed [69].

Although these measurements are technically demanding, they do provide additional information on how interchromophoric coupling in conjugated polymers can alter both the electronic and the vibrational observables of these large macromolecules [69].

10.6
Influence of Initial Excitation Energy on Energy Transfer

In the above discussion of energy transfer, and indeed quite generally in the consideration of resonant dipole–dipole coupling, the exchange of excited state energy is thought to occur independently of the donor's original excitation energy; internal conversion, nonradiative dissipation of energy, precedes energy transfer [57, 82]. However, through-bond energy transfer, as apparently occurs in endcapped conjugated polymers, is observed even in the absence of any significant spectral overlap between donor and acceptor [38]. Indeed, the observation of this efficient coupling in PEC-PIFTEH suggests that the through-bond transfer processes may also be relevant to interchromophoric coupling on the backbone in general. As this form of energy transfer, which is mediated by charge delocalization between chromophores, is expected to be extremely fast, the excitation may not have sufficient time to thermalize on one chromophore before it is passed onto the next. The question is then whether the initial excitation energy of the chromophore, the energy of the photon absorbed, impacts in any way on the energy transfer process or the fate of the ultimate emissive state.

In order to investigate the effect of the initial excitation energy on energy transfer, the emission from the perylene endcap of PEC-PIFTEH was monitored under different excitation wavelengths, using a broadly tunable frequency-doubled Ti:sapphire laser. This experiment can be thought of as a broadband photoluminescence excitation measurement. It is similar to the more conventional photocurrent action spectroscopy in which the photocurrent in a device is probed as a function of excitation wavelength [139]. In this case, however, it is the exciton "current" which is probed – that is, the flux of excitations along the polymer back-

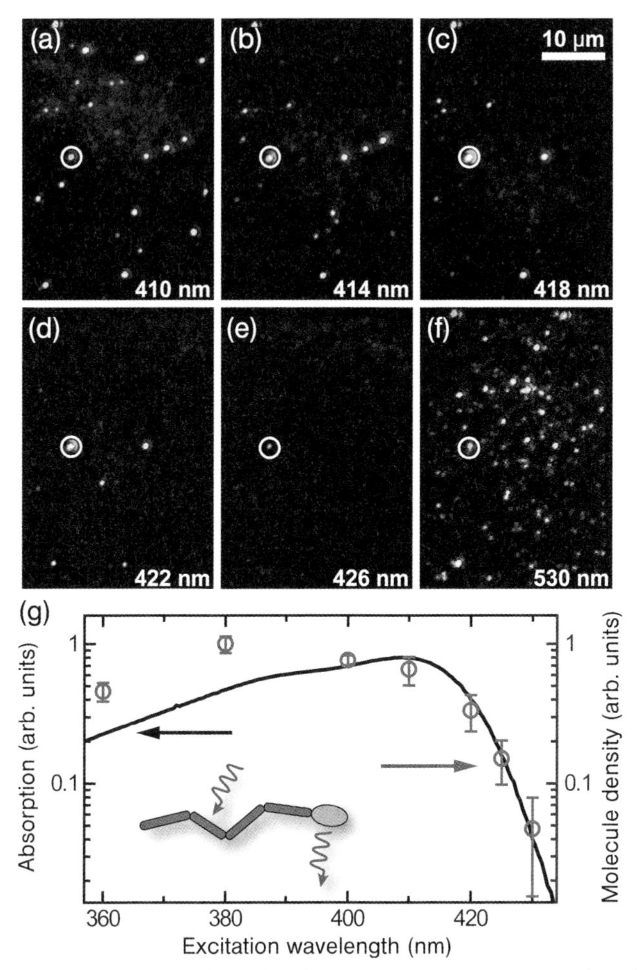

Figure 10.16 Single-chain photoluminescence excitation spectroscopy at 5 K of the PEC-PIFTEH system. Only the emission of the endcap is considered. (a–e) Fluorescence microscopy images of the same molecules for different excitation wavelengths, shown on the same intensity scale at constant excitation power of 2 mW. (f) Direct endcap excitation reveals the highest molecular density (intensity scale reduced by a factor of 2.5). (g) The density of observable molecules (○ grey) characterizes the probability of exciting the polymer backbone and transferring the energy to the endcap. This probability scales directly with the (solution) absorption spectrum of the polymer backbone (black). Adapted from Ref. [71].

bone to the emissive endcap. Figure 10.16 shows fluorescence microscopy images of the same sample region (at 5 K) under excitation at different wavelengths. The absorption of the backbone and the endcap are well separated spectrally, so that the two species can be excited independently. First should be considered the case of energy transfer, where the incident radiation is absorbed by the backbone and only the emission of the endcap is monitored through a 545 nm long-pass filter.

Figure 10.16a shows the spatial density of emitting endcaps under excitation at 410 nm. However, as the excitation is tuned to longer wavelengths, an increasing number of single molecules disappears from the images (Figure 10.16b–e). The highest density of spots is observed under direct excitation of the endcap at 530 nm (Figure 10.16f), due to the fact that approximately 80% of all chains do not display energy transfer to the endcap [37, 38]. Remarkably, it is found that if a particular molecule absorbs at long wavelengths, it will also absorb at shorter wavelengths, which suggests that the intrinsic chromophore absorption spectrum is very broad. An example of this effect is indicated by the molecule circled in the images. It should be noted that, because of the light-harvesting properties of the polymer chain, the direct excitation in Figure 10.16f leads to a lower emission intensity per molecule [71].

Automation of the laser wavelength scanning can be used to record true single-chain photoluminescence excitation spectra. Whereas many spectra do show the expected mirror image relation in terms of linewidth and vibronic progression with the emission spectra, in some cases a broad excitation continuum is superimposed on the narrower molecular resonances. Even at low temperatures, where molecular electronic transitions should be narrow, it is possible to excite the polymer over a wide range of laser wavelengths. It is speculated that this dramatic spectral broadening in excitation arises from an ultrafast dephasing of the excited state due to the exceptionally efficient through-bond coupling of the final polymer chromophore on the PIF chain and the covalently bound perylene endcap. Interestingly, photoluminescence excitation spectroscopy of MEH-PPV has revealed both very broad spectral features along with ultranarrow transitions [66], which are far below the resolution of this broad-range set-up. The coexistence of broad and narrow excitation features is expected to arise from the fact that some chromophores will always exhibit energy transfer to an adjacent unit, whereas others – at least at low temperatures – remain isolated and uncoupled. In addition, as discussed above, the overall physical shape of the chromophore determines spectral diffusion, which leads to line broadening. Narrow lines will only be observed in MEH-PPV in the case of geometrically straight chromophores [65, 70], from which no energy transfer can occur due to lack of a suitable acceptor.

While the average emission intensity per molecule remains approximately constant, the number of molecules detected as a function of excitation wavelength corresponds directly to the ensemble absorption as shown in Figure 10.16g for a total of 1698 single chains. It is not the overall strength of the single molecule absorption which scales with the ensemble absorption, but rather the probability of being able to excite a single molecule. The shorter the excitation wavelength, the more polymer chains have a chromophore which overlaps spectrally with the excitation. The actual absorption process itself is more of a digital nature: either the polymer chain absorbs, or it does not [71].

Although the PEC-PIFTEH polymer is nominally rather rigid, individual chromophores on the polymer can differ significantly in orientation, leading to a reduction in polarization anisotropy in excitation [37]. If shorter excitation wavelengths imply a greater probability of exciting a single chain which exhibits energy transfer, as inferred by Figure 10.16, it is to be expected that an increasing number of chromophores will become involved in the absorption process as the excitation wavelength is reduced. The polarization anisotropy should consequently decrease as the excitation wavelength is reduced. It is found that histograms of the polarization anisotropy under backbone excitation and detection of the emission from the endcap (i.e., the fluorescence resonant energy transfer (FRET) case) display a marked dependence on excitation wavelength [71]. As the excitation wavelength is reduced, more chromophores – the resonance energies of which scatter statistically – absorb radiation of a given polarization. The polarization anisotropy therefore decreases as the higher states of the inhomogeneously broadened absorption spectrum are probed. As this is a strictly statistical process, the decrease in anisotropy is witnessed in a shift of the histogram of polarization anisotropy to lower anisotropy values at shorter excitation wavelengths [71].

The involvement of multiple chromophores in the light absorption process implies that excitation energy is passed between individual chromophores before it reaches the fluorescent endcap. Intuitively, this cascade process would not be expected to exert any influence on the final emissive state of the perylene endcap. Remarkably, however, the initial excitation energy does reveal a subtle correlation with the ultimate emission–a consequence of intramolecular energy dissipation on the polymer backbone during the cascade process, which leads to a very slight rearrangement of the endcap's local environment. In effect, this constitutes a physically remote thermodynamic process, considering that the polymer chain length at over 100 nm is more than two orders of magnitude greater than the extension of the perylene dye from which emission occurs. Basically, a careful consideration of the spectral dynamics in acceptor emission following a change of excitation wavelength of the backbone should reveal subtle spectral diffusion driven by the excess photon energy in excitation. Whenever a red-shift in excitation leads to a blue-shift in emission, the shift is accompanied by spectral narrowing: the less energy dissipated on the backbone, the narrower the endcap emission spectrum [71]. The link between endcap emission and backbone excitation can be rationalized by a simple two-level system model. The occurrence of spectral diffusion in molecular compounds is often interpreted in terms of transitions occurring between different local energetic minima [26, 58]. Excitation at low photon energy only probes the emission of one local minimum. As the excitation energy is raised, the higher photon energies lead to the creation of a further, lower-energy potential minimum, for example by polarization of the environment of the emitting molecule. A random averaging over emission from these different states gives rise to both spectral broadening and an apparent red-shift in emission. Lowering the photon energy in excitation then leads to a blue-shift in emission, accompanied by spectral narrowing [71].

The photophysics of organic semiconductors is generally interpreted in the framework of thermalized, tightly-bound excitons [57, 82]. In a photoluminescence excitation experiment, an excitation wavelength dependence of the emitting species would not be expected. Thus, the present case of endcapped polymers is rather intriguing since, as the backbone excitation energy is varied, the energy difference is deposited on the backbone. The energetic gap between backbone and endcap, the energy dissipated in the final step of energy transfer, remains virtually constant. Evidently, an increased energy dissipation on the polymer backbone drives spectral diffusion of the spatially and energetically remote endcap. This observation is significant in terms of understanding the thermodynamics of nanoscale systems – and not least biological light absorbers – in general. Although the absorbing and emitting units on the polymer chain may be separated by tens of nanometers in the present case, energy dissipation in the absorber can significantly influence the overall energetics of the emitter. This form of remote interchromophoric coupling offers an additional conceptional subtlety to developing a complete microscopic understanding of intramolecular interactions in macromolecular aggregates [71].

10.7 Conclusions

Conjugated polymers display a vast array of interchromophoric coupling phenomena, and represent exciting model systems for acquiring knowledge of the very nature of energy transfer in macromolecular aggregates. As experimental techniques improve, we would expect to continue to learn about the influence of physical shape on the formation of chromophores on the polymer chain, and on the interaction between chromophores. On the one hand, polymers appear poorly defined, disordered – if not "dirty" – systems, yet on the other hand the diversity in form and function is often reminiscent of purely biological complexes, with the benefit of the facility of structural and chemical engineering availed by organic chemistry. As the contribution of single-molecule spectroscopic techniques to materials science becomes ever more apparent – not least in an ability to identify low concentrations of structural or chemical defects that ultimately may lead to catastrophic material failure through cascaded amplification during device operation – the question arises as to how the unique optical and electronic properties of elongated macromolecules can be exploited in new generations of devices. In this respect, two goals are apparent, and will define the symbiotic relationship between single-molecule spectroscopy and polymer electronics for years to come:

- First, synthetic methods must evolve to provide a precise control over molecular chain conformation, in order to ultimately simulate intermolecular behavior in the single-molecular entity. For example, the controlled clustering of single polymer strands is conceivable by macromolecular templating, using, for example, large macrocycles [96]. The key idea would then be to induce intermolecular function within the single molecule.

- Second, techniques need to be developed to address and manipulate individual molecules, for instance by reversibly switching the polymer conformation. Previously, we have explored the use of static electric fields in tuning molecular resonances, as well as dipole–dipole coupling [64, 105] – an approach which led to the creation of the single-molecule energy transfer gate, in which the migration of excitation energy could be switched electrically [105]. Polymeric macromolecules offer the prospect of combining different functional elements to interface the molecule with the outside world, as well as offering a variety of functions, such as logic operations or sensory utility, within the smallest conceivable volume.

Acknowledgments

The authors are indebted to Drs Florian Schindler, Klaus Becker, Jürgen Müller, Enrico Da Como, Prof. Jochen Feldmann, and Mr Nicholas Borys for their many contributions to the study of energy transfer in single conjugated polymer chains as discussed in this chapter. They are also most grateful to Professors Ullrich Scherf, Klaus Müllen, and Sigurd Höger for the kind provision of materials used in the experiments described. Financial support by the Volkswagen Foundation, the DFG (Sonderforschungsbereich 486), the NSF (grant No. CHE-ASC 0748473) and the Petroleum Research Fund (grant No. 46795) is greatly appreciated.

References

1 Pope, M. and Swenberg, C.E. (1999) *Electronic Processes in Organic Crystals and Polymers*, 2nd edn, Oxford University Press, Oxford.

2 Forrest, S.R. (2004) *Nature*, **428**, 911.

3 Friend, R.H., Gymer, R.W., Holmes, A.B., Burroughes, J.H., Marks, R.N., Taliani, C., Bradley, D.D.C., Dos Santos, D.A., Brédas, J.L., Lögdlund, M., and Salaneck, W.R. (1999) *Nature*, **397**, 121.

4 Malliaras, G. and Friend, R. (2005) *Phys. Today*, **58**, 53.

5 Lupton, J.M. (2008) *Nature*, **453**, 459.

6 McQuade, D.T., Pullen, A.E., and Swager, T.M. (2000) *Chem. Rev.*, **100**, 2537.

7 Sariciftci, N.S. (1997) *Primary Photoexcitations in Conjugated Polymers: Molecular Exciton versus Semiconductor Band Model*, World Scientific, Singapore.

8 Rauscher, U., Bässler, H., Bradley, D.D.C., and Hennecke, M. (1990) *Phys. Rev. B*, **42**, 9830.

9 Leng, J.M., Jeglinski, S., Wei, X., Benner, R.E., Vardeny, Z.V., Guo, F., and Mazumdar, S. (1994) *Phys. Rev. Lett.*, **72**, 156.

10 Scholes, G.D. and Rumbles, G. (2006) *Nat. Mater.*, **5**, 683.

11 Hu, D.H., Yu, J., Wong, K., Bagchi, B., Rossky, P.J., and Barbara, P.F. (2000) *Nature*, **405**, 1030.

12 Hagler, T.W., Pakbaz, K., and Heeger, A.J. (1994) *Phys. Rev. B*, **49**, 10968.

13 Beenken, W.J.D. and Pullerits, T. (2004) *J. Phys. Chem. B*, **108**, 6164.

14 Schindler, F., Lupton, J.M., Feldmann, J., and Scherf, U. (2004) *Proc. Natl Acad. Sci. USA*, **101**, 14695.

15 Schindler, F., Jacob, J., Grimsdale, A.C., Scherf, U., Müllen, K., Lupton, J.M., and Feldmann, J. (2005) *Angew. Chem., Int. Ed.*, **44**, 1520.

16 Grimm, S., Tabatabai, A., Scherer, A., Michaelis, J., and Frank, I. (2007) *J. Phys. Chem. B*, **111**, 12053.

17 Schwartz, B.J. (2008) *Nat. Mater.*, **7**, 427.

18 Sartori, S.S., De Feyter, S., Hofkens, J., Van der Auweraer, M., De Schryver, F., Brunner, K., and Hofstraat, J.W. (2003) *Macromolecules*, **36**, 500.

19 Liang, J.J., White, J.D., Chen, Y.C., Wang, C.F., Hsiang, J.C., Lim, T.S., Sun, W.Y., Hsu, J.H., Hsu, C.P., Hayashi, M.,

Fann, W.S., Peng, K.Y., and Chen, S.A. (2006) *Phys. Rev. B*, **74**, 085209.

20 Lammi, R.K. and Barbara, P.F. (2005) *Photochem. Photobiol. Sci.*, **4**, 95.

21 Hennebicq, E., Deleener, C., Brédas, J.L., Scholes, G.D., and Beljonne, D. (2006) *J. Chem. Phys.*, **125**, 054901.

22 Van Averbeke, B., Beljonne, D., and Hennebicq, E. (2008) *Adv. Funct. Mater.*, **18**, 492.

23 Sundström, V., Pullerits, T., and van Grondelle, R. (1999) *J. Phys. Chem. B*, **103**, 2327.

24 van Oijen, A.M., Ketelaars, M., Köhler, J., Aartsma, T.J., and Schmidt, J. (1999) *Science*, **285**, 400.

25 Fleming, G.R. and Scholes, G.D. (2004) *Nature*, **431**, 256.

26 Hofmann, C., Aartsma, T.J., and Köhler, J. (2004) *Chem. Phys. Lett.*, **395**, 373.

27 Jang, S.J., Newton, M.D., and Silbey, R.J. (2004) *Phys. Rev. Lett.*, **92**, 218301.

28 Hofmann, C., Aartsma, T.J., Michel, H., and Köhler, J. (2003) *Proc. Natl Acad. Sci. USA*, **100**, 15534.

29 Schwartz, B.J. (2003) *Annu. Rev. Phys. Chem.*, **54**, 141.

30 List, E.J.W., Creely, C., Leising, G., Schulte, N., Schlüter, A.D., Scherf, U., Müllen, K., and Graupner, W. (2000) *Chem. Phys. Lett.*, **325**, 132.

31 Vanden Bout, D.A., Yip, W.T., Hu, D.H., Fu, D.K., Swager, T.M., and Barbara, P.F. (1997) *Science*, **277**, 1074.

32 Yu, J., Hu, D.H., and Barbara, P.F. (2000) *Science*, **289**, 1327.

33 Becker, S., Ego, C., Grimsdale, A.C., List, E.J.W., Marsitzky, D., Pogantsch, A., Setayesh, S., Leising, G., and Müllen, K. (2001) *Synth. Met.*, **125**, 73.

34 Ego, C., Marsitzky, D., Becker, S., Zhang, J., Grimsdale, A.C., Müllen, K., MacKenzie, J.D., Silva, C., and Friend, R.H. (2003) *J. Am. Chem. Soc.*, **125**, 437.

35 Beljonne, D., Pourtois, G., Silva, C., Hennebicq, E., Herz, L.M., Friend, R.H., Scholes, G.D., Setayesh, S., Müllen, K., and Brédas, J.L. (2002) *Proc. Natl Acad. Sci. USA*, **99**, 10982.

36 Hennebicq, E., Pourtois, G., Scholes, G.D., Herz, L.M., Russell, D.M., Silva, C., Setayesh, S., Grimsdale, A.C., Müllen, K., Brédas, J.L., and Beljonne, D. (2005) *J. Am. Chem. Soc.*, **127**, 4744.

37 Becker, K. and Lupton, J.M. (2006) *J. Am. Chem. Soc.*, **128**, 6468.

38 Becker, K., Lupton, J.M., Feldmann, J., Setayesh, S., Grimsdale, A.C., and Müllen, K. (2006) *J. Am. Chem. Soc.*, **128**, 680.

39 Huser, T., Yan, M., and Rothberg, L.J. (2000) *Proc. Natl Acad. Sci. USA*, **97**, 11187.

40 Gaylord, B.S., Heeger, A.J., and Bazan, G.C. (2002) *Proc. Natl Acad. Sci. USA*, **99**, 10954.

41 Currie, M.J., Mapel, J.K., Heidel, T.D., Goffri, S., and Baldo, M.A. (2008) *Science*, **321**, 226.

42 Kersting, R., Lemmer, U., Mahrt, R.F., Leo, K., Kurz, H., Bässler, H., and Göbel, E.O. (1993) *Phys. Rev. Lett.*, **70**, 3820.

43 Kennedy, S.P., Garro, N., and Phillips, R.T. (2001) *Phys. Rev. Lett.*, **86**, 4148.

44 Basché, T., Kummer, S., and Bräuchle, C. (1995) *Nature*, **373**, 132.

45 Dickson, R.M., Cubitt, A.B., Tsien, R.Y., and Moerner, W.E. (1997) *Nature*, **388**, 355.

46 Lu, H.P. and Xie, X.S. (1997) *Nature*, **385**, 143.

47 Moerner, W.E. and Orrit, M. (1999) *Science*, **283**, 1670.

48 Barbara, P.F., Gesquiere, A.J., Park, S.J., and Lee, Y.J. (2005) *Acc. Chem. Res.*, **38**, 602.

49 Hofkens, J., Maus, M., Gensch, T., Vosch, T., Cotlet, M., Köhn, F., Herrmann, A., Müllen, K., and De Schryver, F. (2000) *J. Am. Chem. Soc.*, **122**, 9278.

50 Guillet, T., Berréhar, J., Grousson, R., Kovensky, J., Lapersonne-Meyer, C., Schott, M., and Voliotis, V. (2001) *Phys. Rev. Lett.*, **87**, 087401.

51 Hofkens, J., Schroeyers, W., Loos, D., Cotlet, M., Köhn, F., Vosch, T., Maus, M., Herrmann, A., Müllen, K., Gensch, T., and De Schryver, F.C. (2001) *Spectrochim. Acta A*, **57**, 2093.

52 Hettich, C., Schmitt, C., Zitzmann, J., Kühn, S., Gerhardt, I., and Sandoghdar, V. (2002) *Science*, **298**, 385.

53 Jung, Y.J., Barkai, E., and Silbey, R.J. (2002) *J. Chem. Phys.*, **117**, 10980.

54 Cotlet, M., Gronheid, R., Habuchi, S., Stefan, A., Barbafina, A., Müllen, K.,

Hofkens, J., and De Schryver, F.C. (2003) *J. Am. Chem. Soc.*, **125**, 13609.

55 Kiraz, A., Ehrl, M., Bräuchle, C., and Zumbusch, A. (2003) *J. Chem. Phys.*, **118**, 10821.

56 Müller, J.G., Lemmer, U., Raschke, G., Anni, M., Scherf, U., Lupton, J.M., and Feldmann, J. (2003) *Phys. Rev. Lett.*, **91**, 267403.

57 Wang, C.F., White, J.D., Lim, T.L., Hsu, J.H., Yang, S.C., Fann, W.S., Peng, K.Y., and Chen, S.A. (2003) *Phys. Rev. B*, **67**, 035202.

58 Barkai, E., Jung, Y.J., and Silbey, R. (2004) *Annu. Rev. Phys. Chem.*, **55**, 457.

59 Müller, J.G., Anni, M., Scherf, U., Lupton, J.M., and Feldmann, J. (2004) *Phys. Rev. B*, **70**, 035205.

60 Müller, J.G., Lupton, J.M., Feldmann, J., Lemmer, U., and Scherf, U. (2004) *Appl. Phys. Lett.*, **84**, 1183.

61 Rønne, C., Trägårdh, J., Hessman, D., and Sundström, V. (2004) *Chem. Phys. Lett.*, **388**, 40.

62 Becker, K. and Lupton, J.M., (2005) *J. Am. Chem. Soc.*, **127**, 7306.

63 De Schryver, F.C., Vosch, T., Cotlet, M., Van der Auweraer, M., Müllen, K., and Hofkens, J. (2005) *Acc. Chem. Res.*, **38**, 514.

64 Schindler, F., Lupton, J.M., Müller, J., Feldmann, J., and Scherf, U. (2006) *Nat. Mater.*, **5**, 141.

65 Da Como, E., Becker, K., Feldmann, J., and Lupton, J.M. (2007) *Nano Lett.*, **7**, 2993.

66 Feist, F.A., Tommaseo, G., and Basché, T. (2007) *Phys. Rev. Lett.*, **98**, 208301.

67 Hildner, R., Lemmer, U., Scherf, U., van Heel, M., and Köhler, J. (2007) *Adv. Mater.*, **19**, 1978.

68 Métivier, R., Nolde, F., Müllen, K., and Basché, T. (2007) *Phys. Rev. Lett.*, **98**, 047802.

69 Walter, M.J., Lupton, J.M., Becker, K., Feldmann, J., Gaefke, G., and Höger, S. (2007) *Phys. Rev. Lett.*, **98**, 137401.

70 Becker, K., Da Como, E., Feldmann, J., Scheliga, F., Csányi, E.T., Tretiak, S., and Lupton, J.M. (2008) *J. Phys. Chem. B*, **112**, 4859.

71 Walter, M.J., Borys, N.J., van Schooten, K.J., and Lupton, J.M. (2008) *Nano Lett.*, **8**, 3330–5.

72 Orrit, M. and Bernard, J. (1990) *Phys. Rev. Lett.*, **65**, 2716.

73 Ha, T., Enderle, T., Chemla, D.S., Selvin, P.R., and Weiss, S. (1996) *Phys. Rev. Lett.*, **77**, 3979.

74 Ruseckas, A., Wood, P., Samuel, I.D.W., Webster, G.R., Mitchell, W.J., Burn, P.L., and Sundström, V. (2005) *Phys. Rev. B*, **72**, 115214.

75 Forster, M., Thomsson, D., Hania, P.R., and Scheblykin, I.G. (2007) *Phys. Chem. Chem. Phys*, **9**, 761.

76 White, J.D., Hsu, J.H., Fann, W.S., Yang, S.C., Pern, G.Y., and Chen, S.A. (2001) *Chem. Phys. Lett.*, **338**, 263.

77 Christ, T., Kulzer, F., Weil, T., Müllen, K., and Basché, T. (2003) *Chem. Phys. Lett.*, **372**, 878.

78 Lemmer, U., Mahrt, R.F., Wada, Y., Greiner, A., Bässler, H., and Göbel, E.O. (1993) *Chem. Phys. Lett.*, **209**, 243.

79 Grage, M.M.L., Wood, P.W., Ruseckas, A., Pullerits, T., Mitchell, W., Burn, P.L., Samuel, I.D.W., and Sundström, V. (2003) *J. Chem. Phys.*, **118**, 7644.

80 Scholes, G.D., Larsen, D.S., Fleming, G.R., Rumbles, G., and Burn, P.L. (2000) *Phys. Rev. B*, **61**, 13670.

81 Schindler, F. and Lupton, J.M. (2005) *ChemPhysChem*, **6**, 926.

82 Bässler, H. and Schweitzer, B. (1999) *Acc. Chem. Res.*, **32**, 173.

83 Heun, S., Mahrt, R.F., Greiner, A., Lemmer, U., Bässler, H., Halliday, D.A., Bradley, D.D.C., Burn, P.L., and Holmes, A.B. (1993) *J. Phys. Condens. Matter*, **5**, 247.

84 Romanovskii, Y.V., Bässler, H., and Scherf, U. (2004) *Chem. Phys. Lett.*, **383**, 89.

85 Empedocles, S.A., Norris, D.J., and Bawendi, M.G. (1996) *Phys. Rev. Lett.*, **77**, 3873.

86 Neuhauser, R.G., Shimizu, K.T., Woo, W.K., Empedocles, S.A., and Bawendi, M.G. (2000) *Phys. Rev. Lett.*, **85**, 3301.

87 Schindler, F. (2006) Molekulare Optoelektronik mit einzelnen konjugierten Polymermolekülen, PhD thesis. Ludwig-Maximilians-Universität München.

88 Grell, M., Bradley, D.D.C., Ungar, G., Hill, J., and Whitehead, K.S. (1999) *Macromolecules*, **32**, 5810.

89 Dubin, F., Melet, R., Barisien, T., Grousson, R., Legrand, L., Schott, M., and Voliotist, V. (2006) *Nat. Phys.*, **2**, 32.

90 Shand, M.L., Chance, R.R., Le Postollec, M., and Schott, M. (1982) *Phys. Rev. B*, **25**, 4431.

91 Rossi, G., Chance, R.R., and Silbey, R. (1989) *J. Chem. Phys.*, **90**, 7594.

92 Müllen, K. and Wegner, G. (1998) *Electronic Materials: The Oligomer Approach*, Wiley-VCH Verlag, Weinheim.

93 Rohlfing, M. and Louie, S.G. (1999) *Phys. Rev. Lett.*, **82**, 1959.

94 Nguyen, T.Q., Wu, J.J., Doan, V., Schwartz, B.J., and Tolbert, S.H. (2000) *Science*, **288**, 652.

95 Brédas, J.L., Beljonne, D., Coropceanu, V., and Cornil, J. (2004) *Chem. Rev.*, **104**, 4971.

96 Becker, K., Lagoudakis, P.G., Gaefke, G., Höger, S., and Lupton, J.M. (2007) *Angew. Chem., Int. Ed.*, **46**, 3450.

97 Becker, K., Fritzsche, M., Höger, S., and Lupton, J.M. (2008) *J. Phys. Chem. B*, **112**, 4849.

98 Ha, T.J., Ting, A.Y., Liang, J., Deniz, A.A., Chemla, D.S., Schultz, P.G., and Weiss, S. (1999) *Chem. Phys.*, **247**, 107.

99 Ha, T. (2001) *Methods*, **25**, 78.

100 Jäckel, F., De Feyter, S., Hofkens, J., Köhn, F., De Schryver, F.C., Ego, C., Grimsdale, A., and Müllen, K. (2002) *Chem. Phys. Lett.*, **362**, 534.

101 Hofkens, J., Cotlet, M., Vosch, T., Tinnefeld, P., Weston, K.D., Ego, C., Grimsdale, A., Müllen, K., Beljonne, D., Brédas, J.L., Jordens, S., Schweitzer, G., Sauer, M., and De Schryver, F. (2003) *Proc. Natl Acad. Sci. USA*, **100**, 13146.

102 Hübner, C.G., Zumofen, G., Renn, A., Herrmann, A., Müllen, K., and Basché, T. (2003) *Phys. Rev. Lett.*, **91**, 093903.

103 Heilemann, M., Tinnefeld, P., Mosteiro, G.S., Parajo, M.G., Van Hulst, N.F., and Sauer, M. (2004) *J. Am. Chem. Soc.*, **126**, 6514.

104 Scholes, G.D. (2003) *Annu. Rev. Phys. Chem.*, **54**, 57.

105 Becker, K., Lupton, J.M., Müller, J., Rogach, A.L., Talapin, D.V., Weller, H., and Feldmann, J. (2006) *Nat. Mater.*, **5**, 777.

106 Melnikov, S.M., Yeow, E.K.L., Uji-i, H., Cotlet, M., Müllen, K., De Schryver, F.C., Enderlein, J., and Hofkens, J. (2007) *J. Phys. Chem. B*, **111**, 708.

107 Curutchet, C., Mennucci, B., Scholes, G.D., and Beljonne, D. (2008) *J. Phys. Chem. B*, **112**, 3759.

108 Lin, H.Z., Tabaei, S.R., Thomsson, D., Mirzov, O., Larsson, P.O., and Scheblykin, I.G. (2008) *J. Am. Chem. Soc.*, **130**, 7042.

109 Mirzov, O., Cichos, F., von Borczyskowski, C., and Scheblykin, I.G. (2004) *Chem. Phys. Lett.*, **386**, 286.

110 Hirsch, J.E., Huberman, B.A., and Scalapino, D.J. (1982) *Phys. Rev. A*, **25**, 519.

111 Platt, N., Spiegel, E.A., and Tresser, C. (1993) *Phys. Rev. Lett.*, **70**, 279.

112 Heagy, J.F., Platt, N., and Hammel, S.M. (1994) *Phys. Rev. E*, **49**, 1140.

113 Efros, A.L. and Rosen, M. (1997) *Phys. Rev. Lett.*, **78**, 1110.

114 Chung, I.H. and Bawendi, M.G. (2004) *Phys. Rev. B*, **70**, 165304.

115 Müller, J., Lupton, J.M., Rogach, A.L., Feldmann, J., Talapin, D.V., and Weller, H. (2005) *Phys. Rev. B*, **72**, 205339.

116 Cichos, F., von Borczyskowski, C., and Orrit, M. (2007) *Curr. Opin. Colloid Interface Sci.*, **12**, 272.

117 Zondervan, R., Kulzer, F., Orlinskii, S.B., and Orrit, M. (2003) *J. Phys. Chem. A*, **107**, 6770.

118 Kraus, R.M., Lagoudakis, P.G., Müller, J., Rogach, A.L., Lupton, J.M., Feldmann, J., Talapin, D.V., and Weller, H. (2005) *J. Phys. Chem. B*, **109**, 18214.

119 List, E.J.W., Kim, C.H., Naik, A.K., Scherf, U., Leising, G., Graupner, W., and Shinar, J. (2001) *Phys. Rev. B*, **64**, 155204.

120 Bell, T.D.M., Jacob, J., Angeles-Izquierdo, M., Fron, E., Nolde, F., Hofkens, J., Müllen, K., and De Schryver, F.C. (2005) *Chem. Commun.*, 4973.

121 Reufer, M., Walter, M.J., Lagoudakis, P.G., Hummel, B., Kolb, J.S., Roskos, H.G., Scherf, U., and Lupton, J.M. (2005) *Nat. Mater.*, **4**, 340.

122 Walter, M.J. and Lupton, J.M. (2007) Spin correlations in organic light-emitting diodes, in *Highly Efficient OLEDs with Phosphorescent Materials* (ed. H. Yersin), Wiley-VCH Verlag, Weinheim, pp. 99–129.

123 Hertel, D., Setayesh, S., Nothofer, H.G., Scherf, U., Müllen, K., and Bässler, H. (2001) *Adv. Mater.*, **13**, 65.

124 Müller, J., Lupton, J.M., Rogach, A.L., Feldmann, J., Talapin, D.V., and Weller, H. (2004) *Phys. Rev. Lett.*, **93**, 167402.

125 Tretiak, S., Saxena, A., Martin, R.L., and Bishop, A.R. (2002) *Phys. Rev. Lett.*, **89**, 097402.

126 Becker, K., Lupton, J.M., Feldmann, J., Nehls, B.S., Galbrecht, F., Gao, D.Q., and Scherf, U. (2006) *Adv. Funct. Mater.*, **16**, 364.

127 Tinnefeld, P., Weston, K.D., Vosch, T., Cotlet, M., Weil, T., Hofkens, J., Müllen, K., De Schryver, F.C., and Sauer, M. (2002) *J. Am. Chem. Soc.*, **124**, 14310.

128 Hollars, C.W., Lane, S.M., and Huser, T. (2003) *Chem. Phys. Lett.*, **370**, 393.

129 Kumar, P., Lee, T.H., Mehta, A., Sumpter, B.G., Dickson, R.M., and Barnes, M.D. (2004) *J. Am. Chem. Soc.*, **126**, 3376.

130 Park, S.J., Gesquiere, A.J., Yu, J., and Barbara, P.F. (2004) *J. Am. Chem. Soc.*, **126**, 4116.

131 Palacios, R.E., Fan, F.R.F., Grey, J.K., Suk, J., Bard, A.J., and Barbara, P.F. (2007) *Nat. Mater.*, **6**, 680.

132 Palacios, R.E., Fan, F.R.F., Bard, A.J., and Barbara, P.F. (2006) *J. Am. Chem. Soc.*, **128**, 9028.

133 Mukamel, S., Tretiak, S., Wagersreiter, T., and Chernyak, V. (1997) *Science*, **277**, 781.

134 Muls, B., Uji-i, H., Melnikov, S., Moussa, A., Verheijen, W., Soumillion, J.P., Josemon, J., Müllen, K., and Hofkens, J. (2005) *ChemPhysChem*, **6**, 2286.

135 Nie, S.M. and Emory, S.R. (1997) *Science*, **275**, 1102.

136 Kneipp, K., Wang, Y., Kneipp, H., Perelman, L.T., Itzkan, I., Dasari, R., and Feld, M.S. (1997) *Phys. Rev. Lett.*, **78**, 1667.

137 Pieczonka, N.P.W. and Aroca, R.F. (2005) *ChemPhysChem*, **6**, 2473.

138 Wang, Z.J., Pan, S.L., Krauss, T.D., Du, H., and Rothberg, L.J. (2003) *Proc. Natl Acad. Sci. USA*, **100**, 8638.

139 Köhler, A., dos Santos, D.A., Beljonne, D., Shuai, Z., Brédas, J.L., Holmes, A.B., Kraus, A., Müllen, K., and Friend, R.H. (1998) *Nature*, **392**, 903.

11
Reactions at the Single-Molecule Level

Maarten B. J. Roeffaers, Gert De Cremer, Bert F. Sels, Dirk E. De Vos, and Johan Hofkens

11.1
Introduction

The ever-improving spatiotemporal resolution and detection sensitivity of fluorescence microscopy offer unique opportunities to deepen our insights into the function of chemical and biological catalysts [1, 2]. Because single-molecule microscopy allows for the counting of single turnover events, the distribution of the catalytic activities of different sites in solid heterogeneous catalysts can be mapped [3], or time-dependent activity fluctuations of individual sites in enzymes or chemical catalysts studied [4]. By experimentally monitoring individuals rather than populations, the origin of complex behavior – for example, in kinetics or in deactivation processes – can be successfully elucidated. In this chapter, an overview is presented of the application of single-molecule fluorescence microscopy for studying individual catalysts and single reaction events. An initial discussion of the advances in the field of biocatalysis is followed by descriptions of some key studies on single enzymatic reactions, detailing the consequences of the results obtained in helping to understand the complex dynamic behavior of these biological catalysts. A second section provides information on chemocatalysts, such as catalytic crystals. Here, in addition to observing (single) reaction events, the different types of dynamic processes such as adsorption, desorption, and diffusion – all of which play a crucial role in catalysis – are discussed in more detail.

11.2
Biocatalysis at the Single-Molecule Level

Nearly all biochemical processes are mediated by the action of enzymes which, over billions of years, have been developed and optimized by Nature to a point where they catalyze hundreds of different types of reactions. As a consequence of this, the application of enzymes in organic synthesis has recently attracted much interest, the aim being to obtain drastic increases in both reaction rates and selec-

Single Particle Tracking and Single Molecule Energy Transfer
Edited by Christoph Bräuchle, Don C. Lamb, and Jens Michaelis

ISBN: 978-3-527-32296-1

tivities. Today, gene technology allows enzymes to be expressed on-demand in a variety of microorganisms, such that a huge portfolio of mutants can be created within a reasonably short time span. As a result, the "library" of available enzymes continues to expand at a rapid rate. Unfortunately, the structural complexities of these poly-amino acid strands, when folded into particular three-dimensional (3-D) structures, hampers our in-depth knowledge of the function of the different reactive groups within the protein structure. In this section, an overview is provided as to how fluorescence-based techniques, notably single-molecule fluorescence spectroscopy (SMFS), have contributed to these investigations into enzyme activity.

11.2.1
Kinetics of Single Biocatalysts

Enzyme kinetics are typically described by the Michaelis–Menten model (Equation 11.1) [5]. First, a substrate molecule (S) reversibly forms a complex (ES) with the enzyme's active site (E), after which the actual enzymatic reaction proceeds irreversibly, yielding the free enzyme (E) and a product molecule (P).

$$\mathrm{E}+\mathrm{S} \underset{k_{-1}}{\overset{k_1}{\rightleftharpoons}} \mathrm{ES} \xrightarrow{k_2} \mathrm{E}+\mathrm{P} \tag{11.1}$$

Briggs and Haldane found that, under steady-state conditions, the initial reaction rate (V_0) was given by:

$$V_0 = \frac{k_2 [\mathrm{E}]_{total} [\mathrm{S}]}{[\mathrm{S}] + K_\mathrm{M}} \tag{11.2}$$

$$\text{where } K_\mathrm{M} = \frac{k_{-1} + k_2}{k_1} = \frac{[\mathrm{E}][\mathrm{S}]}{[\mathrm{ES}]} \tag{11.3}$$

This implies that, at low substrate concentrations, the reaction rate is almost directly proportional to the substrate concentration, whereas at high substrate concentrations the enzyme becomes saturated. In the latter case the system is under reaction control, and the rate is independent of any further increase in [S]; thus, a plateau level for the reaction rate is reached. The parameter K_M, which represents the substrate concentration at which the reaction rate equals half the maximum rate, can therefore be regarded as an indicator of whether the system is under reaction control, or under diffusion/complex formation control. In many cases, complicated enzymatic mechanisms with several elementary steps in the catalytic cycle can be simplified to this basic kinetic scheme.

It is important to note that this model is meant to be used for traditional bulk kinetic assays, in which an ensemble-averaged activity value is obtained for the whole enzyme population. So, does the Michaelis–Menten equation also hold when examining individual enzymes [6]? As noted above, enzymes may have extremely complex 3-D structures, in addition to which their structures can demonstrate pronounced conformational dynamics. It is not surprising therefore, that

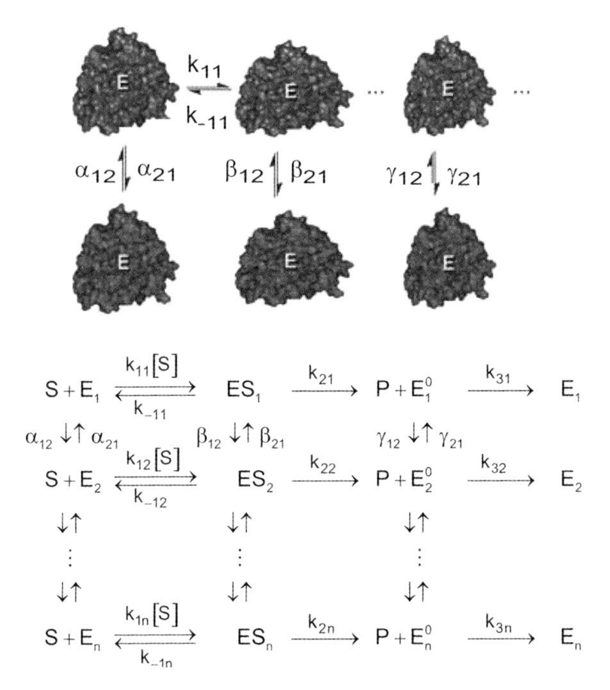

Figure 11.1 The fluctuating enzyme model. The thermodynamic equilibrium between several conformational substates of an enzyme, each with their own specific catalytic parameters, gives rise to time-dependent fluctuations in the enzymatic activity. These fluctuations can be implemented in the classical Michaelis–Menten model by adding an extra dimension that describes the interconversion between the enzymatic substates. Reproduced with permission from Ref. [9].

within a seemingly homogeneous enzyme population, different activities can be found for the individual enzymes, depending on the exact conformational state in which they reside at the moment of assay. This spread of single enzyme activity is known as *static disorder* [7, 8]. Moreover, depending on the timescales of such dynamics in comparison with the experimental time-resolution, time-dependent fluctuations in the activity of one individual enzyme can be found; this is termed *dynamic disorder*. Related to this concept is the so-called *memory effect*, which means that the probability for a turnover to occur is greater shortly after a previous successful enzymatic turnover. In an attempt to take into account the influence of the conformational dynamics on enzymatic activity, a single enzyme variant of the Michaelis–Menten model has been proposed; this is known as the *fluctuating enzyme model* (Figure 11.1) [9]. In this model, an extra dimension is added to the traditional Michaelis–Menten system that describes the thermodynamic equilibrium between the different conformational substates.

In the following sections, a nonexhaustive overview is provided of the key studies that have been conducted with single enzymes and which have proved the added value of single-molecule assays over traditional ensemble assays:

- In *time-averaged single enzyme studies*, the individual enzymes – together with a suitable fluorogenic substrate – are spatially confined in a form of microreactor. A fluorescence-based detection system is then used to quantify the concentration of the product liberated by the enzyme, as a function of time. When, typically, a wide-field fluorescence microscope is used for this purpose, it allows the activities of multiple individual enzymes to be followed simultaneously, albeit with a limited time resolution. Unfortunately, when using this approach effects such as dynamic disorder are difficult to identify.
- In *single-turnover detection*, the activities of the individual enzymes are highlighted. In this case, the turnovers of each enzyme are probed one by one, allowing time-resolved fluctuations such as dynamic disorder within a single enzyme's activity to be analyzed.

11.2.1.1 **Single Enzyme Studies**

Long before single-molecule detection methods were made readily available for *in situ* studies, many research groups had already investigated the *outcome* of a single enzyme's action. By using a fluorogenic substrate, the effect of an individual enzyme could be "amplified" by monitoring the accumulation of liberated fluorescent product molecules. For this, the heavily diluted enzyme solution was confined within a small volume (typically in the picoliter to femtoliter range), which served as a microreactor. The first such attempt was made by Rotman in 1961 [10], who used micrometer-sized droplets of a water-in-oil emulsion as microreactors to study the thermal denaturation of individual β-D-galactosidase enzymes. By starting from an extremely diluted enzyme concentration (subnanomolar) to prepare the emulsion, Rotman was able to assure statistically that most droplets contained, at maximum, one enzyme molecule. The fluorogenic probe used, 6-hydroxyfluoran-β-D-galactose (6HFG), is converted by the enzyme into the fluorescent 6-hydroxyfluoran (Figure 11.2). Following incubation at elevated temperatures, the time-based evolution of the fluorescence intensity of individual droplets was monitored, using wide-field fluorescence microscopy. In this way, Rotman unearthed evidence that heat-induced denaturation caused only a fraction of the enzymes to completely lose their activity, while another fraction retained full activity, rather than there being a gradual activity decrease in the activity of *all* individual enzymes. By using a similar approach, Lee and Brody recently discovered static disorder effects in α-chymotrypsin enzymes [11].

One major drawback of the emulsion-based method is the nonuniformity of the water droplets, as an unknown droplet size may cause the relationship between the intensity and amount of molecules produced by the enzyme to be more complicated. By using photolithographic procedures, small (femtoliter-sized) wells with a very homogeneous size distribution were produced in a polydimethylsiloxane (PDMS) polymer matrix [12]. By sandwiching a solution containing the fluorogenic fluorescein-di-β-D-galactopyranoside probe and β-D-galactosidase enzymes at very low concentration, the activity of the individual enzymes could then be monitored using wide-field fluorescence microscopy (see Figure 11.3). In order to prevent any

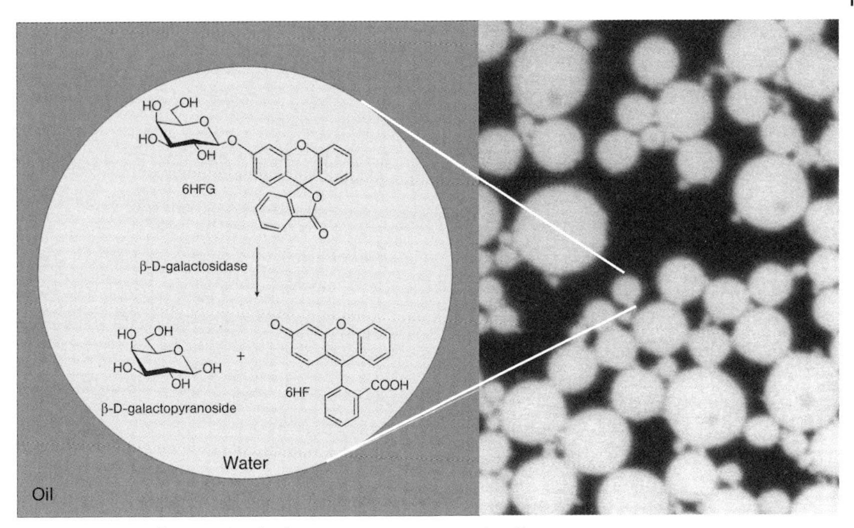

Figure 11.2 Studying individual enzymes using water-in-oil emulsions. The enzyme is confined together with its fluorogenic substrate within the water droplets of an oil-in-water emulsion. Upon enzymatic action, the fluorescent products will accumulate in the droplets; this can be visualized as an increase in the emission intensity.

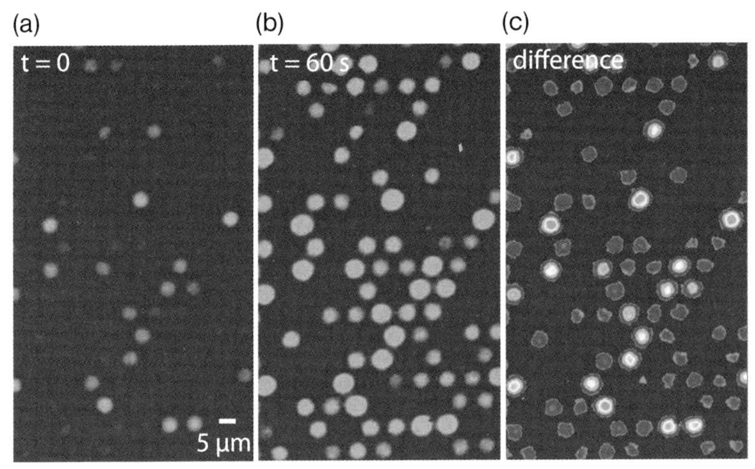

Figure 11.3 Wide-field images of β-D-galactosidase activity in femtoliter chambers. Upon enzymatic action, resorufin accumulates in the chambers, yielding a high emission that originates from chambers containing at least one active enzyme. (a) The initial fluorescence of the chambers immediately after closing. (b) Emission after 1 min of reaction. (c) The intensity difference between panels (a) and (b). This intensity difference is proportional to the enzyme activity. Reproduced with permission from Ref. [12]; © 2005, Macmillan.

nonspecific adsorption of the enzymes onto the PDMS polymers, the chamber walls were precoated with a layer of bovine serum albumin (BSA).

Alternatively, the group of Walt proved that a single β-D-galactosidase enzyme could be captured together with its substrate, resorufin-β-D-galactopyranoside, in an array of 5×10^4 ultrasmall reaction chambers located at the end of an optical fiber bundle [13, 14]. The fiber could then serve as both a light carrier and a micro-reactor, thereby assuring that each microreactor would be well illuminated. This method was used to study the kinetics of inhibitor release in the tetrameric β-D-galactosidase enzymes. First, the enzyme was deactivated by saturating the four active sites with the inhibitor, D-galactal. The galactal concentration was greatly reduced by dilution immediately before starting the measurements, and the activity–evolution then monitored. The sudden dilution served as a driving force to release the inhibitor molecules from the enzyme's active sites. Based on the discrete changes in a single enzyme's activity over time, the dissociation kinetics of the inhibitor could be derived such that a stochastic cooperative inhibitor release model was proposed. This inferred that the release of an inhibitor from one of the tetrameric enzyme's subunits would lead to a sudden release of inhibitor from the other subunits.

11.2.1.2 Individual Enzymatic Turnover Studies

Since the 1990s, several strategies have been developed to detect individual turnovers in enzymes, based on SMFS. Although an overview of these strategies will be highlighted below, attention will initially be focused on the implications of single-turnover experiments in relation to the concept of enzyme kinetics. Whereas, in bulk experiments kinetics are expressed as changes in reactant or product concentrations over time, the concept of *concentration* becomes meaningless at the single-turnover level. Here, turnovers are *stochastic events*, and kinetics are therefore expressed as the *probability* for a turnover to occur. The inverse of the average time span between two successive turnovers (the average waiting time, $\langle\tau\rangle$, then acts as a measure of this probability. Consequently, the Michaelis–Menten formula under single-turnover conditions for a certain enzyme at a certain conformational state becomes:

$$\frac{1}{\langle\tau\rangle} = \frac{k_2 \cdot [\mathrm{S}]}{[\mathrm{S}] + K_M} \tag{11.4}$$

In the absence of static disorder the ergodic principle is valid, which means that averaging over time for one individual is equivalent to averaging over a whole population at a certain moment in time. In this case, Equation 11.4 should be equivalent to Equation 11.2, and thus $1/\tau$ will be equal to $V_0/[\mathrm{E}]_{\mathrm{total}}$. If k_2 is a static (nonfluctuating) first-order rate constant, then the histogram of the waiting times, which represents the probability density function (pdf)[1] of the waiting times,

1) The pdf or probability density function $f(x)$ of the waiting times is the continuous form of the waiting time histogram, normalized such that $\int_{-\infty}^{+\infty} f(x)dx = 1$. The probability that a waiting time has a value between a and b is then given by $\int_a^b f(x)dx$.

would be expected to follow a single-exponential decay. However, due to the complicated dynamics between different conformational substates – known as the "energy landscape" of an individual enzyme – k_2 is continuously changing and thus the pdf should decay according to a multiexponential decay with an infinite number of components. This would result in a very broad decay over many orders of magnitude. It has been proven that this can be simplified mathematically to a *stretched exponential function* [15]. Thus, the pdf for the waiting times becomes:

$$P_{\text{waiting times}} = \int_0^\infty A(\tau)\exp\left(-\frac{t}{\tau}\right)d\tau \approx A\cdot\exp\left[-\left(\frac{t}{\tau_{\text{waiting}}}\right)^\alpha\right] \quad (11.5)$$

The factor α, which has a value of between 0 and 1, determines over how many orders of magnitude the decay is stretched as a result of dynamic disorder. For $\alpha = 1$, there is no stretching and the pdf equals a monoexponential decay. However, the more α approaches zero, the more the decay is stretched.

Attention is now turned to the different approaches that have been used to identify such activity fluctuations by visualizing individual turnover events in biocatalysis. The basis of such investigations is the immobilization of an enzyme on a glass coverslip at low surface concentrations; this is carried out in order to guarantee that the individual enzymes are separated by distances larger than the diffraction limit, which is the resolution limit of optical microscopy. The turnovers can then be probed either by using fluorescent or fluorogenic substrate molecules, or by monitoring the fluctuations of a fluorescently labeled enzyme itself during the catalytic cycle. The latter approach, however, implies that the time over which the fluorescent biocatalyst can be investigated is limited by the photostability of the used reporter dye; once the dye bleaches it is no longer possible to monitor the single catalytic events. However, this limitation does not exist when using a fluorogenic substrate, since for every turnover a freshly formed fluorescent molecule is probed. In this scheme, the most crucial factor is to find a suitable combination of enzyme–profluorescent substrate.

On the majority of occasions, confocal fluorescence microscopy is used for single-turnover experiments on enzymes. This technique is employed not only because of the better time-resolution of point-detectors compared to the CCD cameras used in wide-field microscopy, but also because of the improved resolution along the optical axis. The high resolution is due to the presence of the pinhole, which greatly reduces any fluorescence background from the impurities in the bulk solution. Nevertheless, wide-field techniques have also proven useful.

The first true single-turnover experiments date back to 1995, when Funatsu *et al.* visualized the binding of Cy3-labeled ATP on an immobilized myosin protein [16]. Myosin, as a molecular motor, converts the energy released by the enzymatic cleavage of ATP into mechanic movement. In order to distinguish between Cy3-ATP in solution and Cy3-ATP bound to myosin, total internal reflection illumination was applied, using wide-field fluorescence microscopy. In this way, only the fluorophores in the first few hundred nanometers of the sample (i.e., the fluorophores bound to myosin) were excited. Binding of the emissive substrate to the

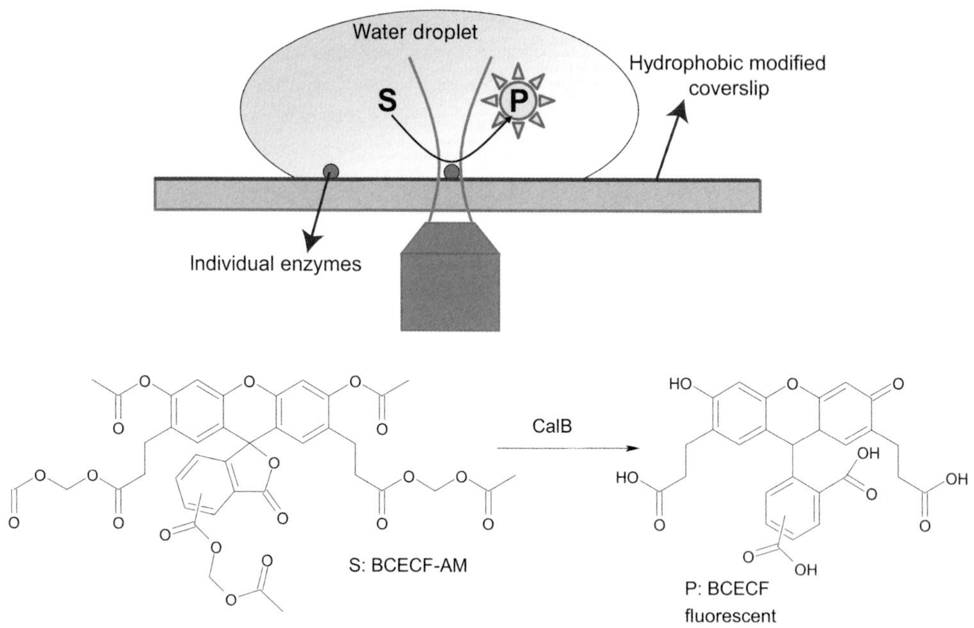

Figure 11.4 Scheme of the single-turnover study of the *Candida antarctica* lipase B (CalB)-catalyzed hydrolysis of BCECF-AM. By using confocal fluorescence microscopy, the formation of individual fluorophores from a nonfluorescent substrate in the enzyme can be monitored.

enzyme, followed by its release, would then give rise to a flash of light that originated from the position of the enzyme. The major drawback of this approach was that the assumption was made that every substrate binding event would also lead to an enzymatic turnover. However, this condition would not necessarily be fulfilled in other enzymatic systems, and great care was required when drawing conclusions on the measured activity values and using this approach in other systems.

By using a nonfluorescent substrate that is converted by the enzyme into a fluorescent product, the turnover itself can be visualized, rather than simply the substrate binding [4, 9, 15, 17]. Moreover, as the substrate (which is present in relatively high concentrations) is itself nonemissive, the background would be reduced significantly, allowing for higher signal-to-noise ratios (SNRs). This technique was applied by the groups of Hofkens and Nolte in their study of the *Candida antarctica* lipase B (CalB), immobilized by hydrophobic interaction on a modified coverslip. By using a fluorogenic fluorescein derivative, BCECF-AM, the individual turnovers could be visualized as intensity bursts in the emission–intensity time transients (Figures 11.4 and 11.5a). These authors found that the waiting times between successive turnovers followed a stretched exponential decay distribution, with an α-value of about 0.15, which indicated activity fluctuation

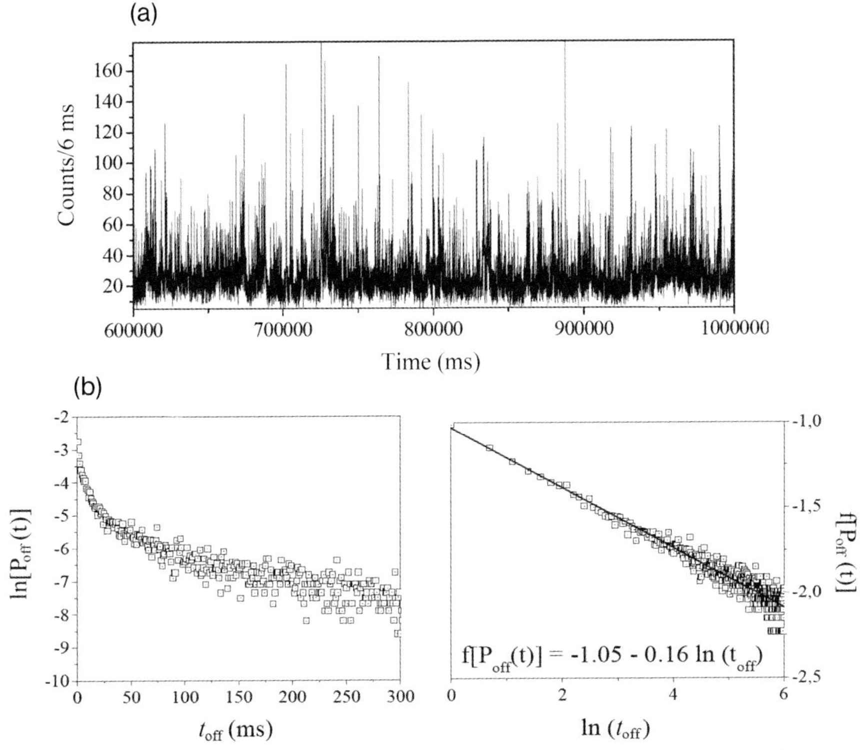

Figure 11.5 (a) Representative part of the fluorescence intensity time trace of a single CalB molecule hydrolyzing BCECF-AM. Every intensity burst corresponds to one enzymatic turnover (formation of one fluorescent BCECF molecule). (b) Distribution of the waiting times $[P_{off}(t)]$ between successive turnovers of the enzyme. Left: The distribution is plotted on a logarithmic scale; the deviation from a straight line indicates that the distribution is not monoexponentially decaying. Right: A stretched exponential function can be linearized by plotting $f(P_{off}(t)) = -\ln[-\ln(P_{off}(t))]$ versus $\ln(t_{off})$.

over time scales spanning several orders of magnitude! An example of this waiting time distribution is given in Figure 11.5.

In a similar study, the group of Xie investigated the activity dynamics at varying substrate concentrations of β-D-galactosidase [17]. The enzyme was first immobilized by coupling the biotinylated enzyme to a streptavidin-coated polystyrene bead that was tethered through a biotin linker to a poly(ethyleneglycol) (PEG)-modified coverslip. When resorufin β-D-galactopyranoside was used as the fluorogenic substrate (Figure 11.6), the waiting times were seen to be exponentially distributed only under substrate-saturating conditions, whereas at low concentrations the distribution followed an increasingly monoexponential decay. This inferred that the conformational fluctuations created dynamic disorder only in k_2 in Equation 11.1 (which represents the rate-limiting step at high substrate concentrations), but (almost) not in the reversible complex formation between enzyme and substrate.

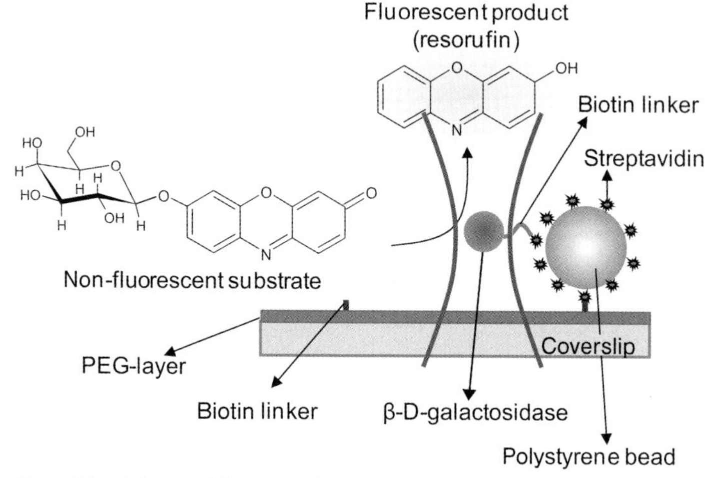

Figure 11.6 Scheme of the set-up for the single-turnover study of β-D-galactosidase [17]. The enzyme is immobilized on a large polymeric bead that can be easily localized by optical microscopy. Confocal fluorescence microscopy allows the detection of single fluorescent products molecule formed inside the enzyme from a nonfluorescent substrate.

The same method also proved to be useful when investigating the transient phases in enzyme denaturation, which are hidden by ensemble averaging in traditional bulk experiments. Recently, De Cremer *et al.* reported the details of a new spontaneous deactivation pathway of chymotrypsin immobilized by entrapment in an agarose polymer. The pathway was examined over extended periods of time, by following single turnovers of the hydrolysis of a fluorogenic peptide derivative of rhodamine 110 [4]. Rather than a sudden "all-or-nothing" deactivation, a transient phase was observed to precede the totally deactivated state. During this transient phase, a reversible conformational change caused the enzyme to switch between active and inactive states, with stepwise inactivation occurring before the enzyme could deactivate irreversibly (Figure 11.7). Although it is still unclear as to which conformational dynamics drive this switching behavior, such results have provided new insights into how transitions occur during deactivation.

A number of other enzymatic systems have been explored with single-turnover precision, by using a fluorogenic substrate to monitor individual catalytic events. In these studies, the choice of a suitable combination of enzyme and profluorescent substrate is clearly crucial; hence, a list of combinations proven to be successful in single turnover monitoring is provided in Table 11.1.

Instead of monitoring the emission from the formed product to identify single catalytic events, it is also possible to monitor the intrinsic fluorescence from the catalyst molecule–provided that changes occur in the fluorescence properties (intensity, spectrum or lifetime) of some of the intermediates within the enzymatic cycle. This concept was extended by Xie and coworkers in 1998, when studying

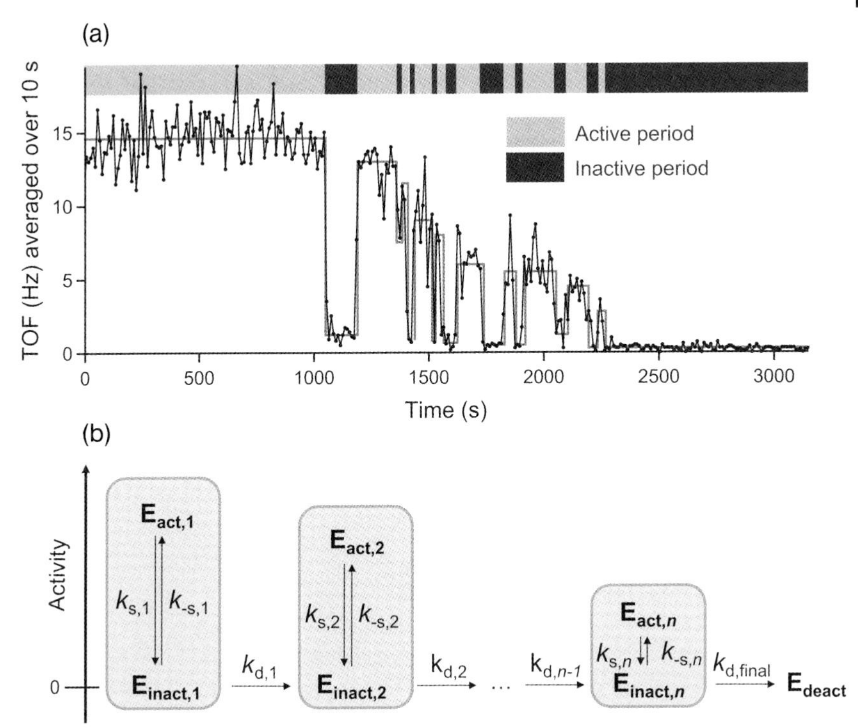

Figure 11.7 Spontaneous deactivation of a single α-chymotrypsin enzyme. (a) Time transient of the turnover frequency (TOF, averaged over 10 s) of a single deactivation chymotrypsin enzyme. The grey line indicates the different states and serves as a guide to the eye. (b) The extended single-molecule deactivation model for enzymes. A reversible conformational change causes the enzyme to switch between active and inactive states. During this equilibrium, stepwise inactivation occurs before the enzyme deactivates irreversibly. Reproduced with permission from Ref. [4]; © 2007, American Chemical Society.

cholesterol oxidase [21]. This enzyme contains a fluorescent flavine adenine dinucleotide (FAD) cofactor that acts as an electron shuttle, and switches between its reduced ($FADH_2$) and oxidized (FAD) states during the catalytic cycle. In contrast to $FADH_2$, FAD is fluorescent and can be monitored at the single-molecule level. Thus, a single on-off cycle in the fluorescence intensity–time trace would represent one catalytic cycle. On analysis, these fluorescence–time traces showed a strong correlation between the duration of individual catalytic cycles (see Figure 11.8). Figure 11.8a shows a two-dimensional (2-D) histogram of the duration of turnover i versus turnover $i + 1$. The strong diagonal trend indicates that a rapid, respectively slow, turnover is likely to be followed by another fast, respectively slow, turnover. This phenomenon is known as the *memory effect of enzymes*, and is assigned to the slow interconversion of the enzyme's structure into different conformational isomers with distinct activities. A similar histogram is shown in Figure 11.8b for turnovers separated by 10 events. Here, no diagonal trend is seen; this means that

Table 11.1 Nonexhaustive overview of suitable combinations of enzyme–profluorescent substrate for single-turnover activity measurements.

Enzyme	Profluorescent substrate	Fluorescent product	Reference
Candida antarctica lipase B (CALB)	BCECF-AM[a]	BCECF[b]	[9, 15]
Thermomyces lanuginosa lipase (TLL)	Carboxyfluorescein diacetate	Fluorescein	[18]
Horseradish peroxidase	Dihydrorhodamine 6G	Rhodamine 6G	[19]
β-D-Galactosidase	Resorufin β-D-galactopyranoside	Resorufin	[17]
α-Chymotrypsin	Rhodamine 110 bis-(suc-Ala-Ala-Pro-Phe)	Rhodamine 110	[4]
Curvularia verruculosa bromoperoxidase	Aminophenyl fluorescein	Fluorescein	[20]

a 2′,7′-Bis-[2-carboxyethyl]-carboxyfluorescein, acctoxymethyl ester.
b 2′,7′-Bis-[2-carboxyethyl]-carboxyfluorescein.

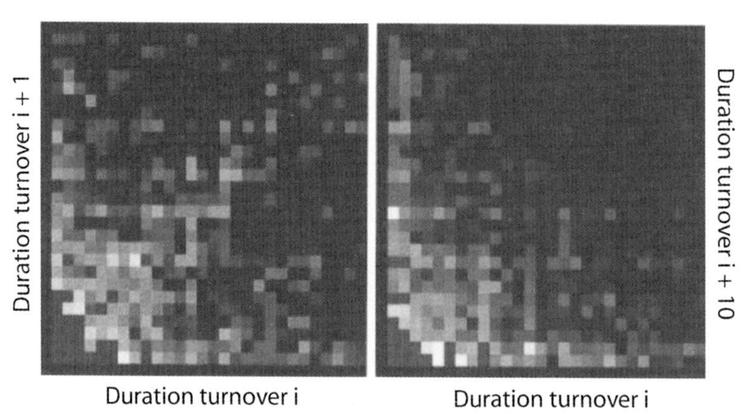

Figure 11.8 Two-dimensional correlation plot of the waiting times between successive turnovers (left) and the waiting times separated by 10 turnovers (right). For successive turnovers a clear correlation is observed, while this correlation is lost over 10 or more turnovers. Reproduced with permission from Ref. [21]. ©, American Association for the Advancement of Science.

no correlation is present, and indicates that the conformational dynamics are occurring on a time scale between one and 10 turnovers.

Recently, Kuznetsova *et al.* proved that, even in the absence of structural rearrangements during the catalytic cycle, fluorescence resonant energy transfer (FRET) can be applied to visualize single turnovers. For example, in the case of nitrite reductase (NiR) from *Alcaligenes faecalis*, when fluorescently labeled with ATTO 655, a copper atom that switches its redox state during the catalytic cycle is responsible for cycling between different FRET efficiencies during catalysis [22]. Although this enzyme is trimeric, the labeling conditions were chosen such that the N-terminus of only one monomer was labeled. Each monomer of the enzyme contains two copper centers that act as a redox shuttle. In their oxidized form, one of the copper centers has a broad absorption spectrum, and thus by a FRET mechanism the excitation energy of the ATTO label can be transferred to an adjacent copper center, which relaxes in a nonradiative manner. However, in its reduced state, the absorption band of the copper complex disappears and FRET is no longer possible. Thus, only when the copper center is reduced can bright fluorescence be seen from the ATTO label (Figure 11.9). In this way, the individual catalytic cycles can be monitored. A careful analysis of the data obtained allowed an identification of the electron flow through the different redox sites of these enzymes during catalysis. Moreover, the rate constants calculated from the single-molecule data were in good agreement with those obtained from traditional ensemble experiments.

The above-described examples clearly demonstrate that several approaches are possible to study single enzymatic reactions, providing new insights into enzyme kinetics and revealing formerly unknown deactivation behaviors in comparison with ensemble averages. Although these new observations have been interpreted

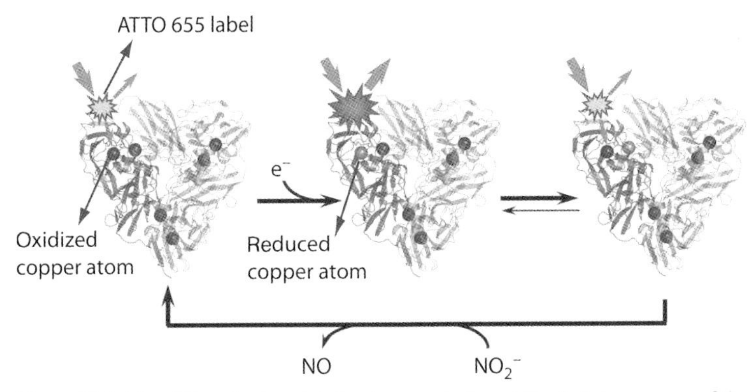

Figure 11.9 Identifying single catalytic steps within nitrite reductase from *Alcaligenes faecalis* by means of FRET. The enzymes are labeled with an ATTO 665 tag. During the catalytic cycle, two copper centers cycle between an oxidized and reduced state. Only in its oxidized state can one of these centers accept the excitation energy of the ATTO 655, and thus quench the emission. Reproduced with permission from Ref. [22]; © 2008, National Academy of Sciences, USA.

within the concept of the conformational dynamics of the protein backbone, the exact nature of the conformational changes involved in enzyme kinetics have not yet been directly identified. Indeed, this issue will become a major challenge in single enzyme research in the future. One potential approach has included the use of a fluorogenic substrate in combination with a FRET pair covalently bound to the enzyme; however, this requires simultaneous three-color detection that will prove to be experimentally challenging.

11.2.2
Unraveling the Modes of Action of Individual Enzymes

Until now, only those enzymes immobilized on a surface, on top of which a solution of freely diffusing substrate molecules can be applied, have been described. Yet, some enzymes are designed to function on a (quasi-) immobilized substrate, in which situation the concept of enzyme kinetics towards such substrates differs totally from the Michaelis–Menten approach. Now, the reaction rate is determined by the diffusion of an enzyme from bulk solution towards the immobilized substrate, by the intrinsic reaction rate, by the movement of the enzyme over the substrate, and also by the dissociation kinetics of the enzyme from the substrate. Movement of the enzyme over a substrate may imply that it can catalyze multiple consecutive reactions before becoming desorbed from that substrate. It would be interesting, therefore, to investigate exactly *how* these enzymes interact with their substrate. In this section, an example is provided of how SMFS has been applied to investigate single enzyme movements over fixed substrates.

11.2.2.1 Phospholipase A1 on Phospholipid Bilayers

Lipases are enzymes that hydrolyze phospholipids. For example, phospholipase Al (PLA1) can dock to a phospholipid bilayer and break down its structure by successive hydrolysis of the individual phospholipids. However, an important question in this process is *where* the enzyme can dock on the bilayers: Does it require edges or defects in the layers in order to reach the ester bonds of the phospholipid, or can the enzyme force its way through an intact layer? Recently, Hofkens *et al.* investigated the hydrolysis of a POPC (1-palmitoyl-2-oleoyl-*sn*-glycero-3-phosphocholine) bilayer by PLA1 by labeling the enzyme with an extremely photostable perylene diimide (PDI) derivative [23, 24]. The bilayer itself was fluorescently marked by the incorporation of a weakly fluorescent 3,3′-dioctadecyloxacarbocyanine perchlorate (DiO) fluorophore. Despite the inherent fluorescent background from this phospholipid bilayer, the individual PDI-labeled enzymes were still bright enough to be distinguished when adsorbed onto the bilayer. The enzymes could be seen to dock preferentially to the edges of the bilayers (Figure 11.10), while some were also adsorbed on top of the intact bilayers, from where they diffused relatively rapidly over the surface. This diffusion appeared as blurry spots (the areas marked with white contour lines in Figure 11.10), as the emission spread over the trajectory of the enzyme within the 50 ms accumulation time of

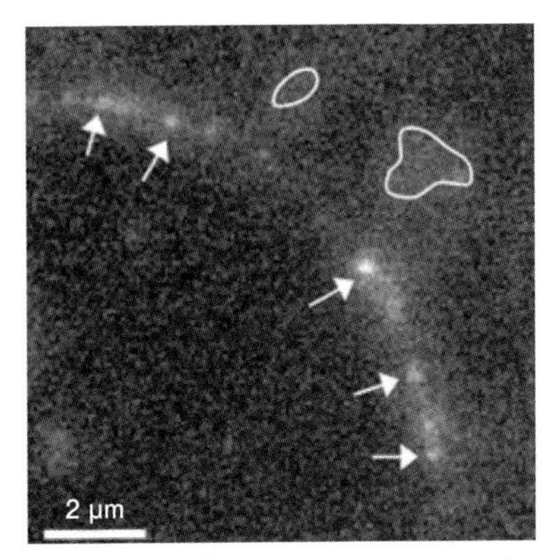

Figure 11.10 Fluorescence microscopy image of individual PDI-labeled phospholipase A1 (PLA1) enzymes on POPC bilayers. The white arrows indicate single enzyme molecules docked to the edge of a bilayer. Because of their relative immobility, they appear in the image as small bright spots. The areas marked with a white contour line represent single enzymes that are rapidly diffusing over an intact bilayer. Because of this rapid diffusion, the emission spot is spread out over the image. Reproduced with permission from Ref. [23].

the CCD camera. The disappearance of the bilayer's weak fluorescence led the authors to conclude that the hydrolysis had occurred almost exclusively from the bilayer's edges, and that the rapidly diffusing enzymes on top of the intact bilayers had not contributed significantly to any disruption of the layers.

11.3 Chemocatalysis at the Single-Molecule Level

The sensitivity of fluorescence microscopy can provide major benefits to catalyst science. Chemocatalysts, in contrast to the point-like biocatalysts discussed so far, often have a 3-D structure throughout which their active sites are distributed. Hence, not only the temporal resolution but also the spatial resolution of the catalyst is of prime importance. This point will now be briefly addressed before attention is focused on the recent application of fluorescence microscopy in the study of chemocatalysts. The complex 3-D structure of a chemocatalyst may imply that the actual chemical transformation is the result of a complex interplay of different dynamic processes, such as adsorption/desorption and diffusion. Hence, it is important to understand these elementary steps and to relate them to the actual chemical transformation. These different dynamic processes can be characterized at the single catalytic particle level down to the single-molecule level by using fluorescence microscopy, in combination with a broad range of fluorescent and profluorescent probes.

11.3.1 Spatial Resolution

As the typical length scales for catalysts and catalytic phenomena range from the subnanometer to the macroscopic scale, an effective characterization technique should be capable of monitoring ultralow quantities, preferentially with single-molecule sensitivity, at these different length scales. Because of light diffraction, problems typically arise at the nanometer length scale, as the maximal resolution at the focal plane is limited to $\lambda/2$, or approximately 250 nm for visible light. Since, in near-field microscopy, light focusing is no longer necessary, any objects at subdiffraction distance can be resolved; consequently, this approach is valuable for the study of 2-D structures, although the depth profile is much more limited than in far-field methods. Far-field methods optimally exploit the optical transparency of many materials to look beyond the outer surface.

Confocal fluorescence microscopy is an often-used far-field method which facilitates 3-D imaging that is very useful in the detailed study of catalyst objects; however, the resolution is improved only by a factor of 2 [2]. The development of stimulated emission depletion (STED) microscopy provided a major step towards fluorescence nanoscopy [25, 26]. Here, the focused excitation laser is overlapped with a donut-shaped beam that serves as a de-excitation beam, forcing the excited molecules back to the ground state. The resultant fluorescence hence stems from a very small central node of down to 15 nm, which is not covered by the depletion beam. Similarly, other reversible saturation processes such as photoswitching can be used to drastically enhance the spatial resolution. Such an approach can be used, for example, to reduce the area over which turnovers are counted [27, 28].

In traditional linear wide-field fluorescence microscopy (WFFM), the location of a single emitter can be determined to almost arbitrarily high accuracy, depending on the SNR that can be achieved. By fitting a 2-D Gaussian to the recorded point spread function (PSF) of a single emitter, the exact location of its center can be determined with nanometer accuracy [29]. However, the simultaneous resolution of multiple emitters in close proximity is not possible. When the detection of these nonresolved molecules can be separated in time by stochastic processes (such as stepwise bleaching, photoswitching, etc.), their relative positions can easily be traced back by fitting consecutive PSFs, such that a high-resolution image can be reconstructed [30–33]. This type of scheme can easily be implemented in the observation of single catalytic cycles based on the formation of fluorescent products. Because of the stochastic nature of reactions at the different sites, the recording of consecutive frames should allow determination of the activity of surface areas down to a few square nanometers. The required SNR can be obtained by using high excitation powers because bleaching is not an issue here. Indeed, it might be possible to visualize differences in activity governed by surface cracks or local defects, or to examine nanometer-sized crystals. Recently, structured illumination wide-field microscopy has also been applied to record super-resolution images [34].

In addition to the direct imaging of objects through their fluorescence intensity, indirect methods such as FRET and electron transfer can be used to study processes at the subnanometer to 10 nm length scale (*vide infra*). FRET is ideally suited to quantify the distances between fluorescent donors and acceptors in the range of 1 to 10 nm [35], while shorter distances (<1 nm) can be probed by monitoring electron transfer between a fluorescent molecule and donors or acceptors in its proximity [36].

11.3.2
Sorption and Diffusion Studies on Chemocatalytic Materials

The combination of time-resolved imaging and spatial resolution of fluorescence microscopy fulfils the requirements for investigating the diffusion of (single) molecules in the pores of individual catalyst particles. Even if the channel dimensions of catalytic materials are well below the diffraction limit, useful information can be retrieved by analyzing the changes in fluorescence intensity, spectra, decay times, and the polarization of guest molecules in the catalyst particles. In the past, fluorescence microscopy has been used at different scales when moving from the bulk to the single-molecule level, including intercrystalline and intracrystalline diffusion at the individual particle level using multiple fluorescent guest molecules, or intracrystalline diffusion at the single-molecule level. Whilst to date, intracrystalline diffusion has been studied using a wide range of techniques, including pulsed-field gradient and other nuclear magnetic resonance (NMR) approaches and quasi-elastic neutron scattering, the direct observation of intercrystalline diffusion has been explored to a much lesser degree.

11.3.2.1 Intercrystalline Diffusion

During intercrystalline diffusion, molecules undergo a cascade of processes that include desorption, diffusion, and readsorption. By exploiting a whole range of photophysical and photochemical processes, such as heavy metal-induced intersystem crossing, charge transfer, and triplet–triplet energy transfer between guest species, intercrystalline migration can be monitored by using fluorescence microscopy [37]. These studies have provided direct evidence of the migration directionality and distances over which diffusion takes place. [37, 38] However, until now intercrystalline migration has been studied only qualitatively using fluorescence microscopy.

11.3.2.2 Intracrystalline Diffusion at the Individual Particle Level

Time-dependent fluorescence intensity profiles from multiple dyes interacting with individual particles allow the visualization and calculation of intracrystalline diffusion. In one example, the sorption and diffusion processes of molecules at the outer surface of layered double hydroxides were monitored using an anionic fluorescent dye [39]. Whereas, the crystal initially showed no observable fluorescence (see Figure 11.11; 0 s), shortly after addition of the dye solution a pronounced increase in fluorescence intensity at the crystal surface was detected (see Figure

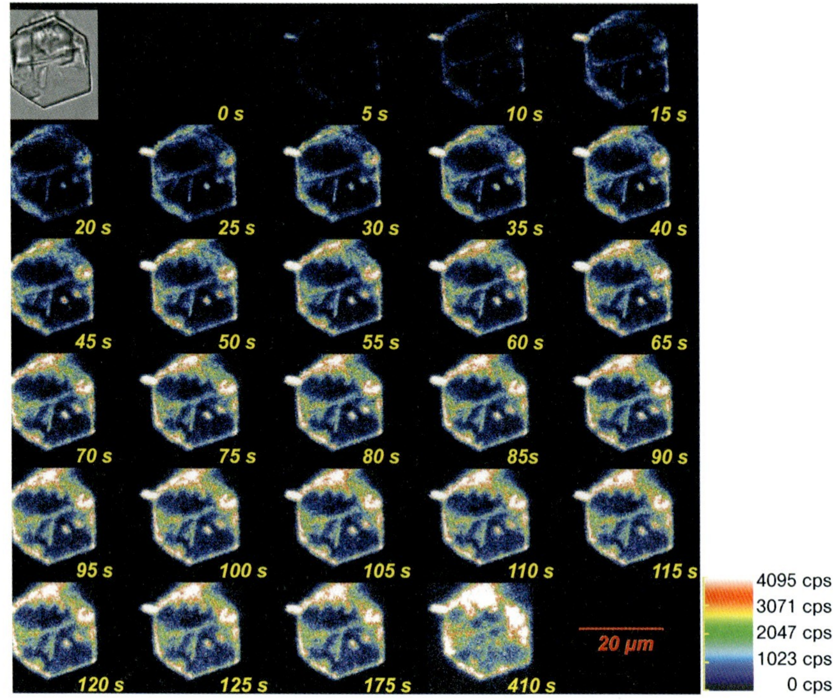

Figure 11.11 Fluorescence microscopy observation of intracrystalline diffusion. After adding the dye solution at 0 s, a gradual increase in fluorescence intensity at the crystal surface can be seen. Initially, the dyes accumulate at the crystal edges, but after 410 s the whole surface has an elevated fluorescence intensity. The fluorescence images are in false color (scale right bottom), and the corresponding optical image is in grayscale. Reproduced with permission from Ref. [39].

11.11). This fluorescence was not distributed homogeneously but was rather concentrated at the crystal rim. However, over time a gradual spread of fluorescence towards the center of the large basal crystal faces was observed.

A linear dependence between fluorescence intensity and local dye concentration was assumed when the dye concentrations were low, while intercalation of the dye was ruled out on the basis of emission spectra. Based on this information, a model was proposed in which the time-dependent fractional coverage of the crystal could be calculated based on the dye concentration in solution via local adsorption and desorption constants, k_a and k_d, respectively, and the local surface diffusion constant, D. Fitting of this model to the experimental data revealed that dye molecules under these conditions only adsorbed at the crystal's edge ($k_a = 27\,m^3\,mol^{-1}\,s^{-1}$), from which occupation of the basal plane via surface diffusion ($D = 310^{-14}\,m^2\,s^{-1}$) takes place. Similar strong interactions between organic molecules and the crystal edges in an aqueous environment have also been reported at the single-molecule level (*vide infra*) [40]. This may possibly be related to stronger and/or better accessible adsorption sites, or to local differences in basicity. In contrast, the data for displacement of the dye molecules by strongly adsorbing CO_3^{2-} revealed that desorption had taken place both from the crystal edge and the basal plane sites ($k_d = 0.04\,s^{-1}$).

11.3.2.3 Intracrystalline Diffusion at the Single-Molecule Level

Although the migration of large amounts of dyes within or between crystals has been discussed at length, for strongly emissive molecules, diffusion inside the host particles can be evaluated at the single-molecule level. Whilst the evaluation of guest diffusion in nanoporous hosts is discussed in Chapter 12, the aim here is to emphasize the possibility of determining the local diffusional processes of dye molecules, based on fluorescence correlation spectroscopy. In this technique, dyes diffusing through a small focused laser beam temporarily generate fluorescence. Consequently, by evaluating these time-dependent intensity fluctuations, using temporal autocorrelation, it is possible to observe dynamic processes that take place over the submillisecond to second time range. Focusing the excitation light inside the host particle results in fluorescence bursts as a function of time, emanating from guest molecules that are diffusing through this illuminated zone. An analysis of many such single-molecule events makes it possible to determine the diffusion coefficient of, for example, rhodamine 6G in mesoporous materials [41]. Such measurements have revealed that, instead of simple free diffusion inside the large 13 nm pores, rhodamine 6G was also sensitive to adsorption/desorption events. In addition to the diffusion coefficient, the fraction of a dye that is temporarily adsorbed at the host structure, as well as the mean desorption time of such events, can be calculated. Similar studies have revealed comparable behaviors of other dyes in different porous host materials [42, 43].

These three different approaches can be used to determine spatially resolved data relating to the nature and size of the diffusing organic molecules. Then, depending on the length and timescale of interest, the dye concentration and the properties, the most suitable technique can be selected.

11.3.3 Chemical Transformation at the Single-Molecule Level

In addition to a detailed characterization of the diffusion and adsorption/desorption processes, the actual catalytic turnover can also be monitored by using fluorescence microscopy. The different measurement schemes used to measure and characterize biocatalytic activity (as described above) can readily be integrated into chemocatalytic research. An initial possibility involves the use of a profluorescent substrate that is converted during the catalytic cycle into an emissive product. This technique was first applied to observe the activity of single catalytic sites by Tan and Yeung [44]. In these experiments, single osmium catalysts converted the nonemissive Ce(IV) to the emissive Ce(III), with the time-dependent conversion of such spatially separated single catalysts being monitored by recording the very slow build up of Ce(III) emission. However, the emissive properties of this reaction product would not allow any observation of the individual catalytic turnovers. Recently, the authors' group succeeded in observing (for the first time) single catalytic turnovers on inorganic catalytic crystals, using the base-catalyzed conversion of fluorescein diacetate to the highly emissive fluorescein products (see Figure 11.12a). This real-time observation of single catalytic turnover events on basic layered double hydroxides was carried out in two different solvents, water and *n*-butanol, which also served as second reactants in the reaction. By filming

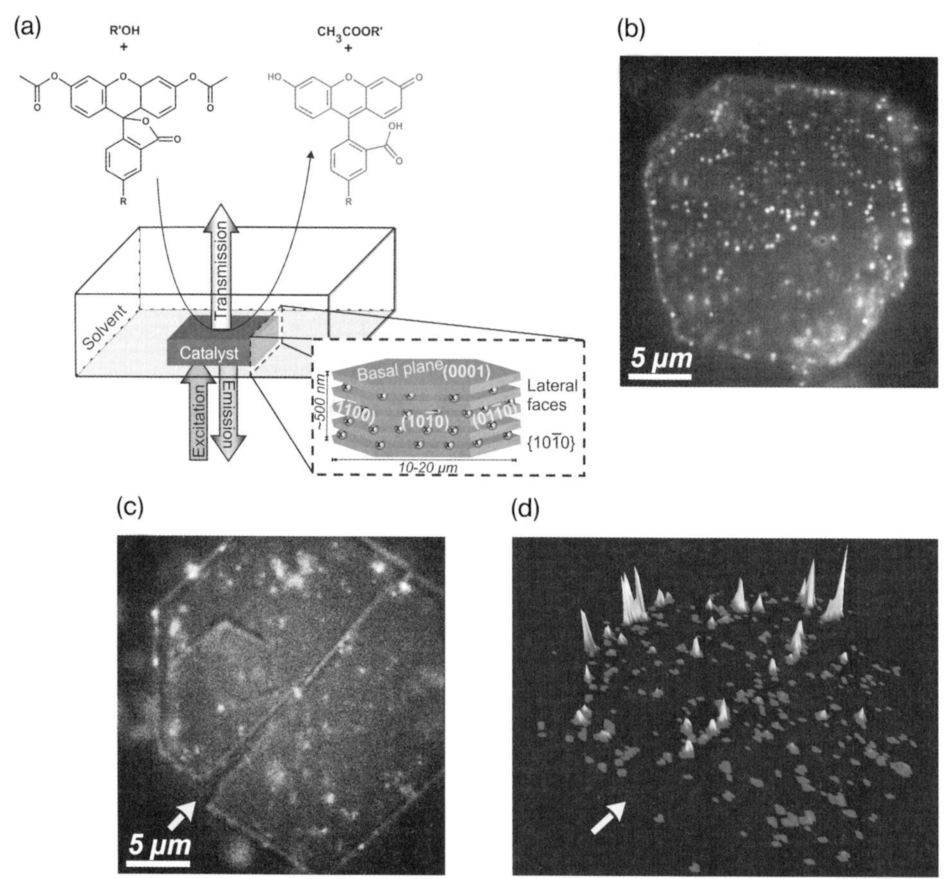

Figure 11.12 Chemical transformation at the single molecule level. (a) Schematic representation of the experimental set-up in which the layered double hydroxides (LDH) catalyst is contained in a measurement cell containing the fluorogenic substrate. Catalytic conversion of the fluorogenic ester yields strongly fluorescent fluorescein molecules. (b) Single transesterification reactions at the LDH surface are visible as fluorescent spots (white dots). These reactions are clearly homogeneously distributed around the outer surface. (c) By changing the solvent from butanol to water, the observed reaction changes from transesterification to a hydrolysis reaction. This reaction is clearly linked to the crystal edges and defects. (d) Accumulated spot intensity on the same crystal over 256 consecutive images.

these single turnover events with time, using wide field microscopy, it was possible to map the spatial distribution of the catalytic activity over the entire crystal, and also to calculate local reaction kinetics (see Figure 11.12b–d). In this way, a pronounced difference between the layered double hydroxides (LDH)-catalyzed transesterification reaction in butanol, and the hydrolysis reaction in water, could be observed. Whereas, the former reaction was seen to take place at the complete outer surface (Figure 11.12b), the latter reaction mainly took place at the crystal edge (Figure 11.12c,d). A simple counting of the turnovers for a local area allowed the calculation of local catalytic rates. For 40 nM of fluorogenic substrate there were, on average, 4.2 transesterification reactions per 100 μm^2 per 96 ms, resulting in approximately 7.4×10^{-13} mol m^{-2} s^{-1}. With a 15-fold increased substrate concentration (600 nM), however, this rate increased almost linearly to 1.1×10^{-11} mol m^{-2} s^{-1}. This approach allowed the measurement of reaction kinetics not only at different substrate concentrations but also for different crystals (Figure 11.13). The linearity of the catalytic activity with increasing concentration, when measured for different catalyst crystals, showed that the reaction was first order in the fluorogenic substrate. Based on the spatial resolution of fluorescence microscopy, and using a similar approach, the contribution of the different crystal parts in the hydrolysis reaction could be calculated, with the crystal edges accounting for >85% of the overall activity (4.7×10^{-12} mol m^{-2} s^{-1}). This was the first direct observation of crystal-face-dependent catalysis using a single-molecule fluorescence approach.

A similar approach was used to observe the catalytic activity in individual ZSM-5 zeolite crystals. Here, the acid-catalyzed conversion of small furfuryl alcohol to strongly fluorescent product oligomers allowed a time-dependent observation of catalytic activity in three dimensions (Figure 11.14a) [45]. Due to the small size of the probe molecule, not only the outer surface but also the extended crystal interior surface became accessible. Under these experimental conditions, the initial reaction was seen to take place at the whole outer surface, visible as a homogeneously

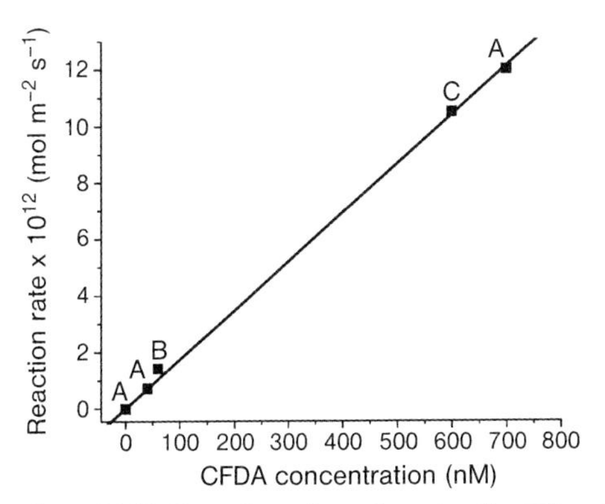

Figure 11.13 Recording of reaction rates for different crystals (A, B, C) with different fluorogenic substrate concentrations in the range of 0 to 700 nM proves that the transesterification reaction is first order in reagent.

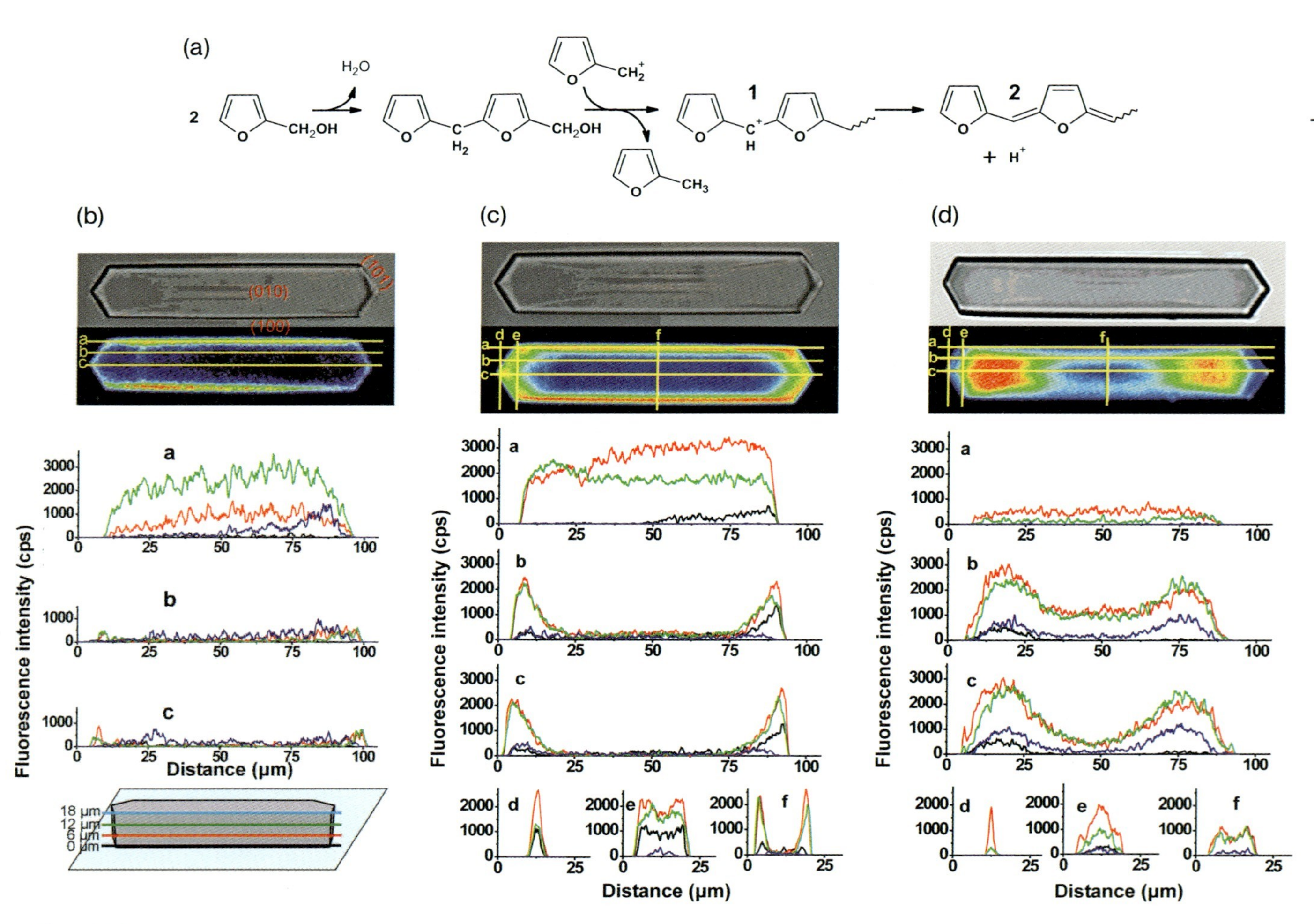

Figure 11.14 (caption see p. 303)

distributed fluorescence intensity at the main crystal faces [(100) and (010); Figure 11.14b]. After some time, the reaction at the crystal tips became more apparent and the reaction spread towards the crystal interior (Figure 11.14c). After very long times, coke formation and an accumulation in the crystal region below the sharp crystal tips became very pronounced (Figure 11.14d). This could only be explained by the presence of internal diffusion barriers between the different crystal components, which contained these longer oligomeric product molecules in the zone below the crystal tips [46]. One unique feature of this reaction was the selectivity at which the different reaction products could be measured; notably, the longer the formed conjugated oligomers, the more red- shifted was the fluorescence emission. Consequently, multicolor fluorescence measurements allowed selective monitoring of the different reaction products. Recently, a wide range of organic reporter substrates has been described, both for microscopic studies of zeolite's catalytic activity [2, 47–49] and for the rapid and sensitive catalyst screening in high-throughput experiments [50–52]. By using functional and structure-sensitive probes, fluorescence microscopy may also be used to thoroughly characterize zeolites' complex crystal structures [53].

Essentially every sufficient bathochromic or hypsochromic shift can be used with fluorescence microscopy to observe chemical reactions. The latter principle was employed by Majima and coworkers to study the remote oxidation of organic molecules by photocatalytically formed singlet oxygen [54]. Following the irradiation of TiO_2 with ultraviolet (UV) light, singlet oxygen was formed. Subsequently, these molecules diffused away from the surface, and induced an oxidation reaction several micrometers to millimeters away from the photocatalyst surface. The oxidation of terrylene diimide (TDI) by singlet oxygen yields a TDI diepoxide product that, when excited at 532 nm, shows a strong fluorescence at 560–650 nm. In contrast, the unreacted TDI shows a weaker and much more red-shifted fluorescence at >650 nm. The reaction-induced hypsochromic shift allows the spectral separation of both fluorescent molecules at the single-molecule level. During the UV irradiation of TiO_2, other reactive oxygen species (ROS) were also formed, for which specific fluorogenic substrates have been developed [55–58]. As the transient concentrations of oxidative species are very low, this single-molecule approach may represent a very useful technique. Subsequently, by varying the distance

Figure 11.14 Confocal fluorescence microscopy imaging of acid-catalyzed furfuryl alcohol oligomerizations. (a) Mechanism of chromophore (**1** and **2**) formation. (b) Homogeneous fluorescence intensity at the (100) face after 10 min. The fluorescence intensities along selected lines (a,b,c) were measured at various depths along the crystal (see schematic representation). Similarly, at the (010) face a strong fluorescence is measured (not shown; see Ref. [45]). (c) After 16 h the slower reaction in the crystal tip zones becomes increasingly apparent. Again, the fluorescence intensities along selected lines (a–f) are shown at different depths in the crystal. (d) After 50 days, a strong accumulation of reaction products in the zones below the crystal tips appears as intense fluorescent zones. This proves the presence of diffusion barriers inside the crystal. Reproduced with permission from Ref. [45].

between the TiO_2 film and the layer of reporter molecules, the mobility of the ROS could be monitored, thus proving the remote oxidative power of photocatalytically formed oxidative oxygen species.

Rather than observing the catalytic formation of fluorescent molecules, a second widely applicable approach utilizes fluorescent catalysts. An excellent example of this is the switching between nonfluorescent and fluorescent states of FAD as the cofactor of cholesterol oxidase during the catalytic cycle (see Section 11.2.1.2) [21]. Changes in emissive properties such as intensity, lifetime, or the spectrum of metal complexes when changing in redox state, or the more general switching of a fluorescent reporter group due to energy or electron transfer to/from the catalytic site, can be used to visualize chemocatalytic turnovers. These ideas have already been widely used in the design of molecular switches, and of fluorescent, cryptand-based indicators for Na^+ and Ca^{2+} [59, 60]. Recently, the group of Herten explored a similar scheme to explore Cu^{2+} complexation at the single-molecule level [61, 62]. For this, the reversible complexation of dissolved copper ions by bipyridine chelates was monitored using a fluorescent reporter group in the vicinity of the complexating group. In the absence of copper, the fluorescence reporter group exhibited a strong fluorescence which, following the complexation of a copper ion, was almost completely quenched. Hence, such a reversible association–dissociation process may result in an "on-off" switching of the fluorescence reporter group, which can be related to the complexation dynamics at the single-molecule level.

Whereas, most biocatalytic reactions occur at room temperature under atmospheric conditions, chemocatalysts often operate under more severe conditions (e.g., high temperature and pressure). By using fluorescence microscopy, single molecules have been visualized at elevated temperatures of up to 373 K, indicating that the technique is not limited to ambient temperature [63, 64].

11.3.4 Conclusions

In this chapter, the potential of fluorescence microscopy for the *in situ* study of both chemocatalysts and biocatalysts has been reviewed. In addition to highlighting previously reported examples, new concepts aimed at broadening the scope of fluorescence microscopy in the catalytic sciences have been discussed. Future advances in spatiotemporal resolution and sensitivity will undoubtedly further enhance the impact of this technique. Clearly, the synergistic application of (*in situ*) fluorescence microscopy and other (*ex situ*) characterization techniques, should result in an explosion of new insights in the coming years [1].

Acknowledgments

M.B.J.R. thanks the Institute for the Promotion of Innovation through Science and Technology in Flanders (IWT-Vlaanderen) for a fellowship. M.B.J.R and G.D.C. thank F.W.O. (Research Foundation Flanders) for financial support. These studies were conducted within the framework of the IAP-V-03 program "Supramolecular Chemistry and Catalysis" of the Belgian Federal government and of GOA sponsoring. The authors also gratefully acknowledge support from the K.U. Leuven in the frame of the Center of Excellence CECAT.

References

1 Roeffaers, M.B.J., De Cremer, G., Uji-i, H., Muls, B., Sels, B.F., Jacobs, P.A., De Schryver, F.C., De Vos, D.E., and Hofkens, J. (2007) Single-molecule fluorescence spectroscopy in (bio) catalysis. *Proc. Natl Acad. Sci. USA*, **104** (31), 12603–12609.

2 Roeffaers, M.B.J., Hofkens, J., De Cremer, G., De Schryver, F.C., Jacobs, P.A., De Vos, D.E., and Sels, B.F. (2007) Fluorescence microscopy: bridging the phase gap in catalysis. *Catal. Today*, **126** (1–2), 44–53.

3 Roeffaers, M.B.J., Sels, B.F., Uji-i, H., De Schryver, F.C., Jacobs, P.A., De Vos, D.E., and Hofkens, J. (2006) Spatially resolved observation of crystal-face-dependent catalysis by single turnover counting. *Nature*, **439** (7076), 572–575.

4 De Cremer, G., Roeffaers, M.B.J., Baruah, M., Sliwa, M., Sels, B.F., Hofkens, J., and De Vos, D.E. (2007) Dynamic disorder and stepwise deactivation in a chymotrypsin catalyzed hydrolysis reaction. *J. Am. Chem. Soc.*, **129** (50), 15458–15459.

5 Berg, J.M., Tymockzo, J.L., and Stryer, L. (2002) *Biochemistry International Edition*, 5th edn, Freeman.

6 Min, W., Gopich, I.V., English, B.P., Kou, S.C., Xie, X.S., and Szabo, A. (2006) When does the Michaelis-Menten equation hold for fluctuating enzymes? *J. Phys. Chem. B*, **110** (41), 20093–20097.

7 Xie, X.S. and Lu, H.P. (1999) Single-molecule enzymology. *J. Biol. Chem.*, **274** (23), 15967–15970.

8 Engelkamp, H., Hatzakis, N.S., Hofkens, J., De Schryver, F.C., Nolte, R.J.M., and Rowan, A.E. (2006) Do enzymes sleep and work? *Chem. Commun.*, (9), 935–940.

9 Velonia, K., Flomenbom, O., Loos, D., Masuo, S., Cotlet, M., Engelborghs, Y., Hofkens, J., Rowan, A.E., Klafter, J., Nolte, R.J.M., and De Schryver, F.C. (2005) Single-enzyme kinetics of CALB-catalyzed hydrolysis. *Angew. Chem., Int. Ed.*, **44** (4), 560–564.

10 Rotman, B. (1961) Measurement of activity of single molecules of β-D-galactosidase. *Proc. Natl Acad. Sci. USA*, **47**, 1981–1991.

11 Lee, A.I. and Brody, J.P. (2005) Single-molecule enzymology of chymotrypsin using water-in-oil emulsion. *Biophys. J.*, **88** (6), 4303–4311.

12 Rondelez, Y., Tresset, G., Tabata, K.V., Arata, H., Fujita, H., Takeuchi, S., and Noji, H. (2005) Microfabricated arrays of femtoliter chambers allow single molecule enzymology. *Nat. Biotechnol.*, **23** (3), 361–365.

13 Rissin, D.M., Gorris, H.H., and Walt, D.R. (2008) Distinct and long-lived activity states of single enzyme molecules. *J. Am. Chem. Soc.*, **130** (15), 5349–5353.

14 Gorris, H.H., Rissin, D.M., and Walt, D.R. (2007) Stochastic inhibitor release and binding from single-enzyme molecules. *Proc. Natl Acad. Sci. USA*, **104** (45), 17680–17685.

15 Flomenbom, O., Velonia, K., Loos, D., Masuo, S., Cotlet, M., Engelborghs, Y., Hofkens, J., Rowan, A.E., Nolte, R.J.M., Van der Auweraer, M., De Schryver, F.C., and Klafter, J. (2005) Stretched exponential decay and correlations in the catalytic activity of fluctuating single lipase molecules. *Proc. Natl Acad. Sci. USA*, **102** (7), 2368–2372.

16 Funatsu, T., Harada, Y., Tokunaga, M., Saito, K., and Yanagida, T. (1995) Imaging of single fluorescent molecules and individual ATP turnovers by single myosin molecules in aqueous-solution. *Nature*, **374** (6522), 555–559.

17 English, B.P., Min, W., van Oijen, A.M., Lee, K.T., Luo, G.B., Sun, H.Y., Cherayil,

B.J., Kou, S.C., and Xie, X.S. (2006) Ever-fluctuating single enzyme molecules: Michaelis-Menten equation revisited. *Nat. Chem. Biol.*, **2** (2), 87–94.
18 Hatzakis, N.S., Engelkamp, H., Velonia, K., Hofkens, J., Christianen, P.C.M., Svendsen, A., Patkar, S.A., Vind, J., Maan, J.C., Rowan, A.E., and Nolte, R.J.M. (2006) Synthesis and single enzyme activity of a clicked lipase-BSA heterodimer. *Chem. Commun.*, (19), 2012–2014.
19 Edman, L., Foldes-Papp, Z., Wennmalm, S., and Rigler, R. (1999) The fluctuating enzyme: a single molecule approach. *Chem. Phys.*, **247** (1), 11–22.
20 Martínez, V.M., De Cremer, G., Roeffaers, M.B.J., Sliwa, M., Baruah, M., De Vos, D.E., Hofkens, J., and Sels, B.F. (2008) Exploration of single molecule events in a haloperoxidase and its biomimic: Localization of halogenation activity. *J. Am. Chem. Soc.*, **130** (40), 13192–13193.
21 Lu, H.P., Xun, L.Y., and Xie, X.S. (1998) Single-molecule enzymatic dynamics. *Science*, **282** (5395), 1877–1882.
22 Kuznetsova, S., Zauner, G., Aartsma, T.J., Engelkamp, H., Hatzakis, N., Rowan, A.E., Nolte, R.J.M., Christianen, P.C.M., and Canters, G.W. (2008) The enzyme mechanism of nitrite reductase studied at single-molecule level. *Proc. Natl Acad. Sci. USA*, **105** (9), 3250–3255.
23 Peneva, K., Mihov, G., Nolde, F., Rocha, S., Hotta, J., Braeckmans, K., Hofkens, J., Uji-i, H., Herrmann, A., and Müllen, K. (2008) Water-soluble monofunctional perylene and terrylene dyes: powerful labels for single-enzyme tracking. *Angew. Chem., Int. Ed.*, **47**, 3372–3375.
24 Rocha, S., Hutchison, J.A., Peneva, K., Herrmann, A., Müllen, K., Skjøt, M., Jørgensen, C.I., Svendsen, A., De Schryver, F.C., Hofkens, J., and Uji-i, H. (2009) Linking phospholipase mobility to activity by single molecule wide-field microscopy. *ChemPhysChem*, **10** (1), 151–161.
25 Donnert, G., Keller, J., Medda, R., Andrei, M.A., Rizzoli, S.O., Lurmann, R., Jahn, R., Eggeling, C., and Hell, S.W. (2006) Macromolecular-scale resolution in biological fluorescence microscopy. *Proc. Natl Acad. Sci. USA*, **103** (31), 11440–11445.
26 Hell, S.W. (2003) Toward fluorescence nanoscopy. *Nat. Biotechnol.*, **21** (11), 1347–1355.
27 Hell, S.W. (2007) Far-field optical nanoscopy. *Science*, **316** (5828), 1153–1158.
28 Dedecker, P., Hotta, J., Flors, C., Sliwa, M., Uji-i, H., Roeffaers, M.B.J., Ando, R., Mizuno, H., Miyawaki, A., and Hofkens, J. (2007) Subdiffraction imaging through the selective donut-mode depletion of thermally stable photoswitchable fluorophores: numerical analysis and application to the fluorescent protein dronpa. *J. Am. Chem. Soc.*, **129** (51), 16132–16141.
29 Yildiz, A. and Selvin, P.R. (2005) Fluorescence imaging with one manometer accuracy: application to molecular motors. *Acc. Chem. Res.*, **38** (7), 574–582.
30 Muls, B., Uji-i, H., Melnikov, S., Moussa, A., Verheijen, W., Soumillion, J.P., Josemon, J., Mullen, K., and Hofkens, J. (2005) Direct measurement of the end-to-end distance of individual polyfluorene polymer chains. *ChemPhysChem*, **6** (11), 2286–2294.
31 Betzig, E., Patterson, G.H., Sougrat, R., Lindwasser, O.W., Olenych, S., Bonifacino, J.S., Davidson, M.W., Lippincott-Schwartz, J., and Hess, H.F. (2006) Imaging intracellular fluorescent proteins at nanometer resolution. *Science*, **313** (5793), 1642–1645.
32 Rust, M.J., Bates, M., and Zhuang, X.W. (2006) Sub-diffraction-limit imaging by stochastic optical reconstruction microscopy (storm). *Nat. Methods*, **3** (10), 793–795.
33 Flors, C., Hotta, J., Uji-i, H., Dedecker, P., Ando, R., Mizuno, H., Miyawaki, A., and Hofkens, J. (2007) A stroboscopic approach for fast photoactivation-localization microscopy with dronpa mutants. *J. Am. Chem. Soc.*, **129** (45), 13970–13977.
34 Gustafsson, M.G.L. (2005) Nonlinear structured-illumination microscopy: wide-field fluorescence imaging with theoretically unlimited resolution. *Proc. Natl Acad. Sci. USA*, **102** (37), 13081–13086.

35 Zhuang, X.W., Kim, H., Pereira, M.J.B., Babcock, H.P., Walter, N.G., and Chu, S. (2002) Correlating structural dynamics and function in single ribozyme molecules. *Science*, **296** (5572), 1473–1476.

36 Yang, H., Luo, G.B., Karnchanaphanurach, P., Louie, T.M., Rech, I., Cova, S., Xun, L.Y., and Xie, X.S. (2003) Protein conformational dynamics probed by single-molecule electron transfer. *Science*, **302** (5643), 262–266.

37 Hashimoto, S. and Kiuchi, J. (2003) Visual and spectroscopic demonstration of intercrystalline migration and resultant photochemical reactions of aromatic molecules adsorbed in zeolites. *J. Phys. Chem. B*, **107** (36), 9763–9773.

38 Hashimoto, S. and Yamashita, S. (2004) Visual observation of contact-induced inter-crystalline migration of aromatic species adsorbed in zeolites by fluorescence microscopy. *ChemPhysChem*, **5** (10), 1585–1591.

39 Roeffaers, M.B.J., Sels, B.F., Loos, D., Kohl, C., Mullen, K., Jacobs, P.A., Hofkens, J., and De Vos, D.E. (2005) In situ space- and time-resolved sorption kinetics of anionic dyes on individual LDH crystals. *ChemPhysChem*, **6** (11), 2295–2299.

40 Roeffaers, M.B.J., Sels, B.F., Uji-i, H., De Schryver, F.C., Jacobs, P.A., De Vos, D.E., and Hofkens, J. (2006) Spatially resolved observation of crystal-face-dependent catalysis by single turnover counting. *Nature*, **439** (7076), 572–575.

41 Mahurin, S.M., Dai, S., and Barnes, M.D. (2003) Probing the diffusion of a dilute dye solution in mesoporous glass with fluorescence correlation spectroscopy. *J. Phys. Chem. B*, **107** (48), 13336–13340.

42 Fu, Y., Ye, F.M., Sanders, W.G., Collinson, M.M., and Higgins, D.A. (2006) Single molecule spectroscopy studies of diffusion in mesoporous silica thin films. *J. Phys. Chem. B*, **110** (18), 9164–9170.

43 Ye, F.M., Higgins, D.A., and Collinson, M.M. (2007) Probing chemical interactions at the single-molecule level in mesoporous silica thin films. *J. Phys. Chem. C*, **111** (18), 6772–6780.

44 Tan, W.H. and Yeung, E.S. (1997) Monitoring the reactions of single enzyme molecules and single metal ions. *Anal. Chem.*, **69** (20), 4242–4248.

45 Roeffaers, M.B.J., Sels, B.F., Uji-i, H., Blanpain, B., L'Hoest, P., Jacobs, P.A., De Schryver, F.C., Hofkens, J., and De Vos, D.E. (2007) Space- and time-resolved visualization of acid catalysis in ZSM-5 crystals by fluorescence microscopy. *Angew. Chem., Int. Ed.*, **46** (10), 1706–1709.

46 Geier, O., Vasenkov, S., Lehmann, E., Karger, J., Schemmert, U., Rakoczy, R.A., and Weitkamp, J. (2001) Interference microscopy investigation of the influence of regular intergrowth effects in mfi-type zeolites on molecular uptake. *J. Phys. Chem. B*, **105** (42), 10217–10222.

47 Roeffaers, M.B.J., Sels, B.F., De Schryver, F.C., Jacobs, P.A., Hofkens, J., and De Vos, D.E. (2007) In situ filming of reactions inside individual zeolite crystals using fluorescence microscopy. *Stud. Surf. Sci. Catal.*, **170**, 717–723.

48 Kox, M.H.F., Stavitski, E., and Weckhuysen, B.M. (2007) Nonuniform catalytic behavior of zeolite crystals as revealed by in situ optical microspectroscopy. *Angew. Chem., Int. Ed.*, **46** (20), 3652–3655.

49 Stavitski, E., Kox, M.H.F., and Weckhuysen, B.M. (2007) Revealing shape selectivity and catalytic activity trends within the pores of h-ZSM-5 crystals by time- and space-resolved optical and fluorescence microspectroscopy. *Chem. Eur. J.*, **13** (25), 7057–7065.

50 Shaughnessy, K.H., Kim, P., and Hartwig, J.F. (1999) A fluorescence-based assay for high-throughput screening of coupling reactions. application to heck chemistry. *J. Am. Chem. Soc.*, **121** (10), 2123–2132.

51 Su, H. and Yeung, E.S. (2000) High-throughput screening of heterogeneous catalysts by laser-induced fluorescence imaging. *J. Am. Chem. Soc.*, **122** (30), 7422–7423.

52 Goddard, J.P. and Reymond, J.L. (2004) Enzyme assays for high-throughput screening. *Curr. Opin. Biotechnol.*, **15** (4), 314–322.

53 Roeffaers, M.B.J., Ameloot, R., Baruah, M., Uji-i, H., Bulut, M., De Cremer, G., Müller, U., Jacobs, P.A., Hofkens, J., Sels, B.E., and De Vos, D.E. (2008) Morphology of large ZSM-5 crystals unraveled by fluorescence microscopy. *J. Am. Chem. Soc.*, **130** (17), 5763–5772.

54 Naito, K., Tachikawa, T., Cui, S.C., Sugimoto, A., Fujitsuka, M., and Majima, T. (2006) Single-molecule detection of airborne singlet oxygen. *J. Am. Chem. Soc.*, **128** (51), 16430–16431.

55 Naito, K., Tachikawa, T., Fujitsuka, M., and Majima, T. (2005) Single-molecule fluorescence imaging of the remote TiO_2 photocatalytic oxidation. *J. Phys. Chem. B*, **109** (49), 23138–23140.

56 Tachikawa, T., Fujitsuka, M., and Majima, T. (2007) Mechanistic insight into the TiO_2 photocatalytic reactions: Design of new photocatalysts. *J. Phys. Chem. C*, **111** (14), 5259–5275.

57 Tachikawa, T., Cui, S.C., Top, S., Fujitsuka, M., and Majima, T. (2007) Nanoscopic heterogeneities in adsorption and electron transfer processes of perylene diimide dye on TiO_2 nanoparticles studied by single-molecule fluorescence spectroscopy. *Chem. Phys. Lett.*, **443** (4–6), 313–318.

58 Naito, K., Tachikawa, T., Fujitsuka, M., and Majima, T. (2008) Real-time single-molecule imaging of the spatial and temporal distribution of reactive oxygen species with fluorescent probes: applications to TiO_2 photocatalysts. *J. Phys. Chem. C*, **112** (4), 1048–1059.

59 Rurack, K. (2001) Flipping the light switch "on" – the design of sensor molecules that show cation-induced fluorescence enhancement with heavy and transition metal ions. *Spectrochim. Acta A: Mol. Biomol. Spectrosc.*, **57** (11), 2161–2195.

60 Fabbrizzi, L., Licchelli, M., and Pallavicini, P. (1999) Transition metals as switches. *Acc. Chem. Res.*, **32** (10), 846–853.

61 Kiel, A., Kovacs, J., Mokhir, A., Kramer, R., and Herten, D.P. (2007) Direct monitoring of formation and dissociation of individual metal complexes by single-molecule fluorescence spectroscopy. *Angew. Chem., Int. Ed.*, **46** (18), 3363–3366.

62 Kiel, A., Jarve, A., Kovacs, J., Mokhir, A., Kramer, R., and Herten, D.P. (2007) Single-molecule studies on individual metal complexes. *Proc. SPIE*, **6444**, 64440C.

63 Vallee, R.A.L., Tomczak, N., Kuipers, L., Vancso, G.J., and van Hulst, N.F. (2003) Single molecule lifetime fluctuations reveal segmental dynamics in polymers. *Phys. Rev. Lett.*, **91** (3), 038301-1–038301-4.

64 Mason, M.D., Ray, K., Grober, R.D., Pohlers, G., and Cameron, J. (2004) Single molecule acid-base kinetics and thermodynamics. *Phys. Rev. Lett.*, **93** (7), 073004-1–073004-4.

12
Visualizing Single-Molecule Diffusion in Nanochannel Systems

Christophe Jung and Christoph Bräuchle

12.1
Introduction

Molecular movement in confined spaces is of broad scientific and technological importance in areas ranging from molecular sieving and membrane separation to active transport along intracellular networks. By viewing the movie of a single fluorescent dye molecule moving within the nanostructured channel system of a mesoporous silica host, incredible details can be seen of molecular motions, be they translation, rotation, trapping at specific sites, lateral motion between defect ("leaky") channels, or bouncing back from disordered regions, to mention only a few examples. In this way, single-molecule tracking experiments create a new quality of understanding the dynamics and interactions of molecules in nanoporous or mesoporous silica structures [1–6]. This is of major importance for many applications of these attractive nanomaterials, which have been used as hosts for numerous molecular and cluster-based catalysts [7], for molecular sieving and chromatography [8], for the stabilization of conducting nanowires [9–11], as a matrix for ultrasmall dye lasers [12], and for novel drug-delivery systems [13], to name but a few examples. In many of these cases, a complete characterization of the transport and of the dynamics of guest molecules in the channels is crucial for the successful functionality of these materials. They can be formed through the cooperative self-assembly of surfactants and framework building blocks [14] with widely tunable properties that include channel diameters (2–50 nm), topologies (hexagonal, cubic or lamellar), and functionalized walls.

In this chapter we describe how single molecules can be used to investigate nanoporous materials, with such methodology being helpful in the design of nanoscopic devices where the extreme control of molecular dynamics is required, an example being new silica-based drug-delivery systems. First, the details are presented of unique combined transmission electron microscopy (TEM) mapping and optical single-molecule tracking (SMT) experiments. The use of single dye molecules as nanoscale probes for mapping the structures of mesoporous silica materials with different phase topologies is followed by a presentation of a newly

Single Particle Tracking and Single Molecule Energy Transfer
Edited by Christoph Bräuchle, Don C. Lamb, and Jens Michaelis

ISBN: 978-3-527-32296-1

developed analytical method, as applied to a typical single-molecule trajectory. Details are then provided of polarization-dependent studies revealing simultaneous orientational motion and translational movement, and of single-molecule measurements with extreme positioning accuracy, down to the single-channel limit. Some investigations of molecular diffusion in calcined samples, using fluorescence correlation spectroscopy (FCS), are then presented. In a further step, the inner channels of mesoporous materials were functionalized with organic moieties, and the ability to fine-tune molecular diffusion is demonstrated. There follow descriptions of investigations with the anti-cancer drug doxorubicin in living HeLa cells, following its release from mesoporous host thin films. The chapter concludes by outlining some potential uses for mesoporous nanoparticles with functionalized pore walls as novel drug-delivery systems.

12.2 Correlation of Structural and Dynamic Properties Using TEM and SMT

Mesoporous structures are commonly characterized by using diffraction and electron microscopy methods [15], as well as gas sorption techniques. The ensemble diffusion behavior of small molecules has been examined with pulsed-field gradient nuclear magnetic resonance (NMR) spectroscopy [16] and neutron scattering [17]. Here, the major interest is in those techniques that provide a more direct access to the real structure of the mesoporous host, and to the dynamics on a single molecule basis. In this way, they can reveal those structural and dynamic features that are not obscured by ensemble or statistical averaging, as is the case with conventional techniques.

High-resolution TEM (HRTEM) offers a distinct means of directly visualizing the channel structure of a mesoporous host [18]. Figure 12.1a shows a typical HRTEM image of a thin film (M41S) that exhibits a hexagonal phase (see Figure 12.1b). This film was prepared by spin-coating a mixture of a silica precursor (TEOS: tetra-ethyl-*ortho*-silicate), a template (Brij56: polyoxyethylene-10-cetylether) and the probe dye molecule, a terrylendiimide (TDI) derivative [19], in an acidic water–ethanol solution; the resultant thin film (<100 nm thickness) was produced via evaporation-induced self-assembly [14]. Clearly, such an image can provide a detailed "landscape" of the channels in which the dye molecule can move, and although the image presents only an average view over the thickness of the film (ca. 10–20 layers of channels), it provides the most realistic picture currently obtainable of the host structure. The main drawbacks, however, are that the image size is limited to approximately 200 × 200 nm, and the film must be very thin.

Structural details such as channel diameters on the scale of a few nanometers cannot be directly imaged by using optical methods. However, single-molecule fluorescence microscopy can be used to track the movement of individual dye molecules that have been incorporated as guests into mesoporous silica thin films. This can be achieved in an epifluorescence microscope equipped with a highly

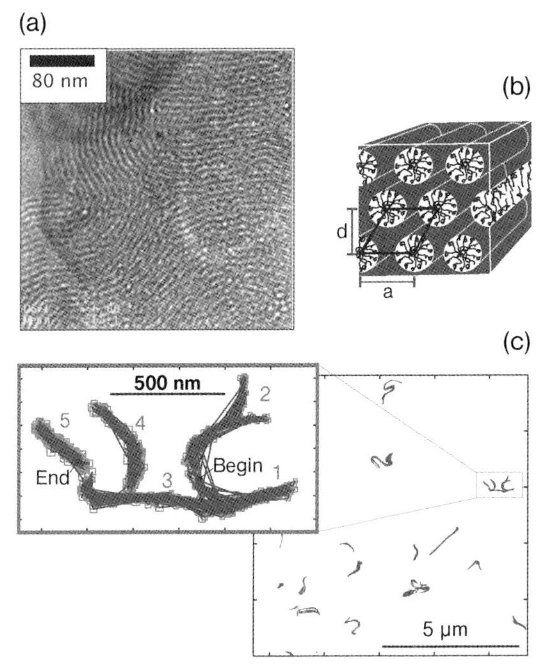

Figure 12.1 Sample system and single-molecule trajectories. (a) High-resolution transmission electron microscopy image showing the landscape of a channelar structure of a hexagonal mesoporous system. (b) Schematic diagram of the hexagonal pore topology. (c) Single-molecule tracking shows the trajectories of the movement of single molecules as guests in the nanoporous host. The enlarged inset shows the trajectory of a molecule as it explores five different domains.

sensitive CCD camera in a wide-field imaging set-up [1, 2]. As the films were much thinner than the focal depth of the microscope objective used, the images were shown to contain data from molecules through the whole film height, and also from the surface of the sample. Series of 1000 images were acquired with a temporal resolution of down to 100 ms per frame. In each movie frame, single molecules showed up as bright spots on a dark background, with their positions being obtained by fitting theoretical diffraction patterns to the spots, such that the positioning accuracy was down to 15 nm. Single-molecule trajectories were then built up by tracking spots from frame to frame; some examples are shown in Figure 12.1c.

The trajectories show that, during the acquisition time of the movie (500 s), the molecules in the hexagonal phase travel in a highly structured manner over distances of several micrometers. The inset in Figure 12.1c shows an enlargement of one of these trajectories, where the molecule travels first along the C-shaped structure on the right (region 1) and, after 65 s, enters the side-arm (region 2). After a further 100 s, it passes into the linear structure at the bottom (region 3). After another 144 s, it enters region 4, where it remains for 69 s before returning to region 3. Ultimately, it passes into region 5, where it moves back and forth for 109 s until the end of the movie (see Movie S1 in the supplementary material of Ref. [1]). It should be noted that the molecule appears to probe the domain boundaries during this process, by repeatedly "bouncing" back from dead ends of the channel regions. This is one of many striking examples showing how a single molecule seems to "explore" the structure of the host. It can be argued that such nonrandom diffusion, which is seen repeatedly in the hexagonal phase, directly maps the alignment of the channels and the domain structure. Moreover, it seems that the structure of the host is "seen" from the viewpoint of the molecule; that is, information is provided regarding the accessibility of the channels and the connectivity of the domains for the molecule in an unprecedented way, which is not possible with any other method. Concise proof, however, that the molecule really follows the channel system can only be provided by a proper overlay of the structure of the channels (as obtained by TEM) with the trajectories of the single molecules (as observed by SMT). Such a correlation would clearly illuminate all of the highly interesting aspects mentioned above, which can be summarized in one general question: How do the structural elements correlate with and influence the dynamics of the molecules in the nanoporous channels? Because the molecular movement in the pore system is the most important and defining characteristic of nanoporous materials and their various applications, knowledge of this behavior as function of the local structure is of major interest.

In order to overlay the optical and electron microscopic images, an approximately 50 nm-thick spin-coated film was prepared directly onto the 30 nm-thick Si_3N_4 membrane of the TEM sample holder [2]. Single polystyrene beads were then incorporated into the film which could be imaged using both TEM and optical microscopy. By first recording the trajectories with the optical wide-field microscope, and then measuring TEM images of the same sample region, a correct correlation of both images could be achieved by overlapping the same pattern of the polystyrene beads. This tedious procedure, which is described in detail in Ref. [2], provides an overlay accuracy of between 4 nm and 30 nm, depending on the number of beads in the images. Figure 12.2a shows the overlay of a single-molecule trajectory with a HRTEM map (×40 000 magnification). The map was obtained from many HRTEM images (as shown in Figure 12.1a), where adjacent square regions of 133 × 133 nm comprise the whole map. Within each square region, a fast Fourier transformation (FFT) resulted in an FFT director that depicted the average orientation of the pores, while the line thickness provided a measure of the degree of structural order in that region. These directors serve as a good guide for the eye with regards to the orientation of the channels, and also provide an overview of the domain sizes.

Figure 12.2a shows an example of a molecule faithfully following the pores and mapping out specific elements of the host structure (see Movie 3 in the supplementary material of Ref. [2]). The perfect overlay of the S-shaped trajectory with the direction of the pore system is shown well by the FFT directors. In Figure 12.2b, a specific region of Figure 12.2a is enlarged to show both the channel structure and the trajectories in greater detail. In both figures, the light blue boxes

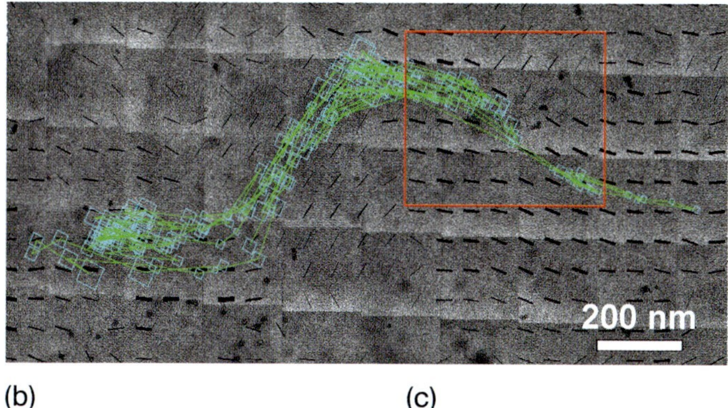

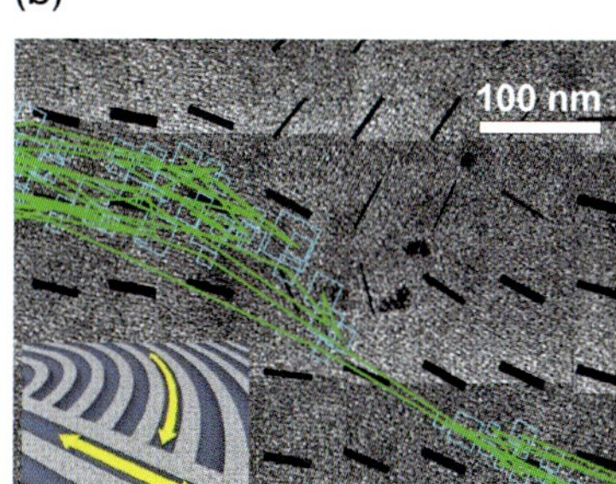

Figure 12.2 Molecular trajectories and structural elements in a hexagonal mesoporous silica film. (a) Overlay of an S-shaped trajectory of a single molecule with a transmission electron microscopy map. The molecule is exploring regions of parallel channels, with strongly curved areas and domain boundaries indicated by the fast Fourier transform directors (black bars). (b) Enlarged area of panel (a), showing part of the trajectory of the single molecule which bounces back repeatedly from a domain boundary formed by channels having different orientations sketched in the inset. The light-blue boxes in panels (a) and (b) depict the positioning accuracy. (c) Sketches of structural elements and molecular movements found in hexagonal mesoporous silica films.

of the trajectory indicate the positional accuracy of the molecular positions determined. As these are in the range of 15–30 nm, it indicates that the molecules' positions can be assigned not to a single channel, but rather to an ensemble of about three to six parallel channels. It should also be borne in mind that the diffusion is being sampled at discrete points in time and space. Thus, the connecting lines simply provide a method of visualizing the trajectories; they do not represent the molecules' exact path. The enlargement in Figure 12.2b, however, shows clearly that the molecule in the upper part bounces back from the domain boundary with channels having different orientation (as sketched in the inset of Figure 12.2b). In the lower part of the trajectory, however, the molecule can find its way through along the main domain which guides the pathway of the S-shaped trajectory. Many more structural elements (such as those shown in Figure 12.2c) are found and can be correlated with the dynamic behavior of single molecules.

In summary, a combination of the two techniques has provided the first direct proof that the molecular diffusion pathway through the pore system correlates with the pore orientation of the hexagonal structure. In addition, the influence of specific structural features of the host on the diffusion behavior of the guest molecules can be clearly seen.

With this approach, it should be possible to determine, in unprecedented detail, how a single fluorescent dye molecule travels through linear or strongly curved sections of the hexagonal channel system, how it changes speed, and how it bounces off a domain boundary with a different channel orientation. Furthermore, it can be shown how molecular travel is stopped at a less-ordered region, or how lateral motions between "leaky" channels allow a molecule to explore different parallel channels within an otherwise well-ordered periodic structure.

12.3
Phase Mixture

One interesting aspect of mesostructured silicas is that, by varying the molar ratio between the surfactant and the silica oligomers of the precursor solution, materials with different mesopore topologies can be prepared. In this section, the details are presented of a single-molecule investigation of mesoporous silica films in which hexagonal and lamellar mesophases coexist [1]. These topologies were in fact shown to strongly influence the diffusion of the single molecules inside the pores and lamellae of the host [6].

Figure 12.3a shows the first image of a movie (see Movie S5 in supplementary material of Ref. [1]) showing single TDI molecules diffusing in a mixture of a hexagonal and of a mesoporous lamellar phase. Here, Gaussian-shaped and donut-shaped patterns as fluorescence images of single molecules are seen to coexist within the same region. A Gaussian-shaped pattern is obtained for a molecule which rotates, whereas a donut-shaped pattern is assigned to a single molecule with its transition dipole (here, the long molecular axis of TDI) aligned along the optical axis of the microscope [20]. In the present case, this means that molecules in the lamellar phase are oriented perpendicular to the glass substrate, and thus are normal to the silica planes of the lamellar phase. The inset of Figure 12.3a shows magnified images of the two molecules indicated by the arrows. Overall, four populations of molecules can be distinguished on the basis of their different diffraction patterns and diffusive behavior in the movie of this phase mixture. The first type comprises molecules with Gaussian-shaped spots, which are diffusing along distinct structures over a large range of 1 to 5 μm. The second type has characteristic donut-shaped diffraction patterns that exhibit unstructured diffusion in two dimensions. Another, much smaller population consists of molecules that diffuse much faster, without showing any particular structure in their trajectories. An additional important observation is that multiple changes between these three

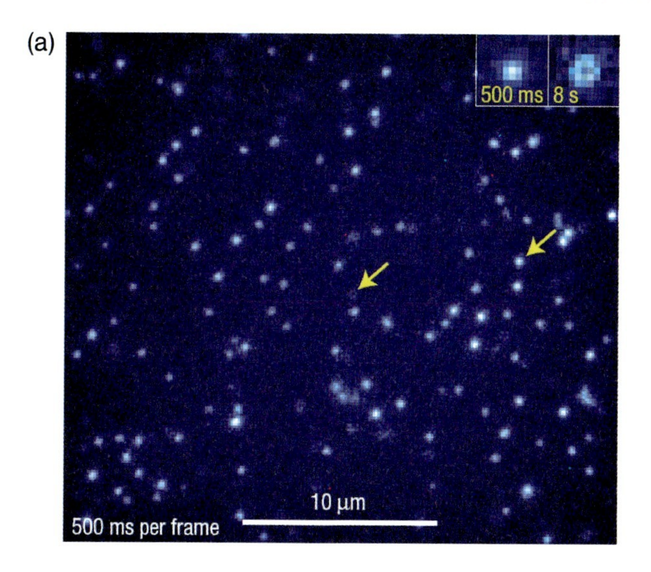

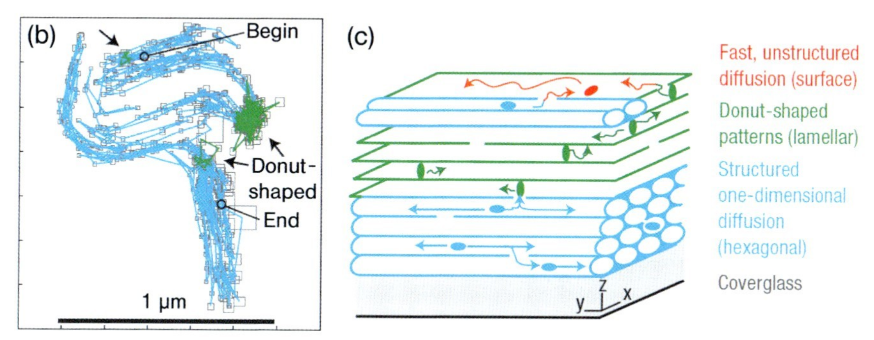

Figure 12.3 Diffusion of single molecules in the phase mixture. (a) A single-molecule image of the phase mixture contains both Gaussian- and donut-shaped diffraction patterns. Magnified images of the molecules indicated by the arrows are shown in the upper right corner. (b) An individual molecule undergoing several changes between the hexagonal (blue parts) and the lamellar phases (green parts, indicated by the arrows). (c) Schematic diagram of the diverse diffusion modes observed in the wide-field movies of the phase mixture.

types of mobility were observed for many of the molecules. Finally, very few immobile molecules were observed.

In accordance with the observations in the pure phases, the structured trajectories of population 1 were assigned to molecules in regions with a hexagonal arrangement of pores, while the molecules with donut-shaped patterns of population 2, that were diffusing very slowly and randomly, were assigned to other regions with lamellar structure, present simultaneously within this sample. The transport behavior of the two remaining populations did not correlate with the pore topologies in the sample. In fact, the fast molecules of population 3, with unstructured trajectories, could be removed by washing the surface of the sample with water, which clearly indicated that the molecules were on the surface of the film.

The diffusion behavior in this phase mixture fitted remarkably well with that in the pure phases. Interestingly, some molecules could be seen migrating between the phases. This observation demonstrates clearly that the two phases are actually interconnected.

A specific example is shown in Figure 12.3b. Again, as in the pure hexagonal phase, the shape of the trajectory explored by the Gaussian pattern clearly reflects the underlying pore structure of the hexagonal phase. This molecule changed three times from a Gaussian spot to a donut and back, with different residence times in each phase (see Movie S6 in supplementary material of Ref. [1]). Such switching phenomena clearly show that the two phases are actually connected, most likely via structural defects at the phase boundaries. Interestingly, other cases were also observed where the molecule switched several times from a Gaussian to a donut-shaped pattern at exactly the same position. This showed that, on occasion, the molecules pass repeatedly through the same defect region between phases.

To conclude, it has been shown that the structure of the trajectories, the diffusivities, and the orientation of single molecules are clearly distinctive for molecules traveling in the lamellar and hexagonal mesophases. A general schematic diagram of the different phases present in the film and the migration within, as well as between, the phases is shown in Figure 12.3c.

12.4
Heterogeneous Dynamics of a Single Molecule

In this section, a demonstration is provided of how the data obtained by SMT can be evaluated. It should be noted in particular that the detailed analysis of a typical single molecule trajectory provides information concerning the heterogeneities of the silica host nanostructure.

Figure 12.4 shows such a case for an individual molecule of population 1 in the phase mixture described in Section 12.3. The molecule in Figure 12.4a moves in a distinct structure and explores at least three different domains, indicated as A, B, and C (see Movie S8 in supplementary material of Ref. [1]). Similar to the molecule in the pure hexagonal phase, this trajectory provides a lucid picture not only of the channel structure but also of the connectivity and accessibility of channels between different domains.

For the molecule moving in this structure, a detailed analysis of the diffusion behavior was made by plotting the cumulative probability $P(r^2,t)$ of the squared displacements r^2 for different time lags t [21, 22]. The analysis of probability distributions instead of histograms allows for a more precise analysis, and avoids any loss of information due to binning of the histogram. The data were fitted with multiexponential decay functions:

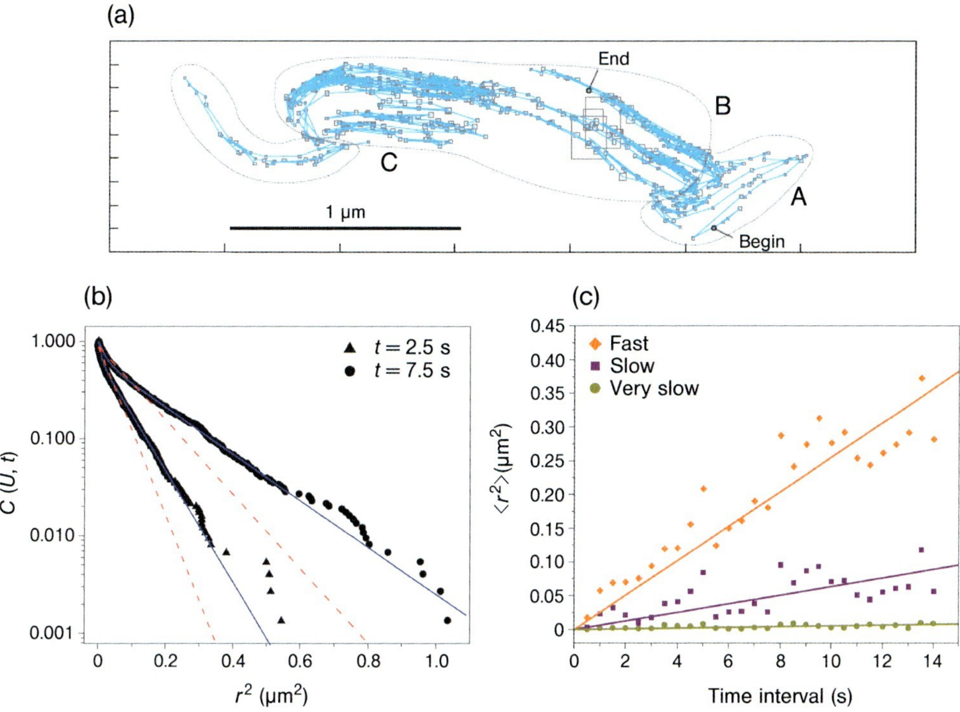

Figure 12.4 Diffusion analysis of an individual trajectory in the hexagonal phase of the mixture. (a) Trajectory of a molecule diffusing in a structured manner in different domains (A, B, C). (b) Plot of cumulative probability of the square displacement r^2 for two sample time intervals (t = 2.5 s,7.5 s). Monoexponential fits (red dashed line) and triexponential fits (blue line) are given. (c) Plot of the mean square displacement $<r^2>$ against the time intervals. Fits are according to $<r^2> = 2Dt$ for the three different characteristic $<r^2>$ distributions.

$$P(r^2, t) = \sum_{i=1}^{n} a_i \exp\left(-\frac{r^2}{\langle r_i(t)^2 \rangle}\right) \qquad (12.1)$$

where a_i is the amplitude of the different components and $<r_i(t)^2>$ the characteristic values for the mean-square displacement (MSD).

Regular diffusion should result in a monoexponential decay (n = 1), giving a characteristic value for the MSD $<r_i(t)^2>$ for each time lag t. Figure 12.4b shows the cumulative probability distributions for two sample time intervals (t = 2.5 s and 7.5 s). Here, the data cannot be fitted with a monoexponential decay function (red dashed lines in Figure 12.4b). Tri-exponential decay functions (n = 3) were found to describe the data best (blue solid lines), giving three characteristic $<r_i(t)^2>$ values for each time lag. These values are plotted against time in Figure 12.4c. The three different sets of $<r_i(t)^2>$ values were fitted with the Einstein–Smoluchowski equation for random diffusion in one dimension:

$$\langle r_i(t)^2 \rangle = 2Dt \quad (12.2)$$

giving values of $D_1 = 1.3 \times 10^{-2}\,\mu m^2 s^{-1}$, $D_2 = 3.2 \times 10^{-3}\,\mu m^2 s^{-1}$, and $D_3 = 2.8 \times 10^{-4}\,\mu m^2 s^{-1}$. These large differences imply that the molecule is diffusing in at least three types of environment. However, it can be shown that the three diffusion regimes are not spatially separated. The step sizes corresponding to these three diffusion modes are equally distributed over all parts of the track, and not segregated in one or other of the domains A, B, or C; the mobility of the molecule does not differ significantly from one domain to the other. Instead, owing to structural heterogeneities, the environment within one channel system changes strongly along the pathway of the molecule. These heterogeneities are revealed by the molecule continuously changing its mode of motion between at least three diffusion coefficients. Therefore, its diffusion cannot be described as a simple Brownian motion. An interpretation of these results could actually be a range of diffusion coefficients due to variations of local environment.

Hence, this example of trajectory analysis demonstrates that the diffusion coefficients vary not only between different phases (as shown in Section 12.3) or between trajectories of individual molecules within one phase, but can also change within the same trajectory of an individual molecule.

12.5 Oriented Single Molecules with Switchable Mobility in Long Unidimensional Nanochannels

By using the surfactant cetyltriethylammonium bromide (CTAB) in the template synthesis of the mesoporous M41S systems instead of Brij56 (which was used to fabricate the films studied in Sections 12.2–12.4), hexagonal arranged channels were obtained where the channel diameter (2–3 nm) was smaller than the length of the TDI molecule (3.2 nm) used as a fluorescent probe. Therefore, rotation of the TDI molecule should be impossible in well-ordered channels.

Polarization-modulated confocal microscopy was performed to monitor, simultaneously, the diffusional and orientational behavior of the TDI molecules in such systems [4]. Figure 12.5a shows three fluorescence images taken at times of 0 min, 2 min, and 4 min, where single TDI dye molecules appeared with a characteristic fluorescence-intensity profile (striped patterns) due to polarization modulation during the scan. From these patterns, both the position of the molecule and the orientation of its transition dipole moment (shown as yellow bars) were computed. By following the molecule in the circle, the trajectory of this single TDI molecule could be obtained, including its position and orientation (see Figure 12.5b). The trajectory showed that the molecule was moving linearly back and forth over a distance of about 2 μm, while it remained remarkably aligned with the direction of the diffusion, which was assigned to the direction of the pores. Figure 12.5c shows the alignment of the fluorophore with respect to the pores, and clearly indicates that free rotation was prevented by the well-ordered channel and the

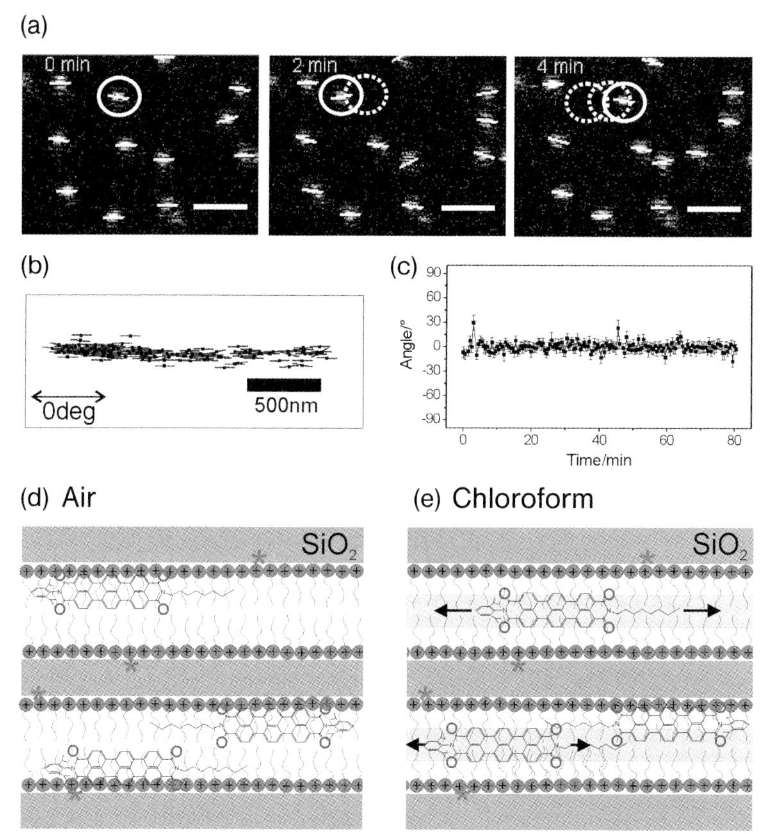

Figure 12.5 Parallel orientation and diffusion of single terrylendiimide (TDI) molecules in a highly ordered domain. (a) Sequence of fluorescence images showing linear diffusion of single TDI molecules in a chloroform atmosphere extracted from a time series. Scale bar – 2 μm. (b) Trajectory extracted from the molecule marked with the white circle in panel (a). (c) Calculated angular time trajectory of the same molecule. (d) Sketch of TDI molecules immobilized in the mesoporous film in an air atmosphere. The asterisks indicate active silanol groups. (e) TDI molecules in the mesoporous film in the presence of chloroform. The solvent provides a lubricant for the molecular movement.

geometrical constraints due to the size of the molecule and the diameter of the channel. Thus, the orientation of single TDI molecules and their trajectories map directly the direction of the channels. Furthermore, the rotational free translation of the single molecule indicated a structurally well-ordered area.

Recently, a method was developed by the present authors which could improve the film preparation to produce highly ordered linear channels in domains up to 100 μm in size [4]. Although, to the authors' knowledge such a high degree of order over long distances has not been reported previously for mesoporous structures, it is however highly desirable for many applications. The observations shown in Figure 12.5a–c were taken with the mesoporous film in a saturated chloroform atmosphere. In this case, the molecules were mobile, but if the atmosphere above the film was exchanged with air (40% relative humidity), the molecular motion ceased immediately. This effect–which proved to be highly reversible–indicated that, by changing the atmosphere around the porous film, the diffusion of TDI guest molecules could be switched on and off reversibly in a very simple manner (see Movies 1 and 2 in the supplementary material of Ref. [4]). These observations led to a model of the (im)mobility of the TDI molecules in the mesoporous host, where the TDI molecule have a very hydrophobic core and four oxygen atoms pointing to the side (Figure 12.5d–e), and with the lone-pair electrons able to interact with the positively charged heads of the CTAB molecules. In addition, interactions are possible with active silanol groups or other defects in the channel walls. These interactions appear to be responsible for the immobilization of the molecule in an air atmosphere (Figure 12.5d). In contrast, when chloroform (which is a good solvent for TDI) was added to the system, it seems likely that the small solvent molecules can form a lubricant-like phase inside the pores (Figure 12.5e), such that the TDI molecules can become solvated and diffuse along the pores [4]. Hence, such solvent exchange can allow an easy control of the diffusional behavior of the guest molecules.

12.6
High Localization Accuracy of Single Molecules Down to the Single Channel Limit

The achievement of an extremely high positioning accuracy in the optical SMT experiments, that was superior to the pore diameter of the mesoporous system, could be achieved with the CTAB-templated M41S system [4]. Here, the TDI molecules moved more slowly than in the Brij56-templated system, whilst by increasing the laser power and improving other parameters of the experiment, it was possible to achieve a spatial resolution of 2–3 nm. This allows the identification of a moving molecule in a specific channel, and the observation of jumps between neighboring channels.

Figure 12.6a shows the trajectory of a TDI molecule (see Movie 6 in the supplementary material of Ref. [4]) which first moves in one channel (black trajectory), and then switches over into the neighboring channel (green trajectory). This can be seen clearly if the movements along the pore, and perpendicular to it, in the $x(t)$ and $y(t)$ graphs are separated (Figure 12.6c). By inspection of the $y(t)$ graph, a jump to a neighboring pore is observable after 103 s. Figure 12.6b displays the histogram of $y(t)$ before (green) and after (black) 103 s; here, the distributions are clearly distinct and can be fitted with two Gaussian curves with a maximum at 0.6 and 6.1 nm and with half-widths (σ) of 2.9 nm and 2.3 nm, respectively. The observed dynamics was attributed to a TDI molecule which switched between two neighboring pores separated by 5–6 nm, as indicated in Figure 12.6d. In other cases, molecules were observed which explored even more distant pores by switching through defects from pore to pore. This seemed to be an important process in order for a molecule to circumvent dead ends in one pore, and to travel over

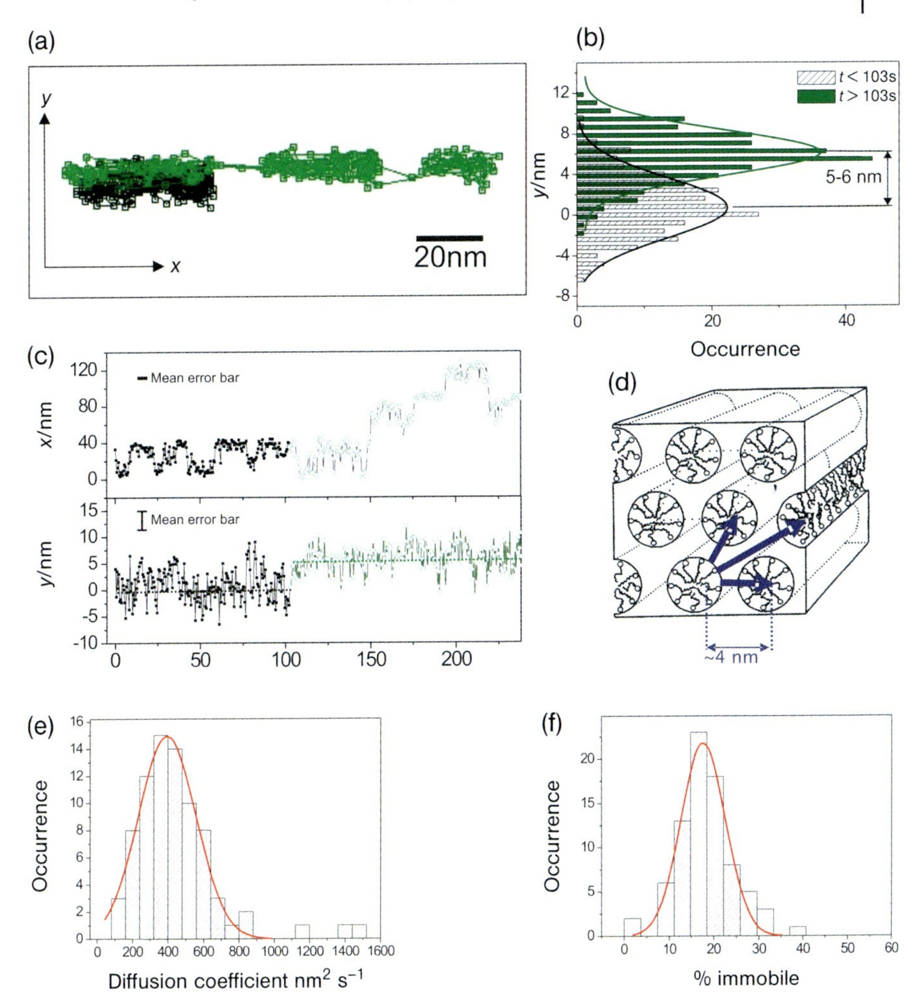

Figure 12.6 Diffusion and switching in two distinct neighboring channels and trapping behavior. (a) Trajectory with high optical resolution of a single terrylendiimide (TDI) molecule switching from one channel (black) to a neighboring channel (green). (b) Histogram with Gaussian distribution of the lateral (y) coordinate for the time intervals before (black) and after (green) the switching to the neighboring channel at time $t = 103$ s. (c) Projected $x(t)$ and $y(t)$ coordinates for the single TDI molecule in panel (a) diffusing in two distinct neighboring pores. $y(t)$ clearly shows the switch into the neighboring channel at the time $t = 103$ s. $x(t)$ shows repeated trapping of the single TDI molecule at the walls. (d) Hexagonal channel system of the CTAB-templated mesoporous host, with arrows indicating a switch to neighboring channels. (e) Histogram of the diffusion coefficients of 80 molecules in a CTAB-templated mesoporous film. (f) Histogram of the percentage of adsorption time per trajectory of 80 molecules in the same system.

larger distances within the mesoporous system. Figure 12.6a also shows an example of this behavior, because the molecule in the first channel was kept between two dead ends and could extend its pathway along the pores simply by switching in the neighboring channel.

The $x(t)$ graph of this molecule (as shown in Figure 12.6c) revealed another interesting property, namely the trapping of a molecule during its passage through the channelar network. Whenever the molecule reached one of the two dead ends of the first channel it became trapped for some seconds. The same was true when it moved in the neighboring channel. A detailed analysis of the trajectories of 80 molecules revealed a Gaussian distribution of the diffusion coefficients (Figure 12.6e) with a mean value of $D = 3.9 \times 10^{-4}\,\mu m^2\,s^{-1}$ and a half-width of $\delta = 10^{-4}\,\mu m^2\,s^{-1}$. In addition, the histogram of the percentage of adsorption time per trajectory (Figure 12.6f) was also Gaussian, with a maximum at 18%. This indicated that a molecule would spend on average 18% of its walk immobilized at an adsorption site. Such kinetic data can provide a very detailed picture of the dynamic behavior of molecules inside the channelar network of a mesoporous system.

12.7
Probing Chemical Interactions in Silica Thin Films Using Fluorescence Correlation Spectroscopy (FCS)

Until now, the mesoporous samples investigated have been as-synthesized; that is, the surfactant used as a structure-directing agent was still present within the pores, and acted as solvent for the guest dye molecules. It would be expected that the host–guest interactions would differ strongly in surfactant-free samples from which the template had been removed with a calcination treatment.

Ye *et al.* [23] used FCS to investigate the mobility and surface adsorption phenomena of dye molecules in both dry and hydrated surfactant-containing and surfactant-free mesoporous silica films. FCS constitutes an interesting method for investigating molecular diffusion in mesoporous materials, as it allows measurements to be made with a high time resolution, which in turn makes possible the observation of rapid processes. In addition, FCS is independent of any photobleaching that would limit the amount of data collected. Unfortunately, however, the trajectories of moving molecules cannot be obtained (i.e., no structural information is provided), and a model for the possible mode of motions is necessary.

Three different dyes, namely Nile red (NR), DiI [1,1′-dihexadecyl-3,3,3′,3′-tetramethylindocarbocyanine perchlorate], and a sulfonated perylene diimide (SPDI), which were selected for their hydrophobicity and charge characteristics, were employed in these studies (for structures, see Figure 12.7a). Both, as-synthesized (with CTAB surfactant as template) and calcined (surfactant-free) mesoporous films were studied, which were either dry or had been hydrated by exposure to a high-humidity environment. FCS was used to assess the relative importance of diffusion and reversible binding in each material – that is, to measure the dye diffusion coefficients and determine the duration of reversible adsorption events.

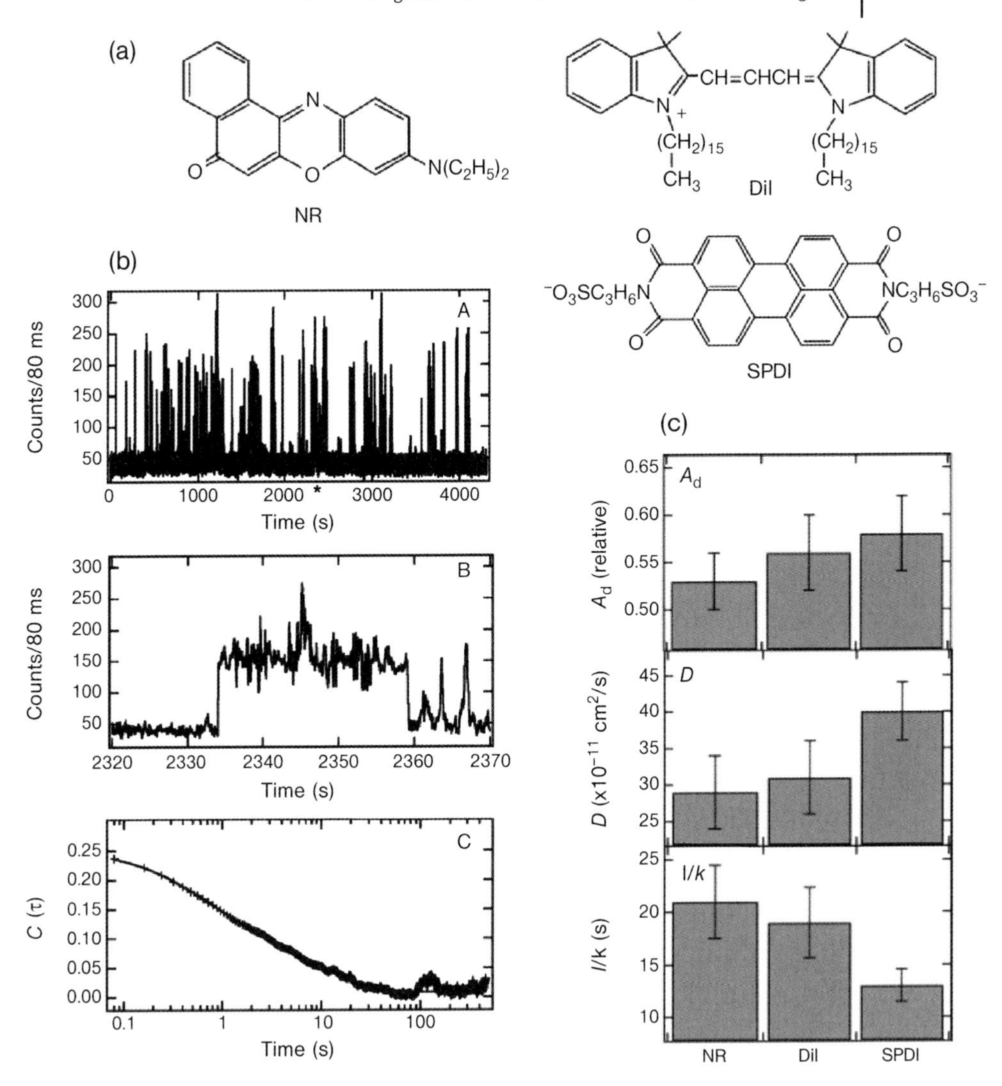

Figure 12.7 Single-molecule diffusion in surfactant-containing and surfactant-free mesoporous silica films. (a) Chemical structures of NR, DiI, and SPDI. (b) A: Representative fluorescence–time transient recorded for DiI in a rehydrated (50% RH) as-synthesized mesoporous silica film. B: Expanded region from part A depicting a long burst event. The expanded region in part A is marked by an asterisk. C: Autocorrelation of the time transient shown in part A (symbols) and its fit to Equation 12.3 (solid line). (c) The measured mean relative amplitude A_d (top), diffusion coefficients D (middle), and mean adsorption time $1/k$ (bottom) values for the three different dyes in rehydrated (50% RH) calcined mesoporous silica films. The error bars show the 95% confidence intervals on each parameter.

Such a STOP and GO behavior due to the presence of adsorption sites within the mesoporous structure has already been reported in Sections 12.5 and 12.6 for as-synthesized CTAB-templated silica films.

Time transients obtained from the different samples studied exhibited distinctly different behaviors that were consistent with a variable level of molecular mobility. A representative time transient obtained from a rehydrated [50% relative humidity (RH)], DiI-doped, as-synthesized film is shown in Figure 12.7b (panel A). The transient exhibited short bursts of fluorescence separated by periods over which only background counts were observed. These bursts continued to occur even over extremely long times (hours), indicating that they had arisen from the translational diffusion of different single molecules into and out of the detection volume of the microscope. On closer inspection of transients (as that shown in Figure 12.7b, panel A), it was clear that they often depicted a mixture of behaviors. For example, an expanded region of the transient in Figure 12.7b (panel A) is shown in Figure 12.7b (panel B); this depicts a fluorescence burst that was too long to be attributed to the translational diffusion of a molecule through the excitation volume. The event shown had a duration of approximately 26 s, while the average burst length due to diffusing molecules in this transient was 0.7 s. Furthermore, the fluorescence signal observed during this event was relatively constant, aside from shot noise fluctuations. As discussed previously for surfactant-containing mesoporous silica materials [3–5, 24, 25], as well as other samples [26–28], all of these events could be attributed to a strong, reversible surface adsorption of the dye molecules. Similar events were observed for all three dyes in different samples, when studied under specified conditions.

The single-point autocorrelation functions, $C(\tau)$, were calculated and subsequently fitted to a model decay function selected for its inclusion of both 2-D diffusion (within the film plane) and reversible surface adsorption [26, 29, 30]:

$$C(\tau)=\frac{A_{\mathrm{d}}}{1+D\tau/s^{2}}+A_{\mathrm{a}}\exp(-\tau k) \tag{12.3}$$

where D is the diffusion coefficient for the dye, s^2 is the beam variance, k is the rate constant for desorption of adsorbed species, and A_d and A_a are the amplitudes of the diffusion and adsorption components of the decay, respectively. The autocorrelation function obtained from the time transient shown in Figure 12.7b (panel A) is shown in Figure 12.7b (panel C). This particular fit yielded a diffusion coefficient of $3.7 \times 10^{-2}\,\mu m^2\,s^{-1}$, a desorption rate constant (k) of $0.080\,s^{-1}$, and a relative A_d of 0.70. This same procedure was used to analyze several hundred time transients obtained from the full range of samples.

The characterization of calcined films provides a means to further explore the mechanisms for dye diffusion and surface adsorption in mesoporous silica films. The water remaining in the calcined films is too little to facilitate molecular diffusion, while the residual surface silanols are still sufficient to impact upon molecular mobility and surface adsorption in hydrated calcined films. All three dyes were observed to be immobile (i.e., they exhibited clear, irreversible photobleaching events) in dry (20% RH) calcined films. Immobility in these samples most likely arises from strong dye–matrix interactions that occur in the absence of solvent (water). This observation also demonstrates the importance of the hydrated surfactant micelles in facilitating molecular mobility in the as-synthesized samples. For the rehydrated (50% RH) calcined samples, all three dyes were found to be mobile. The bar graphs in Figure 12.7c show the average diffusion coefficients, D, the mean relative amplitude A_d, and the mean adsorption time ($1/k$) values obtained. For SPDI, DiI, and NR, the average diffusion coefficients obtained were 4.0×10^{-2}, 3.1×10^{-2}, and $2.9 \times 10^{-2}\,\mu m^2\,s^{-1}$, respectively. The mean A_d (relative) values measured for these three dyes were 0.58, 0.56, and 0.53, respectively, and the mean adsorption times ($1/k$) were 13, 19, and 21 s, respectively. Heterogeneity similar to that observed in the as-synthesized films was also observed here, as reflected by the differences between the mean and most common values of these parameters. Overall, the results showed the SPDI molecules to be the most mobile among the three dyes, and to spend the least time adsorbed to the silica surface. Both, DiI and NR exhibited smaller (but similar) D-values, and spent relatively more time adsorbed to the silica surface. The "water-filled" pores of the hydrated calcined films clearly provide the fluid environment necessary for translational diffusion of the dye molecules.

12.8 Functionalized Mesoporous Silica Structures

For many applications, mesoporous materials are expected to show enhanced properties when their inner channel walls are functionalized with organic moieties to fine-tune the host–guest interactions. Recently, a co-condensation method was developed [31] which enabled the homogeneous incorporation of functional groups (see the sketch in Figure 12.8a), and the diffusion behavior of single TDI dye molecules studied in the functionalized mesoporous films [32]. The films were measured as-synthesized (with Brij-56 as the templating agent) and at 30% RH.

The influence of functionalization density on the D-value of the DIP-TDI was first examined as shown in Figure 12.8b. These data were obtained from propyl- (black line), cyanopropyl- (red line), and phenyl-functionalized (blue line) films with functionalization densities of 2.5, 5, 10, 20, and 30 mol%. The green dot (at 0 mol%) corresponds to an unfunctionalized sample. The mean D-values of the propyl- and cyanopropyl-functionalized samples were increased substantially with increasing functionalization density (sevenfold and fourfold factors, respectively). In contrast to the above samples that were functionalized with flexible chains, the mean D-value for the phenyl-functionalization decreased with increasing functionalization density. Here, the dye inside the film was slowed by almost one order of magnitude, from $D = 6.5 \times 10^{-4}\,\mu m^2\,s^{-1}$ (2.5 mol%) to $D = 8 \times 10^{-5}\,\mu m^2\,s^{-1}$ (30 mol%). These results showed that the diffusion dynamics of guest molecules could be heavily influenced by the introduction of functional groups, and that the mean D-values changed significantly with functionalization density.

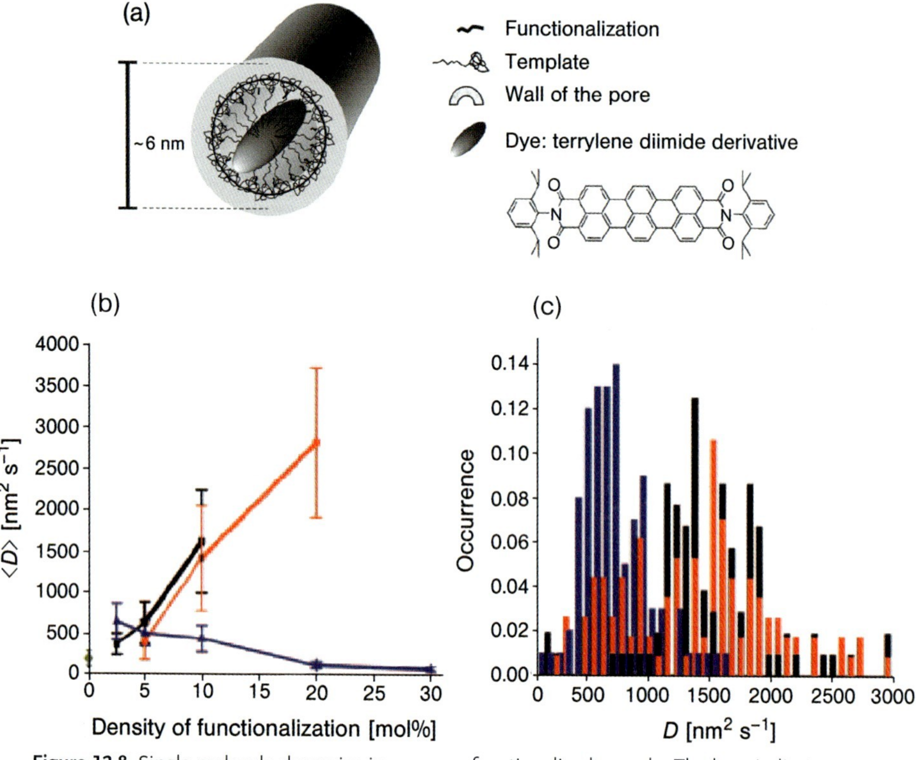

Figure 12.8 Single-molecule dynamics in functionalized mesoporous silica. (a) Sketch of a DIP-TDI dye molecule within one pore. All constituents are drawn to scale. The chemical structure of the dye is displayed on the right. (b) Correlation of the mean diffusion coefficients <D> with the functionalization densities, including data for the unfunctionalized film, given at zero density (black: propyl, red: cyanopropyl, blue: phenyl, green: unfunctionalized). The green dot at 0 mol% corresponds to an unfunctionalized sample. The bars indicate the width of the distribution of the *D*-values due to the heterogeneity of the samples, and not to any error in their determination. (c) Influence of the polarity of the functional groups (red: cyanopropyl, blue: trifluoropropyl, black: propyl) on the diffusion dynamics of the guest molecules. The films were synthesized with 10 mol% functionalization density, and the films measured at 30% RH.

The nature of the functional groups was also varied in order to explore their influence on guest dynamics. As an example, Figure 12.8c displays the influence of functional-group polarity on the *D*-values in a mesoporous film by comparing propyl-, cyanopropyl-, and trifluoropropyl-functionalized films. The strongly polar trifluoropropyl groups decreased the mean *D*-value of the dye to about one-half ($7.4 \times 10^{-4}\,\mu m^2 s^{-1}$) of those of the propyl- and cyanopropyl- functionalized films ($16.2 \times 10^{-4}\,\mu m^2 s^{-1}$ and $14.2 \times 10^{-4}\,\mu m^2 s^{-1}$, respectively). Thus, an increase in the polarity of the functional groups leads to a decrease in dye dynamics.

Taken together, these data demonstrate that the functionalization of the pore walls opens up the opportunity to fine-tune host–guest interactions in these systems. For example, such interactions are of major interest when mesoporous hosts are used for drug-delivery systems, where control of the drug release rate is of paramount importance. Thus, a deceleration in guest dynamics, as was observed for the phenyl-functionalized samples, can lead to the generation of a "depot" effect; in other words, the incorporated drug will be released slowly over a prolonged period of time.

12.9
Single-Molecule Studies of Mesoporous Silica Structures for Drug-Delivery Applications

Novel drug-delivery systems based on nanostructures can provide fundamental progress to many therapies in medicine [33]. Recently, mesoporous thin silica films with nanometer-sized pores have been applied as drug carriers [34], and also used to incorporate the anti-cancer drug, doxorubicin [35–37]. Subsequent measurements conducted on a single-molecule level have provided mechanistic insights into the processes that govern drug diffusion, and revealed the interactions of the drug with the host structure [34]. Drug diffusion inside the nanoporous network is controlled by both pore size and surface modification.

Surprisingly, in the case of an unfunctionalized Brij-56-templated thin film, all molecules were shown to be immobile, after which propyl functional groups were attached covalently to the walls of the silica matrix. As shown in Section 12.8, propyl-functionalization increases the diffusivity of guest molecules inside mesoporous films compared to unfunctionalized films. The single-molecule trajectory of a mobile molecule is displayed in Figure 12.9b; here, the well-structured trajectory can be seen clearly to map the domain structure of the underlying porous network.

Doxorubicin-loaded, Pluronic-templated mesoporous films were then investigated for their applicability as drug-delivery systems. These templated structures were selected on the basis of Pluronic's well-known role as a biocompatible micellar nanocarrier of pharmaceuticals, such as doxorubicin. In this study, a doxorubicin-loaded mesoporous film was immersed into a cell medium containing live HeLa cells (the sample set-up is shown in Figure 12.9c). As the solution entered the pores and triggered the release of doxorubicin from the delivery system, the increased fluorescence intensity of the cell medium was monitored using confocal microscopy (the gray curve in Figure 12.9d). Although no doxorubicin fluorescence was detected within the first 4 min of adding the drug-loaded coverslip to the cell medium (the time taken for the cell medium to flush the pores and trigger drug release), the doxorubicin fluorescence was subsequently rapidly increased. A good agreement between an exponential fit and the experimental data showed the release to follows first-order kinetics, with the majority of the drug having been released over a 10 min period. A similar release kinetics was identified by Cauda *et al.* when the antibiotic vancomycin was loaded into mesoporous silica [38].

(a)

(c) Doxorubicin hydrochloride

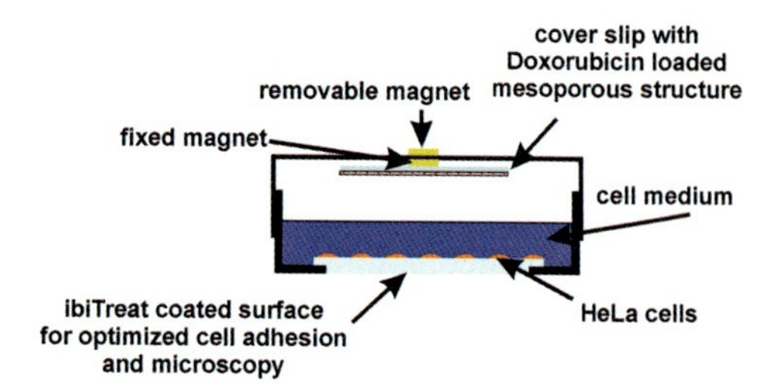

(b)

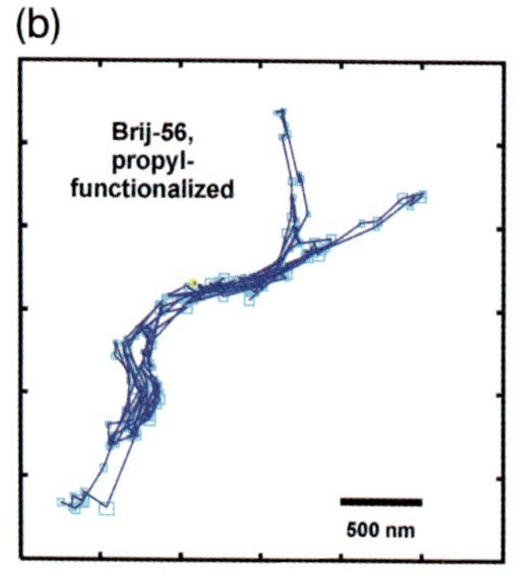

(d)

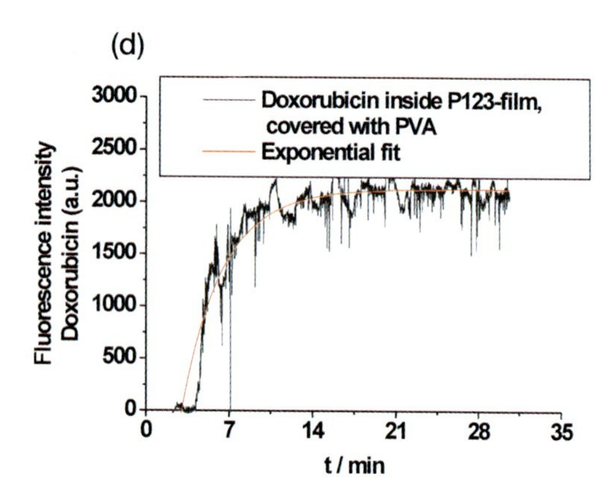

(e)

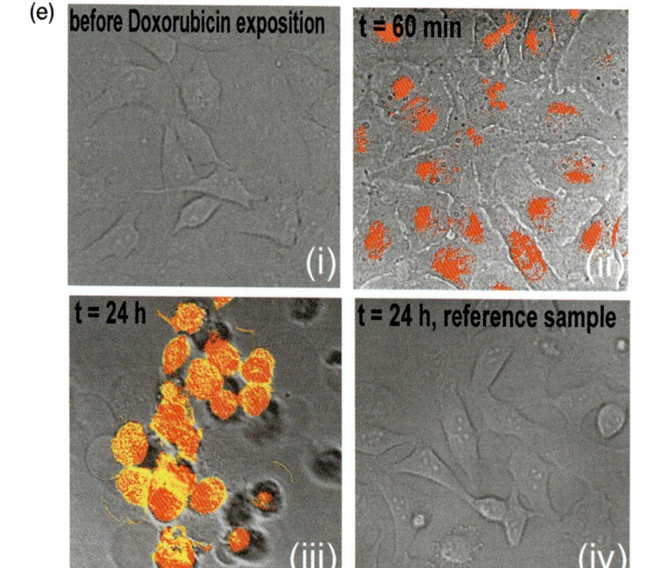

Figure 12.9 (caption see p. 329)

The next step was to investigate the effect of the delivered doxorubicin on HeLa cells. Figure 12.9e shows overlays of confocal transmission images (gray) and fluorescence images (red) of the doxorubicin fluorescence. Based on their shape, the HeLa cells were alive before being exposed to doxorubicin (Figure 12.9e (i)), but after 60 min doxorubicin fluorescence was clearly located in the cell nucleus (Figure 12.9e (ii)). The rationale for this was that the cytostatic properties of doxorubicin arise mainly its direct intercalation into DNA, as well as inhibition of topoisomerase II by interfering with the topoisomerase II–DNA complex [39]. However, the cells had not yet changed their shape. After 24 h, the cells were highly fluorescent (Figure 12.9e (iii)), round in shape, and had become detached from the dish, indicating cell death. The lack of any effect in a control, doxorubicin-free delivery system (Figure 12.9e (iv)) confirmed that cell death had been caused by doxorubicin. It was concluded, therefore, that doxorubicin released from thin films retains its cytostatic properties, and mesoporous films can be applied for drug-delivery purposes.

In summary, the way in which drug interaction with the host-matrix can be influenced on a nanometer-scale, by covalently attaching organic functional groups, has been successfully demonstrated. Such fine-tuning of a host–guest interaction is an essential prerequisite when generating a depot-effect. It has also been shown that the drug can be released from nanochannels in the carrier system, and delivered to the cells.

12.10 Conclusions and Outlook

The studies described in this chapter have demonstrated how single molecules can be used to map out the structure of mesoporous materials. The approach used offers detailed mechanistic insights into the complicated host–guest interplay. The use of mesoporous thin films with nanometer-sized pores as a drug-delivery

Figure 12.9 Mesoporous silica films as novel drug-delivery-systems. (a) Structure of the anticancer drug doxorubicin hydrochloride. (b) Trajectory of a mobile molecule inside a propyl-functionalized Brij-56-templated film. The trajectory reveals the topology of the underlying domain structure of the nanoporous host structure. (c) Sample set-up. The sample consists of a μ-dish with cell medium and HeLa cells adhered to the dish bottom. The coverslip with the doxorubicin-loaded mesoporous structure is held on the upper side of the dish, using magnets. On removing the magnet, the sample is immersed in the cell medium, which is able to flush the pores of the delivery system and trigger drug release. (d) Release kinetics of doxorubicin from a Pluronic P123-templated thin film. The release was monitored via the change in fluorescence intensity of doxorubicin over time (gray curve). The black line shows an exponential fit to the data. (e) Live-cell measurements. Overlay of confocal transmission images (gray) and doxorubicin fluorescence (red). Images were recorded (i) before doxorubicin exposure, and at (ii) 60 min and (iii) 24 h after adding the doxorubicin-loaded delivery system. (iv) Image recorded at 24 h after addition of an unloaded delivery system (control).

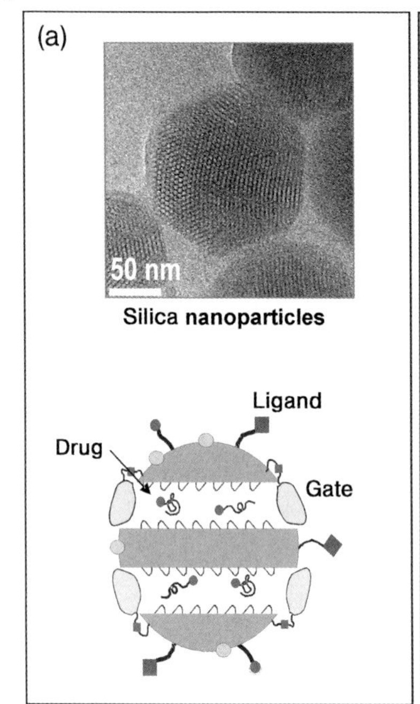

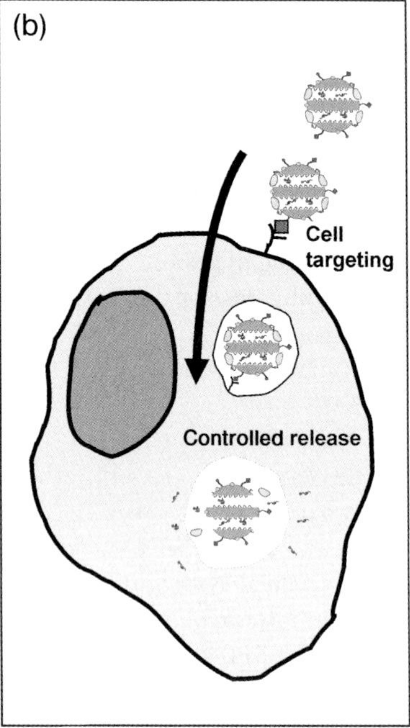

Figure 12.10 Mesoporous silica nanoparticles as novel drug-delivery systems. (a) Upper: Transmission electron microscopy image of mesoporous silica nanoparticles. Lower: Schematic of a novel drug-delivery particle which contains functionalized pores, closed by a gate, and is decorated with ligands for cell targeting. (b) Cell targeting by ligand–receptor interaction at the cell membrane, endosomal uptake, and controlled release after pH change from early to late endosome formation.

system for the cytostatic agent doxorubicin has also been demonstrated. For future applications, mesoporous structures may be employed as nanoparticles for drug-delivery applications, for example in cancer therapy.

Mesoporous nanoparticles of approximately 100 nm diameter can be produced (see Figure 12.10a, upper), the intention being to use these as the basis for a novel drug-delivery system (see Figure 12.10a, lower). This system utilizes several functions for controlled drug release and the targeting of specific cells that include:

- Functionalized pore walls to create a depot effect and controlled release of the drug.
- A gate which closes the channels but can be opened by reducing the pH from 7 to 5, as occurs between early and late endosome formation within the cell.
- Ligands attached to the nanoparticles used for cell-targeting purposes, by using specific receptors at the cell surface.

In this process, the nanoparticle would attach via its ligand to a specific receptor at the cell membrane; detailed live-cell imaging of the uptake and trafficking of polymer-based nanoparticles (polyplexes) as drug-delivery systems [40] targeting to cancer cells has already been demonstrated [41, 42]. Following endocytosis of the nanoparticle, the early endosome would be transformed to its late form by reducing the pH from 7 to 5; this in turn would open the nanoparticle's "gates," releasing the drug in controlled fashion (Figure 12.10b). These drug-delivery systems would be especially effective for the administration of cytotoxic drugs in cancer chemotherapy. In this situation, the delivery system is targeted towards the cancer cell but releases the toxic agent only when inside the cancer cell. Drug-delivery systems using mesoporous nanoparticles with different functionalities are currently undergoing development at the present authors' laboratory.

Acknowledgments

The authors are very grateful to all coworkers and collaborators in their publications, and who have shaped their understanding of molecular dynamics in nanoporous systems and drug delivery in living cells. These studies were supported by the Excellence Clusters Nanosystems Initiative Munich (NIM) and Center for Integrated Protein Science Munich (CIPSM), and also by the collaborative research centers SFB 486 and SFB 749.

References

1 Kirstein, J., Platschek, B., Jung, C., Brown, R., Bein, T., and Bräuchle, C. (2007) Exploration of nanostructured channel systems by single molecule probes. *Nat. Mater.*, **6**, 303–310.

2 Zürner, A., Kirstein, J., Döblinger, M., Bräuchle, C., and Bein, T. (2007) Visualizing single molecule diffusion in mesoporous materials. *Nature*, **450**, 705–708.

3 Jung, C., Hellriegel, C., Platschek, B., Wöhrle, D., Bein, T., Michaelis, J., and Bräuchle, C. (2007) Simultaneous measurement of orientational and spectral dynamics of single molecules in nanostructured host-guest materials. *J. Am. Chem. Soc.*, **129**, 5570–5579.

4 Jung, C., Kirstein, J., Platschek, B., Bein, T., Budde, M., Frank, I., Müllen, K., Michaelis, J., and Bräuchle, C. (2008) Diffusion of oriented single molecules with switchable mobility in networks of long unidimensional nanochannels. *J. Am. Chem. Soc.*, **130**, 1638–1648.

5 Jung, C., Hellriegel, C., Michaelis, J., and Bräuchle, C. (2007) Single molecule traffic in mesoporous materials: translational, orientational and spectral dynamics. *Adv. Mater.*, **19**, 956–960.

6 Feil, F., Jung, C., Kirstein, J., Michaelis, J., Li, C., Nolde, F., Müllen, K., and Bräuchle, C. (2009) Diffusional and orientational dynamics of various single terrylene diimide conjugates in mesoporous materials. *Micropor. Mesopor. Mater.*, in press. doi: 10.1016/j.micromeso.2009.01.024

7 De Vos, D.E., Dams, M., Sels, B.F., and Jacobs, P.A. (2002) Ordered mesoporous and microporous molecular sieves functionalized with transition metal complexes as catalysts for selective organic transformations. *Chem. Rev.*, **102**, 3615–3640.

8 Rebbin, V., Schmidt, R., and Fröba, M. (2006) Spherical particles of phenylene-bridged periodic mesoporous organosilica for high-performance liquid chromatography. *Angew. Chem., Int. Ed.*, **45**, 5210–5214.

9 Cott, D.J., *et al.* (2006) Preparation of oriented mesoporous carbon nano-

filaments within the pores of anodic alumina membranes. *J. Am. Chem. Soc.*, **128**, 3920–3921.

10 Ye, B., Trudeau, M.L., and Antonelli, D.M. (2001) Observation of a double maximum in the dependence of conductivity on oxidation state in potassium fulleride nanowires supported by a mesoporous niobium oxide host lattice. *Adv. Mater.*, **13**, 561–565.

11 Petkov, N., Stock, N., and Bein, T. (2005) Gold electroless reduction in nanosized channels of thiol-modified SBA-15 material. *J. Phys. Chem. B*, **109**, 10737–10743.

12 Braun, I., Ihlein, G., Laeri, F., Nockel, J.U., Schulz-Ekloff, G., Schuth, F., Vietze, U., Weiss, O., and Wohrle, D. (2000) Hexagonal microlasers based on organic dyes in nanoporous crystals. *Appl. Phys. B*, **70**, 335–343.

13 Roy, I., *et al.* (2005) Optical tracking of organically modified silica nanoparticles as DNA carriers: a nonviral nanomedicine approach for gene delivery. *Proc. Natl Acad. Sci. USA*, **102**, 279–284.

14 Brinker, C.J., Lu, Y., Sellinger, A., and Fan, H. (1999) Evaporation-induced self-assembly: nanostructures made easy. *Adv. Mater.*, **11**, 579–585.

15 Terasaki, O. and Ohsuna, T. (2003) Structural study of microporous and mesoporous materials by transmission electron microscopy, in *Handbook of Zeolite Science and Technology* (eds S.M. Auerbach, K.A. Carrado, and P.K. Dutta), Dekker, New York, pp. 291–315.

16 Kukla, V., *et al.* (1996) NMR studies of single-file diffusion in unidimensional channel zeolites. *Science*, **272**, 702–704.

17 Benes, N.E., Jobic, H., and Verweij, H. (2001) Quasi-elastic neutron scattering study of the mobility of methane in microporous silica. *Micropor. Mesopor. Mater.*, **43**, 147–152.

18 Sakamoto, Y., *et al.* (2000) Direct imaging of the pores and cages of three-dimensional mesoporous materials. *Nature*, **408**, 449–453.

19 Jung, C., *et al.* (2006) A new photostable terrylene diimide dye for applications in single molecule studies and membrane labeling. *J. Am. Chem. Soc.*, **128**, 5283–5291.

20 Dickson, R.M., Norri, D.J., and Moerner, W.E. (1998) Simultaneous imaging of individual molecules aligned both parallel and perpendicular to the optic axis. *Phys. Rev. Lett.*, **81**, 5322–5325.

21 Hellriegel, C., *et al.* (2004) Diffusion of single streptocyanine molecules in the nanoporous network of sol-gel glasses. *J. Phys. Chem. B*, **108**, 14699–14709.

22 Hellriegel, C., Kirstein, J., and Bräuchle, C. (2005) Tracking of single molecules as a powerful method to characterise diffusivity of organic species in mesoporous materials. *New J. Phys.*, **7**, 1–14.

23 Ye, F., Higgins, D.A., and Collinson, M.M. (2007) Probing chemical interactions at the single-molecule level in mesoporous silica thin films. *J. Phys. Chem. C*, **111** (18), 6772–6780.

24 Seebacher, C., Hellriegel, C., Deeg, F.W., Bräuchle, C., Altmaier, S., Behrens, P., and Müllen, K. (2002) Observation of translational diffusion of single terrylenediimide molecules in a mesostructured molecular sieve. *J. Phys. Chem. B*, **106**, 5591–5595.

25 Fu, Y., Ye, F., Sanders, W.G., Collinson, M.M., and Higgins, D.A. (2006) Single molecule spectroscopy studies of diffusion in mesoporous silica thin films. *J. Phys. Chem. B*, **110**, 9164–9170.

26 Wirth, M.J. and Swinton, D.J. (1998) Single-molecule probing of mixed-mode adsorption at a chromatographic interface. *Anal. Chem.*, **70**, 5264–5271.

27 Martin-Brown, S.A., Fu, Y., Saroja, G., Collinson, M.M., and Higgins, D.A. (2005) Single-molecule studies of diffusion by oligomer-bound dyes in organically modified sol–gel-derived silicate films. *Anal. Chem.*, **77**, 486–494.

28 Zhong, Z., Lowry, M., Wang, G., and Geng, L. (2005) Probing strong adsorption of solute onto C18-silica gel by fluorescence correlation imaging and single-molecule spectroscopy under RPLC conditions. *Anal. Chem.*, **77**, 2303–2310.

29 Mahurin, S.M., Dai, S., and Barnes, M.D. (2003) Probing the diffusion of a dilute dye solution in mesoporous glass with fluorescence correlation spectroscopy. *J. Phys. Chem. B*, **107**, 13336–13340.

30 Wirth, M.J., Ludes, M.D., and Swinton, D.J. (2001) Analytic solution to the autocorrelation function for lateral diffusion and rare strong adsorption. *Appl. Spectrosc.*, **55**, 663–669.

31 Han, W.S., Kang, Y., Lee, S.J., Lee, H., Do, Y., Lee, Y.-A., and Jung, J.H. (2005) Fabrication of color-tunable luminescent silica nanotubes loaded with functional dyes using a sol-gel cocondensation method. *J. Phys. Chem.*, **109**, 20661–20664.

32 Lebold, T., Mühlstein, L.A., Blechinger, J., Riederer, M., Amenitsch, H., Köhn, R., Peneva, K., Müllen, K., Michaelis, J., Bräuchle, C., and Bein, T. (2008) Tuning single-molecule dynamics in functionalized mesoporous silica. *Chem. Eur. J.*, **15**, 1661–1672.

33 Riehemann, K., Schneider, S.W., Luger, T.A., Godin, B., Ferrari, M., and Fuchs, H. (2009) Nanomedicine–challenge and perspectives. *Angew. Chem., Int. Ed.*, **48**, 872–897.

34 Lebold, T., Jung, C., Michaelis, J., and Bräuchle, C. (2009) Single molecule and cellular studies of nanostructured silica materials as delivery system for the anti-cancer drug doxorubicin. *Nano Lett.* **9**, 2877–2883.

35 Wagner, D., Kern, W.V., and Kern, P. (1994) Liposomal doxorubicin in AIDS-related Kaposi's sarcoma: long-term experiences. *Clin. Invest.*, **72** (6), 417–423.

36 Collins, Y. and Lele, S. (2005) Long-term pegylated liposomal doxorubicin use in recurrent ovarian carcinoma. *J. Natl Med. Assoc.*, **97**, 1414–1416.

37 O'Shaughnessy, J. (2003) Liposomal anthracyclines for breast cancer: overview. *Oncologist*, **8** (90002), 1–2.

38 Cauda, V., Onida, B., Platschek, B., Mühlstein, L., and Bein, T. (2008) Large antibiotic molecule diffusion in confined mesoporous silica with controlled morphology. *J. Mater. Chem.*, **18** (48), 5888–5899.

39 D'Arpa, P. and Liu, L.F. (1989) Topoisomerase-targeting antitumor drugs. *Biochim. Biophys. Acta*, **989** (2), 163–177.

40 Schaffert, D. and Wagner, E. (2008) Gene therapy progress and prospects: synthetic polymer-based systems. *Gene Ther.*, **15**, 1131–1138.

41 Bausinger, R., von Gersdorff, K., Braeckmans, K., Ogris, M., Wagner, E., Brauchle, C., *et al.* (2006) The transport of nanosized gene carriers unraveled by live-cell imaging. *Angew. Chem., Int. Ed.*, **45**, 1568–1572.

42 de Bruin, K., Ruthardt, N., von Gersdorff, K., Bausinger, R., Wagner, E., Ogris, M., and Bräuchle, C. (2007) Cellular dynamics of EGF receptor targeted synthetic viruses. *Mol. Ther.*, **15**, 1297–1305.

Index

Single Particle Tracking and Single Molecule Energy Transfer
Edited by Christoph Bräuchle, Don C. Lamb, and Jens Michaelis

ISBN: 978-3-527-32296-1

t